SCIENCEWORLD

SW·8

> Explore
> Question
> Imagine

Peter Saffin
BSc, BEd, GD Env. Sci

Peter Stannard
BSc, DipEd

Ken Williamson
BSc (Hons), DipEd

VICTORIAN CURRICULUM

First published 2017 by
MACMILLAN EDUCATION AUSTRALIA PTY LTD
15–19 Claremont Street, South Yarra, VIC 3141

Visit our website at www.macmillaneducation.com.au

Associated companies and representatives
throughout the world.

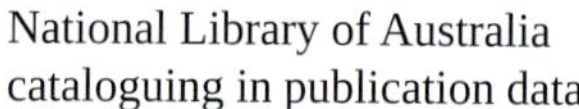

National Library of Australia
cataloguing in publication data

Creator:	Saffin, Peter, author
Title:	Scienceworld Victorian Curriculum 8 / Peter Saffin, Peter Stannard, Ken Williamson.
ISBN:	978 1 4202 3735 1 (paperback)
Target Audience:	For secondary school age
Subjects:	Science--Textbooks.
	Science--Study and teaching (Secondary)--Victoria
Other creators/contributors:	Stannard, Peter, author.
	Williamson, Ken, author.
Dewey number:	500.712945

Publisher: Emma Cooper
Project editors: Barbara Delissen, Diana Saad and Lachlan Smith
Proofreader: Marta Veroni
Illustrators: Andrew Craig, Chris Dent, Brent Hagen, Guy Holt and Nives Porcellato
Cover and text designer: Dimitrios Frangoulis
Production control: Katherine Fullagar
Photo research and permissions clearance: Annthea Lewis and Vanessa Roberts
Typeset in Merriweather Light 9.5/13.5 by Dim Frangoulis
Cover image: NASA/Chris Gunn
Indexer: Karen Gillen

Printed in China

Warning: It is recommended that Aboriginal and Torres Strait Islander peoples exercise caution when viewing this publication as it may contain images of deceased persons.

Contents

Using ScienceWorld

ScienceWorld takes a constructivist approach to learning, helping students explore what they already know, then building on that knowledge as they progress. This guide will show the key features of *ScienceWorld*, and help you understand how best to use it in the classroom.

Each chapter starts with a **Chapter opening page** containing an engaging photo for discussion. It shows the main content and skills to be covered in that chapter.

Students' prior knowledge is explored with a **Get started** activity, each with an **Explore**, **Question** or **Imagine** task that introduces the chapter topic.

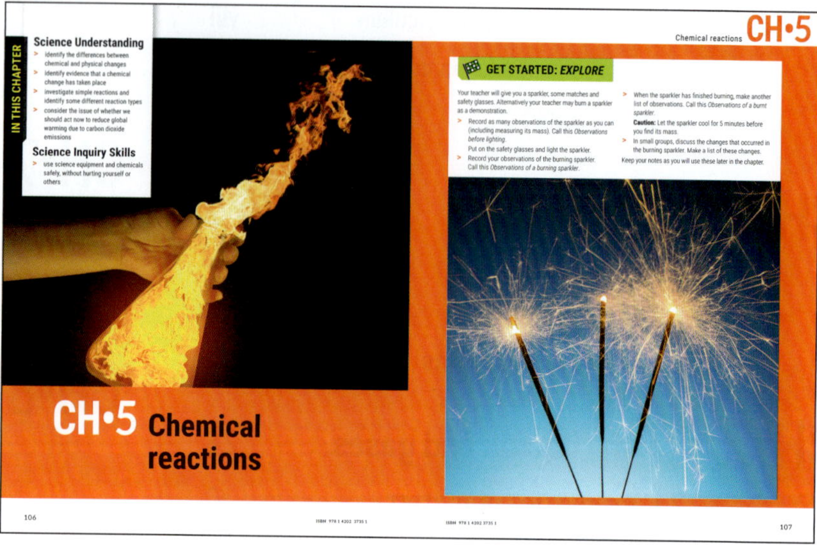

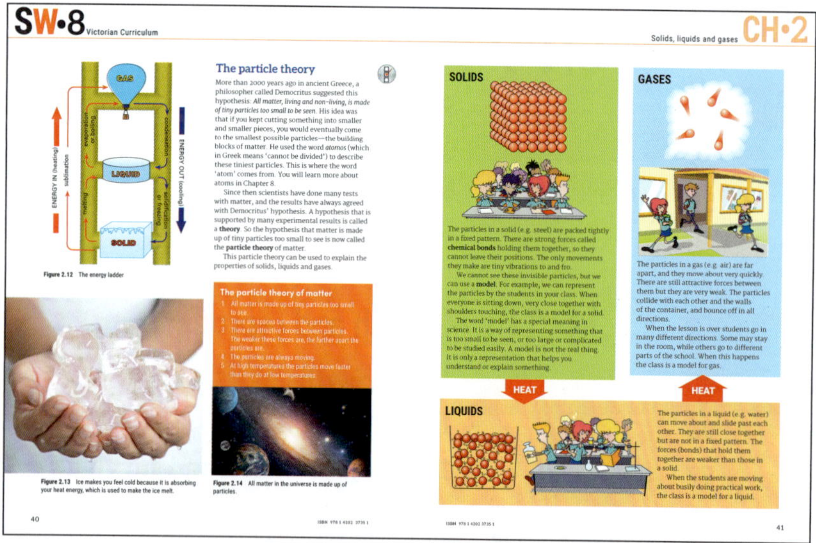

Each chapter is broken into sections focusing on certain aspects of theory and skills covered. *ScienceWorld* uses engaging photos, illustrations, cartoons and contexts to make science accessible to all students.

Activities are dispersed throughout each section to reinforce concepts and provide hands-on learning opportunities for each lesson.

Explore online boxes provide opportunity for wider research.

ISBN 978 1 4202 3735 1

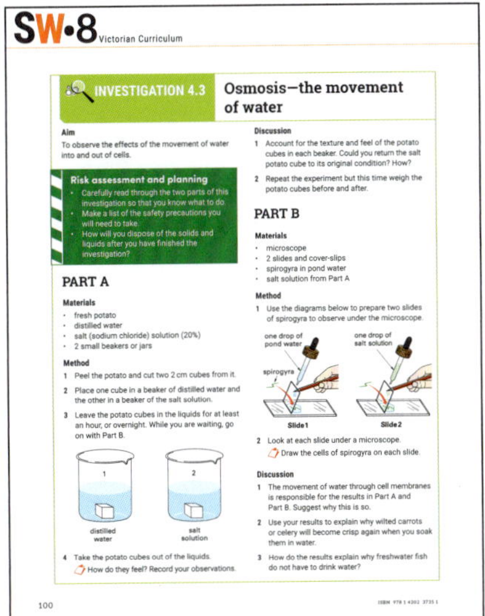

Investigations provide opportunities for students to apply Science Inquiry skills, while exploring key concepts.

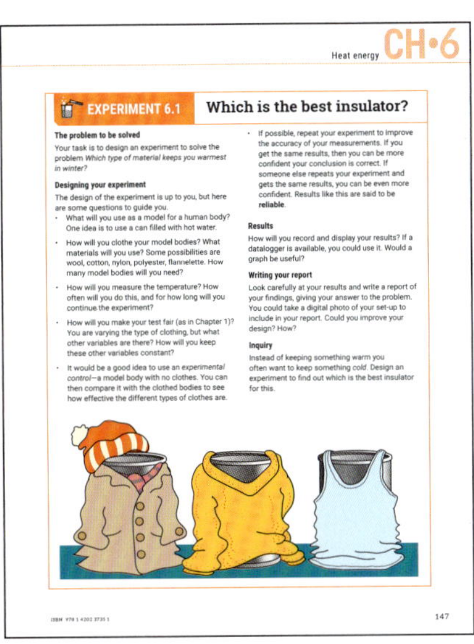

Experiments allow students to design their own experiments and inventions. Students explore and apply science skills and method while solving problems. This allows students to discover and engage with science concepts for themselves.

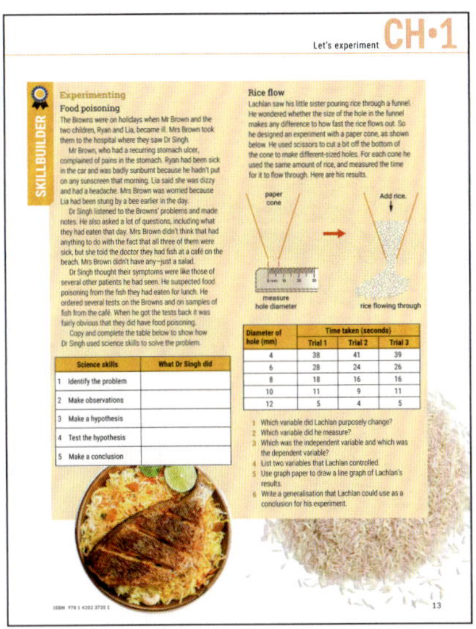

Skillbuilders teach key skills explicitly, supporting a clear progression of skill development throughout the book.

Science as a human endeavour features bring science to life; putting science in context historically, for today, and for the future. These features include activities that allow students to understand the nature and context of science and to imagine the future.

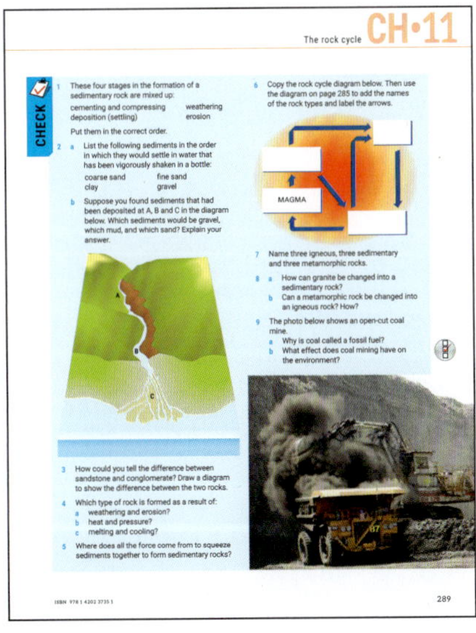

Check questions review students' understanding of concepts for each section.

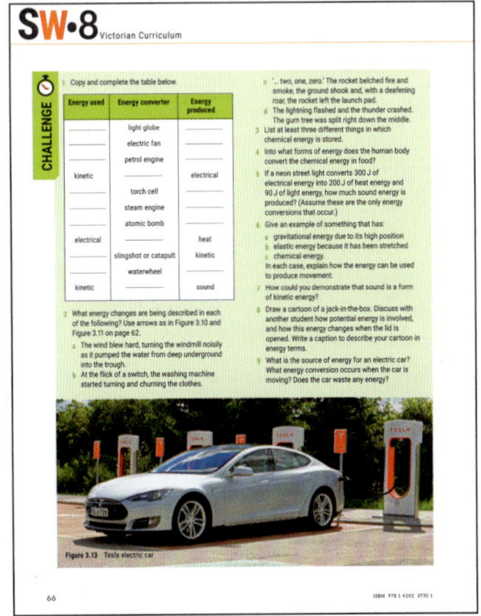

Challenge questions provide an opportunity for students to apply their knowledge and skills in context, and challenge students to think at a higher level.

ISBN 978 1 4202 3735 1

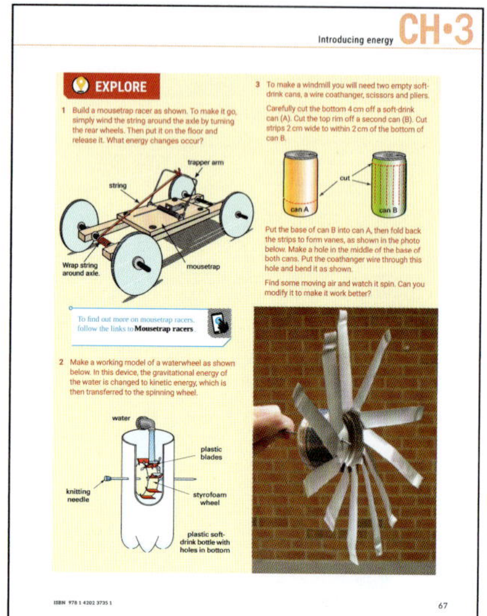

Explore sections allow students to create, invent and inquire into concepts learnt throughout the chapters.

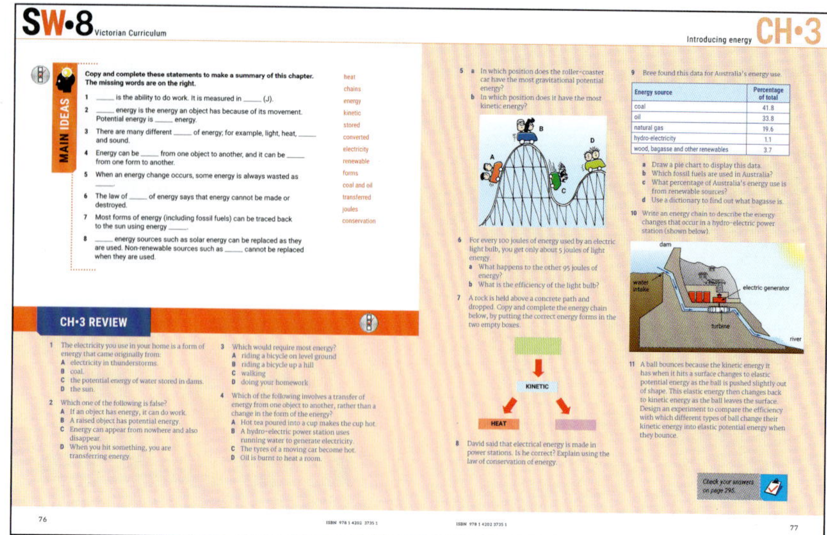

Each chapter ends with a **Main ideas** cloze exercise to test students' understanding of the key chapter concepts through comprehension.

The **Chapter review** then provides an opportunity for students to revise key science knowledge and skills developed through each chapter.

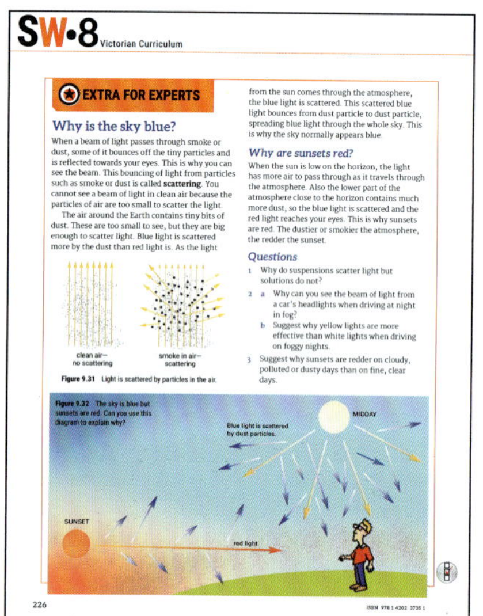

Extra for experts features provide opportunities for students to engage at a higher level, to challenge their scientific thinking and understanding.

ISBN 978 1 4202 3735 1

Foreword

As you probably know, I'm mad about science. Every day I learn something new about the world around me—dark matter and dark energy, living creatures of all shapes and sizes, the amazing irrationalities and untapped abilities of our human brain etc. My *Great Moments in Science* radio series/podcast is one way in which I explore these things and try to make them easier for people to understand. In this book you will explore all these things and learn how to think scientifically, asking questions about the world and imagining new solutions to these questions.

Doing science can lead to so many different and fascinating careers where you can design intelligent robots, use giant telescopes in space, produce food in a world where global warming is real, or go as an astronaut to planet Mars! It's fun to apply your knowledge of science to the real world and let your imagination run riot. After all, it's not the answer that gets you the Nobel Prize, it's the question. You could be the next Elizabeth Blackburn who won a Nobel Prize for her work on the telomere—a previously unexplored section of the human chromosome that gives us new and deep insights into aging.

And remember the words of Richard Feynman, 'Science is a way to not get fooled' ...

Dr Karl

Links to the Victorian Curriculum

This scope and sequence provides an overview of how *ScienceWorld 8* covers the Victorian Curriculum. The focus is on the Science Understanding strand, although only some of the Science as a Human Endeavour content and elaborations are covered in this version of the scope and sequence. Included online in the teacher support are curriculum scope and sequence guides that detail how *ScienceWorld* covers the Victorian Curriculum content descriptions across all four books, and these also include a full mapping of the Science as Human Endeavour sub-strand, and the Science Inquiry Skills.

Abbreviations:
BS: Biological Sciences
CS: Chemical Sciences
ESS: Earth and Space Sciences
PS: Physical Sciences

ScienceWorld 8

Chapter & Unit titles	Science Understanding	Elaborations
1 Let's experiment		
1.1 What is science?	Science Inquiry Skills	
1.2 Experimenting	Science Inquiry Skills	
1.3 Solving problems	Science Inquiry Skills	
2 Solids, liquids and gases		
2.1 Properties of matter	CS: The properties of the different states of matter can be explained in terms of the motion and arrangement of particles (VCSSU096)	
2.2 Solid–liquid–gas	CS: The properties of the different states of matter can be explained in terms of the motion and arrangement of particles (VCSSU096)	• modelling the arrangement of particles in solids, liquids and gases • using the particle model to distinguish between the properties of liquid water, ice and steam
2.3 Using the particle theory	CS: The properties of the different states of matter can be explained in terms of the motion and arrangement of particles (VCSSU096)	• modelling the arrangement of particles in solids, liquids and gases • using the particle model to distinguish between the properties of liquid water, ice and steam
3 Introducing energy		
3.1 What is energy?	PS: Energy appears in different forms including movement (kinetic energy), heat, light, chemical energy and potential energy; devices can change energy from one form to another (VCSSU104)	
3.2 Forms of energy	PS: Energy appears in different forms including movement (kinetic energy), heat, light, chemical energy and potential energy; devices can change energy from one form to another (VCSSU104)	• recognising that kinetic energy is the energy possessed by moving bodies • recognising that potential energy is stored energy, for example, gravitational, chemical and elastic energy • using flow diagrams to illustrate changes between different forms of energy • investigating the energy transformations in devices, for example, a catapult or a water wheel

ISBN 978 1 4202 3735 1

3.3 Energy comes—energy goes	PS: Energy appears in different forms including movement (kinetic energy), heat, light, chemical energy and potential energy; devices can change energy from one form to another (VCSSU104) ESS: Some of Earth's resources are renewable, but others are non-renewable (VCSSU100)	• investigating the energy transformations in devices, for example, a catapult or a water wheel • considering what is meant by the term 'renewable' in relation to the Earth's resources • considering timescales for regeneration of resources
4 Cells of life		
4.1 Cells	BS: Cells are the basic units of living things and have specialised structures and functions (VCSSU092) BS: Multicellular organisms contain systems of organs that carry out specialised functions that enable them to survive and reproduce (VCSSU094)	• examining a variety of cells using a light microscope, by digital technology or by viewing a simulation • distinguishing plant cells from animal and fungal cells • identifying structures within cells and describing their function • recognising that some organisms consist of a single cell • examining the specialised cells and tissues involved in structure and function of particular organs
4.2 Cell processes	BS: Cells are the basic units of living things and have specialised structures and functions (VCSSU092) BS: Multicellular organisms contain systems of organs that carry out specialised functions that enable them to survive and reproduce (VCSSU094)	• examining a variety of cells using a light microscope, by digital technology or by viewing a simulation • distinguishing plant cells from animal and fungal cells • identifying structures within cells and describing their function • examining the specialised cells and tissues involved in structure and function of particular organs
4.3 Investigating cells	BS: Cells are the basic units of living things and have specialised structures and functions (VCSSU092)	
5 Chemical reactions		
5.1 Physical and chemical properties	CS: Chemical change involves substances reacting to form new substances (VCSSU098)	• identifying the differences between chemical and physical changes • identifying evidence that a chemical change has taken place • investigating simple reactions, for example, combining elements to make a compound
5.2 What is a chemical reaction?	CS: Chemical change involves substances reacting to form new substances (VCSSU098)	• identifying the differences between chemical and physical changes • identifying evidence that a chemical change has taken place • investigating simple reactions, for example, combining elements to make a compound
5.3 Some common gases	CS: Chemical change involves substances reacting to form new substances (VCSSU098)	• investigating simple reactions, for example, combining elements to make a compound
6 Heat energy		
6.1 Heat and temperature	PS: Energy appears in different forms including movement (kinetic energy), heat, light, chemical energy and potential energy; devices can change energy from one form to another (VCSSU104)	
6.2 Heat transfer	PS: Energy appears in different forms including movement (kinetic energy), heat, light, chemical energy and potential energy; devices can change energy from one form to another (VCSSU104)	• using flow diagrams to illustrate changes between different forms of energy • investigating the energy transformations in devices, for example, a catapult or a water wheel
6.3 Exploring heat	PS: Energy appears in different forms including movement (kinetic energy), heat, light, chemical energy and potential energy; devices can change energy from one form to another (VCSSU104)	• using flow diagrams to illustrate changes between different forms of energy • investigating the energy transformations in devices, for example, a catapult or a water wheel

ISBN 978 1 4202 3735 1

7 The human body		
7.1 How muscles work	BS: Multicellular organisms contain systems of organs that carry out specialised functions that enable them to survive and reproduce (VCSSU094)	• examining the specialised cells and tissues involved in structure and function of particular organs • describing the structure of each organ in a system and relating its function to the overall function of the system
7.2 Digestion	BS: Multicellular organisms contain systems of organs that carry out specialised functions that enable them to survive and reproduce (VCSSU094)	• examining the specialised cells and tissues involved in structure and function of particular organs • describing the structure of each organ in a system and relating its function to the overall function of the system • identifying the organs and overall function of a system of a multicellular organism in supporting life processes
7.3 Body systems	BS: Multicellular organisms contain systems of organs that carry out specialised functions that enable them to survive and reproduce (VCSSU094)	• examining the specialised cells and tissues involved in structure and function of particular organs • describing the structure of each organ in a system and relating its function to the overall function of the system • identifying the organs and overall function of a system of a multicellular organism in supporting life processes
8 Elements and compounds		
8.1 Atoms and molecules	CS: Differences between elements, compounds and mixtures can be described by using a particle model (VCSSU097)	• modelling the arrangement of particles in elements and compounds
8.2 Elements and compounds	CS: Differences between elements, compounds and mixtures can be described by using a particle model (VCSSU097)	• modelling the arrangement of particles in elements and compounds • recognising that elements and simple compounds can be represented by symbols and formulas
8.3 Chemical reactions	CS: Differences between elements, compounds and mixtures can be described by using a particle model (VCSSU097)	• modelling the arrangement of particles in elements and compounds • recognising that elements and simple compounds can be represented by symbols and formulas • explaining why elements and compounds can be represented by chemical formulas while mixtures cannot
	CS: Chemical change involves substances reacting to form new substances (VCSSU098)	• identifying evidence that a chemical change has taken place • investigating simple reactions, for example, combining elements to make a compound
9 Light and sound		
9.1 Properties of light and sound	PS: Light can form images using the reflective feature of curved mirrors and the refractive feature of lenses, and can disperse to produce a spectrum which is part of a larger spectrum of radiation (VCSSU105) PS: The properties of sound can be explained by a wave model (VCSSU106)	• exploring how images can change when the arrangement of the mirror or lens system is altered • exploring the mechanism of the human eye and corrective technologies
	BS: Multicellular organisms contain systems of organs that carry out specialised functions that enable them to survive and reproduce (VCSSU094)	• examining the specialised cells and tissues involved in structure and function of particular organs • describing the structure of each organ in a system and relating its function to the overall function of the system • identifying the organs and overall function of a system of a multicellular organism in supporting life processes
9.2 Light and colour	PS: Light can form images using the reflective feature of curved mirrors and the refractive feature of lenses, and can disperse to produce a spectrum which is part of a larger spectrum of radiation (VCSSU105)	• observing the spread and order of colours in the visible spectrum

ISBN 978 1 4202 3735 1

9.3 Light and sound as waves	PS: Light can form images using the reflective feature of curved mirrors and the refractive feature of lenses, and can disperse to produce a spectrum which is part of a larger spectrum of radiation (VCSSU105) PS: The properties of sound can be explained by a wave model (VCSSU106)	• describing the different types of radiation in the larger spectrum of radiation • describing how sounds are produced by different musical instruments • measuring the speed of sound • using a wave model to describe the measured properties of sound, wavelength and frequency
9.4 Applications of sound	PS: The properties of sound can be explained by a wave model (VCSSU106) BS: Multicellular organisms contain systems of organs that carry out specialised functions that enable them to survive and reproduce (VCSSU094)	• describing how sounds are produced by different musical instruments • using a wave model to describe the measured properties of sound, wavelength and frequency • examining the specialised cells and tissues involved in structure and function of particular organs • describing the structure of each organ in a system and relating its function to the overall function of the system • identifying the organs and overall function of a system of a multicellular organism in supporting life processes
10 Growth and reproduction		
10.1 Growth	BS: Multicellular organisms contain systems of organs that carry out specialised functions that enable them to survive and reproduce (VCSSU094)	• examining the specialised cells and tissues involved in structure and function of particular organs
10.2 Reproduction	BS: Multicellular organisms contain systems of organs that carry out specialised functions that enable them to survive and reproduce (VCSSU094)	• examining the specialised cells and tissues involved in structure and function of particular organs • describing the structure of each organ in a system and relating its function to the overall function of the system • identifying the organs and overall function of a system of a multicellular organism in supporting life processes • comparing reproductive systems of organisms • comparing similar systems in different organisms, for example, digestive systems in herbivores and carnivores, respiratory systems in fish and mammals
10.3 Reproduction and survival	BS: Multicellular organisms contain systems of organs that carry out specialised functions that enable them to survive and reproduce (VCSSU094)	• examining the specialised cells and tissues involved in structure and function of particular organs • describing the structure of each organ in a system and relating its function to the overall function of the system • identifying the organs and overall function of a system of a multicellular organism in supporting life processes • comparing reproductive systems of organisms
11 The rock cycle		
11.1 Rocks from fire	ESS: Sedimentary, igneous and metamorphic rocks contain minerals and are formed by processes that occur within Earth over a variety of timescales (VCSSU102)	• recognising that rocks are a collection of different minerals • considering the role of forces and energy in the formation of different types of rocks and minerals • identifying a range of common rock types using keys based on observable physical and chemical properties
11.2 Weathering and erosion	ESS: Sedimentary, igneous and metamorphic rocks contain minerals and are formed by processes that occur within Earth over a variety of timescales (VCSSU102)	• recognising that rocks are a collection of different minerals • considering the role of forces and energy in the formation of different types of rocks and minerals • identifying a range of common rock types using keys based on observable physical and chemical properties
11.3 The rock cycle	ESS: Sedimentary, igneous and metamorphic rocks contain minerals and are formed by processes that occur within Earth over a variety of timescales (VCSSU102)	• recognising that rocks are a collection of different minerals • considering the role of forces and energy in the formation of different types of rocks and minerals • identifying a range of common rock types using keys based on observable physical and chemical properties

ISBN 978 1 4202 3735 1

Risk assessment and planning

The best part of science is doing investigations and experiments.

Flick through the book and find:
- where you learn how to design experiments (in Chapter 1)
- where you make some common gases in the laboratory (in Chapter 2)
- where you learn how to use a microscope (in Chapter 6)
- in which chapter you learn how to make soap.

To get the most out of the investigations, you must be well prepared, and you must consider safety issues. This is why most investigations in this book have a **Risk assessment and planning** at the beginning. The first one is on page 14. Before you start an investigation you should follow these steps.

1 Read the investigation carefully and study the diagrams. Make sure you know the *aim* of the investigation, that is why you are doing it.

2 Make sure you know exactly what you will be doing. You will be working in a group most of the time so you will need to sort out who will be doing what.

3 You need to know which materials you will be using. You will also need to know how to use the equipment.

4 You usually need to prepare a *data table* in which to record your results. Sometimes the textbook shows you how to do this and sometimes you have to design it yourselves.

5 Experiments are open-ended investigations where you have to design your own tests to answer a question or solve a problem. Have a quick look at these:

p. 6 How much weight will a paper bridge support?

p. 147 Which type of material keeps you warmest in winter?

6 You will of course need to discuss your design with your teacher before you start.

There are risks involved in doing investigations and experiments, but you can reduce the risks to yourself and others if you follow simple safety procedures.

Make sure you can answer these questions:
- What are the safety rules for the laboratory?
- What safety procedures are necessary when you see these symbols?

a 

b 

c 

- What special precautions are necessary when you use a Bunsen burner?
- When should you use safety glasses?

- What should you do if you get a chemical in your eyes or on your skin?

ISBN 978 1 4202 3735 1

Online resources

Throughout this book you will find links to activities and video or audio files. The activities are for students to practise key skills, or to reinforce learning on key concepts. Activities vary in type and include crosswords, matching, drag and drop, labelling, multiple choice, true and false, and sequencing activities. Students can repeat these activities as revision, and practise them at any time.

Each activity is scored and the teacher can review student progress in the digital mark book. In *ScienceWorld 8* there are approximately 110 activities.

When an activity is available, you will find an icon on the page of the book where it is most relevant to learning.

 **Digital activity**

 **Audio or video material**

 **Supporting documentation**

We hope you enjoy using these activities to improve learning outcomes.

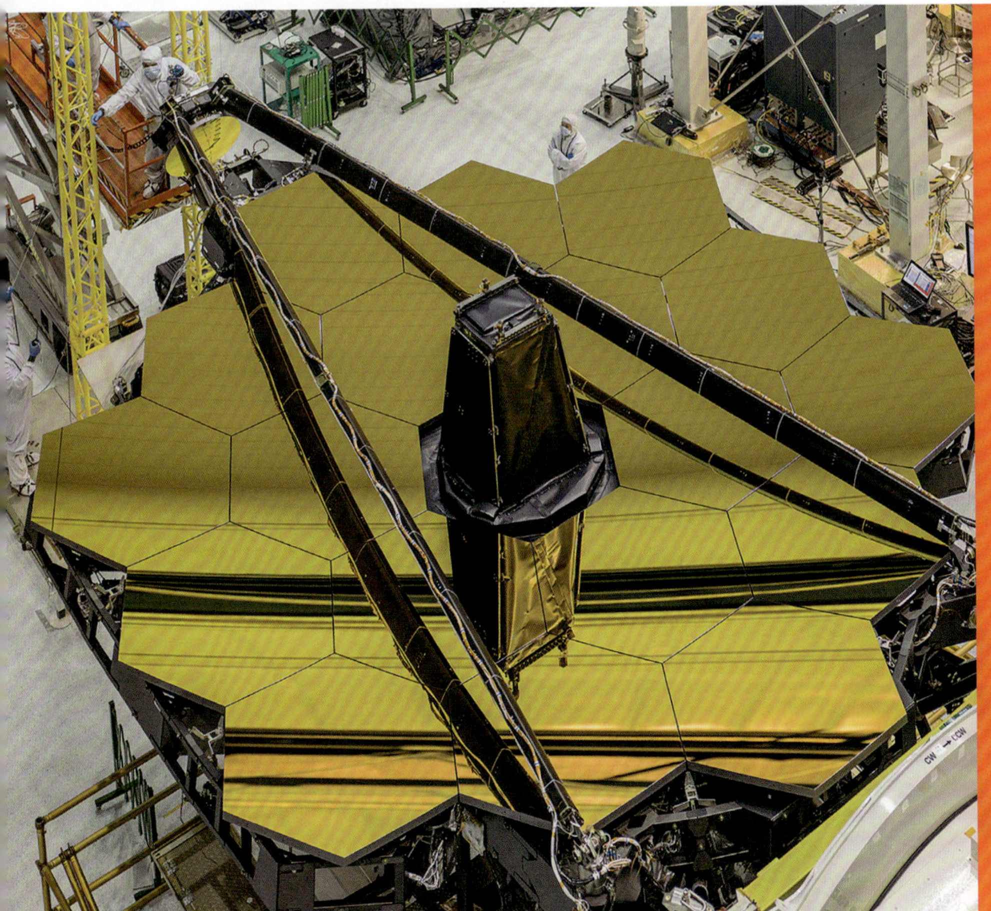

James Webb Space Telescope

The James Webb Space Telescope (JWST) is scheduled for launch in 2018. It will be the largest and most powerful space telescope ever built. The JWST will have a massive 6.5 m gold mirror and will be situated further away from Earth than the Hubble Space Telescope. It will be used to study the solar system, including taking images of the planets. It will also be able to study the history of the universe, taking images of other galaxies, identifying new planets and exploring the black hole at the centre of the Milky Way.

ISBN 978 1 4202 3735 1

IN THIS CHAPTER

Science Understanding

> discuss the work of Australian scientists in solving problems

Science Inquiry Skills

> identify variables to be changed, measured or controlled for a fair test
> draw line graphs, with the independent variable on the horizontal axis and the dependent variable on the vertical axis
> use experiments to test hypotheses and modify them if necessary
> evaluate an experiment and suggest ways of improving it
> work individually or in a group to choose and do a science research project

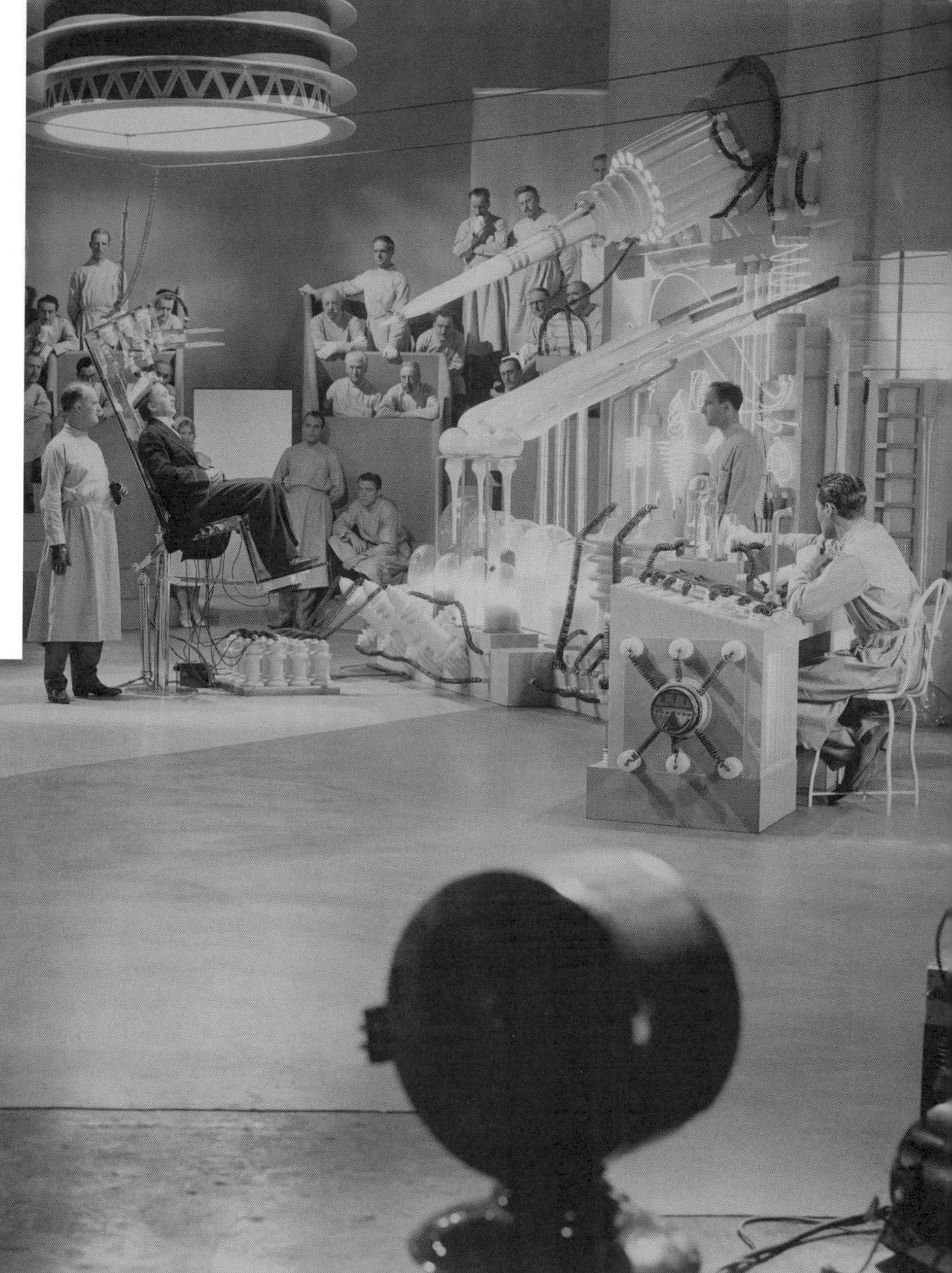

CH•1 Let's experiment

ISBN 978 1 4202 3735 1

GET STARTED: *EXPLORE*

Imagine you are out on a school excursion exploring a local ecosystem, and you discover that some oil has been released into the local waterway. You collect a sample, and now you want to find out what caused the pollution so it can be avoided in the future.

1 Make a list of observations about the water sample and the local area, in order to collect information about the pollution.

2 Make a list of all the possible ways the oil could have leaked and made its way into the water.

3 For each possible way the oil could have got there, discuss how you could test or work out whether that's what actually occurred.

4 Suggest ways to avoid this type of pollution in the future.

ISBN 978 1 4202 3735 1

1.1 What is science?

Finding out why oil was released into your local waterway involves asking questions, testing these questions to see what answers they bring up and, if you have still not uncovered the reason, asking more questions and doing more tests.

Science is all about asking questions, testing, asking more questions and doing more tests. Science is a way of finding out how or why things happen.

You learnt in *ScienceWorld* 7 that an **experiment** is a well thought-out test. The test has a series of steps involving several different skills as shown below.

Observing

Observing is when you use your senses to find out as much as you can about an object or event. Observations can lead to two different types of description. One is recorded in words, and is called **qualitative** (QUAL-i-tate-ive) observation. Other observations involve taking measurements. These are called **quantitative** (QUANT-i-tate-ive) observations. Both observations are called **data**.

Recording

Recording is when you write down your data. You often record this in a *data table*.

Inferring

Inferring is trying to explain your observations. For example, to explain why there was oil in the local waterway you might say that it leaked from the storage drums of a nearby factory. This inference may not be correct, but it could be tested by making further observations.

Predicting

Predicting is the process of making a forecast of what a future observation will be. Predictions are based on your observations and what you already know.

Generalising

Generalising is where you write a statement that seems true in most cases after you have made many observations. Since there may be exceptions to a generalisation, words like 'most' and 'many' are often used.

Often a generalisation links two different factors. For example, when a painter generalises that 'the warmer the day the faster the paint dries', he is linking drying time to temperature.

Figure 1.1 Scientists perform a variety of tasks that require them to use different skills.

ISBN 978 1 4202 3735 1

Investigations and experiments

In *ScienceWorld* 7 you did some laboratory investigations such as filtering and distilling, measuring temperature and testing for gases. In this chapter you will be doing experiments where you have to design tests to answer questions or solve problems.

What's the difference between an investigation and an experiment? The terms mean much the same thing. Both involve carefully planned laboratory or field work. However, an experiment is based on solving a problem or answering a question.

Experiments involve designing tests, observing and recording data, then writing full reports. The Skillbuilder below shows you how to write up a report.

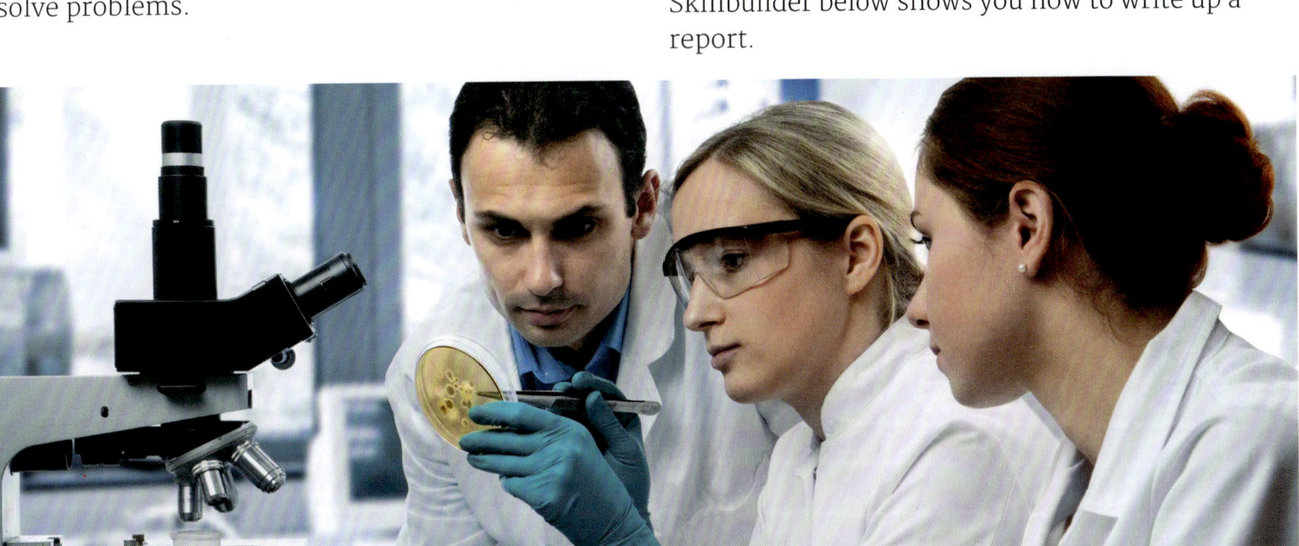

Figure 1.2 Scientists must make observations and record their findings.

SKILLBUILDER

Writing reports

A report is organised using seven headings.

TITLE A very brief description of the investigation, your name and the date.

AIM You say why you did the investigation—sometimes this is a question.

MATERIALS You list the equipment and chemicals you used in the investigation.

METHOD You say what you did in the investigation in numbered steps. Whenever possible include a large, neat diagram of the apparatus.

RESULTS You record the data. Data includes qualitative observations (words) and measurements (numbers). Usually these are recorded in a data table. This makes the data easier to read.

DISCUSSION You try to explain your results, and list any problems that you experienced. You might also explain how you could improve the investigation.

CONCLUSION You answer the question posed in the aim. Often your conclusion will contain a **generalisation**—a statement that seems true in most cases. For example, a student investigating paper bridges concluded: *The more folds the paper bridge has, the more weight it can support.*

ISBN 978 1 4202 3735 1

EXPERIMENT 1.1 Paper bridges

Suppose you suspend a piece of A4 paper between two blocks. How much mass will the paper support? None, you say! Well, look at the paper bridge in the photo.

The paper has been folded many times. It is a paper bridge, and it can support a container with stones in it.

The problem to be solved

If we increase the number of folds in a piece of paper, will it support more mass?

Your task is to work in a small group to design an experiment that will answer this question.

Designing your experiment

1 Discuss what tests you will do to answer the question.

2 Make a list of the equipment you will need.

3 Discuss how you are going to record your observations. Will you make quantitative observations?

4 When you and your teacher are happy with your plan, get started.

Writing your report

1 Write a full report of your experiment, using the headings Title, Aim, Materials, Method, Results, Discussion and Conclusion.

2 Your discussion should contain an inference that tries to explain your observations.

3 Your conclusion should contain a generalisation that links mass and the folds in the paper.

4 You might like to take a digital photo of your set-up and include it in your report.

Extending the experiment

You might like to extend your experiment by testing these predictions:

1 Two layers of folded paper will support twice the mass supported by a single piece of paper.

2 Heavier paper will support more mass than ordinary paper.

3 Dry paper is much stronger than damp paper.

ISBN 978 1 4202 3735 1

1 Use the following words to complete the sentences below.

A generalisation ...
Predicting ...
An experiment ...

a ______ is a scientific test.
b ______ is a statement that is true in most cases.
c ______ is saying what might happen in the future.

2 For each statement below say whether it is an observation, an inference, a prediction or a generalisation.

a It tends to rain more in winter than in summer.
b She must have eaten something that doesn't agree with her.
c There should be a full moon next week.
d The leaves on this plant are turning yellow.
e That colourless liquid must be an acid.

3 You placed a young mouse in a cage with dishes containing three different foods. After observing the mouse for 30 minutes you noticed that it had eaten nothing. What inferences could you make from this?

4 Cameron has a mouse in a cage. The mouse has an exercise wheel with a counter on it. Cameron wrote down the counter reading each morning, but the bottom of his results sheet has been torn off.

a Predict what the counter reading for day 4 should be (approximately).
b Explain how you made this prediction.

day	counter reading
1	49
2	100
3	152
4	

1 Heidi dropped a ball from two different heights and measured how high it bounced each time. She used her data to draw a graph.

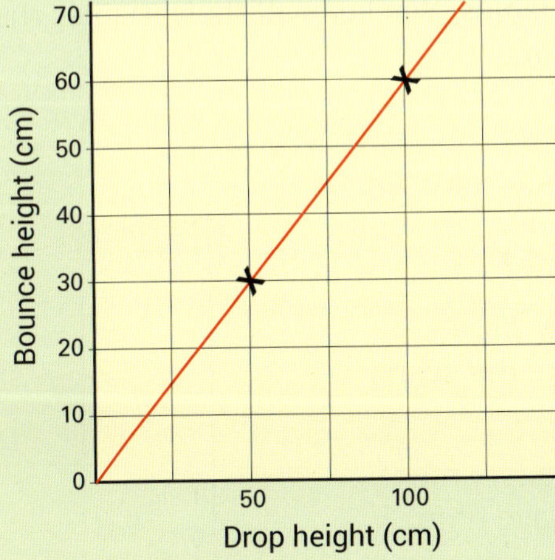

a Predict how high the ball will bounce if she drops it from 75 cm.
b Predict the bounce height for a drop height of 150 cm.

2 Mick peeled a banana for lunch and left it in his bag when he went to play soccer. Later he discovered that the banana had turned brown and soft.

a Pose a question based on Mick's observations.
b Suggest an inference that tries to answer this question.

3 Ask other students in your group these questions:

- Will it rain tomorrow?
- Will it be a full moon tonight?
- How fast can you swim 50 metres freestyle?

a Decide whether the answers they gave you are predictions (based on observations and knowledge), or just guesses.
b What information would you need to turn the guesses into proper predictions?

ISBN 978 1 4202 3735 1

1.2 Experimenting

In the paper bridges experiment, you tried to find out if the number of folds in the paper affected the mass the paper could support. Was your experiment a *fair test*? Did you consider the other factors that might have affected the result?

Figure 1.4 The containers have different shapes. The paper bridge supports the rectangular container better than it supports the circular one.

Controlling variables

There are other factors that could have affected the results of your paper bridge experiment. Look at the photos.

Figure 1.3 One paper bridge is longer than the other. The shorter bridge can support more mass.

There are at least three factors that could affect the results of this experiment:

1 the number of folds in the paper
2 the distance between the supports
3 the shape of the mass container.

These factors that could change the results of an experiment are called **variables**.

You should test only one variable at a time. If you want to increase the number of folds in the paper, then you must keep the other two variables the same: use the same type of container, and keep the length of the bridge the same. This is then called a **fair test**.

The test becomes a fair test when you **control the variables**. You keep all the variables the same, except one.

You can make a pendulum by suspending a steel nut on a paperclip tied by cotton to a metal clamp and stand.

Suppose your group wants to find out whether the mass of a pendulum makes any difference to the time it takes to do a complete swing (from start back to start again).

Use the questions below to design an experiment that will test the statement above.

- What are the variables in this experiment?
- Which variables will we purposely change?
- Which variables will we need to keep the same?
- How will we measure the swing time?

Do the experiment, if you have time.

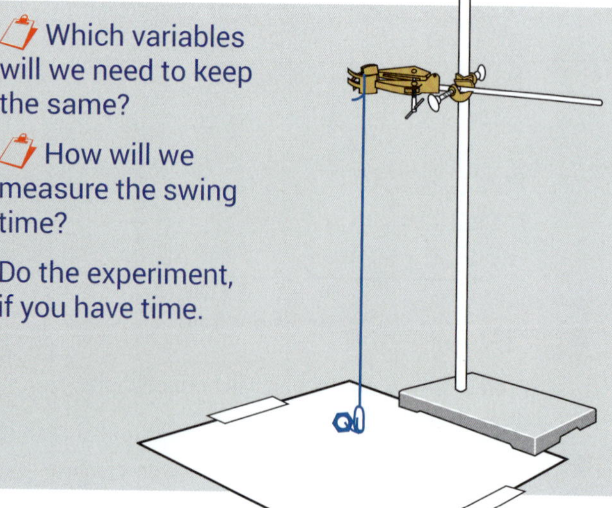

ISBN 978 1 4202 3735 1

Testing a hypothesis

A **hypothesis** (high-POTH-e-sis) is a generalisation that can be tested. It explains a set of observations or gives a possible answer to a question. Note that the plural of hypothesis is hypotheses (high-POTH-e-sees). An example of a hypothesis is given on this page.

1 *Rosco recorded his observations of the effect of a magnet on various materials.*

Material tested	Magnetic (✓) Non-magnetic (✗)
nail	✓
piece of glass	✗
wooden pencil	✗
knife	✓
paperclip	✓

2 *He looked through his results.*

3 *Based on his results he made this generalisation.*

4 *From this generalisation he was able to make a prediction that could be tested.*

5 Rosco experimented further. Because his prediction turned out to be wrong, he had to modify (change) his hypothesis.

ISBN 978 1 4202 3735 1

Which filter?

In *ScienceWorld 7* you learnt how to filter muddy water. In this experiment you will design tests to see whether folding a filter paper in different ways has any effect on the time it takes to filter muddy water.

The problem to be solved

To compare the time it takes to filter muddy water using filter papers folded in different ways. If you can't remember how to fold a filter paper normally, check with you teacher. Follow the instructions below for making a *fluted* (folded many times) filter paper.

Designing the experiment

1. Work in a small group and discuss the tests you will do.
2. Write a hypothesis for the experiment.
3. Make a list of the equipment you will need.
4. Make a list of the safety precautions you will take.
5. Which variables will you control? Which variable are you going to change?
6. Discuss how you are going to record your observations.
7. When you and your teacher are happy with your plan, get started.

Writing your report

1. Write a full report of your experiment, using the headings Title, Aim, Materials, Method, Results, Discussion and Conclusion.
2. Your discussion should contain an inference that tries to explain your observations.
3. Do your results support your hypothesis? If not, write a better hypothesis.

Extending the experiment

You might like to test this prediction: *A sixteen-fold fluted filter paper filters twice as fast as an eight-fold one.*

Making a fluted filter paper

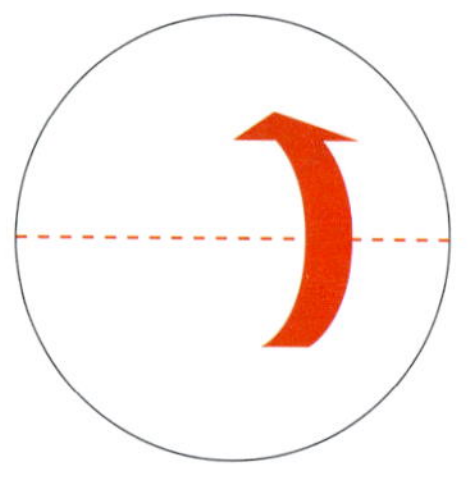

Fold here.

Unfold the filter paper.

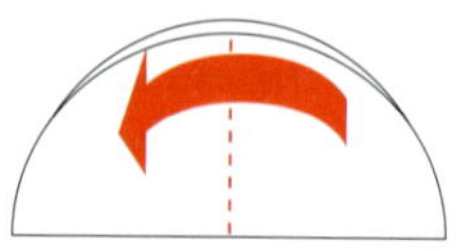

Fold into quarters ...

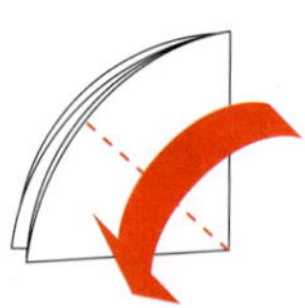

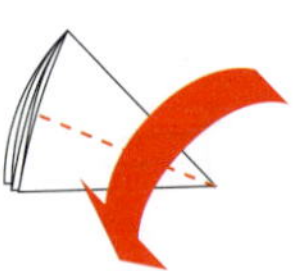

... then fold two more times.

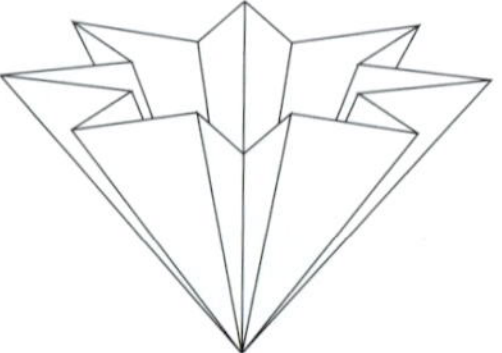

Adjust the folds so that the filter paper forms an eight-pointed star.

ISBN 978 1 4202 3735 1

Graphing

A line graph is a way of displaying data so that it can be interpreted easily. It may be a straight line or a curved line. A line graph shows you the relationship between two variables.

Look at this data from the side of a milk carton. The data was obtained by storing milk at various temperatures and recording the average time before it 'went off'.

Independent and dependent variables

The temperature was changed on purpose. It is called the **independent variable**, because you can select any value for it. The number of days the milk lasted is called the **dependent variable**, because it depends on the temperature. All other variables, e.g. brand of milk and type of container, were *controlled* (kept the same).

SKILLBUILDER

Drawing graphs

1. On a piece of graph paper draw the horizontal axis and the vertical axis.
2. On a line graph, the dependent variable is plotted on the *vertical* axis. The independent variable is plotted on the *horizontal* axis. Label the horizontal axis 'Temperature (°C)'. Label the vertical axis 'Time (days)'.
3. Select suitable scales for the two axes so that the graph fills most of the page.
4. Look at the first pair of numbers in the data table. They are:
   temperature 4 °C
   time 9 days
   In pencil, mark the point where the grid lines meet with a small neat cross. Then do the same with the other pairs of numbers.
5. By looking at the four crosses you have drawn, you can see that this graph is a curved line. Use a pencil to draw a smooth curve through the crosses, as shown. (This may take some practice.) Don't join the crosses with straight lines.
6. Finally, write a title for the graph at the top. This tells others what the graph is about.

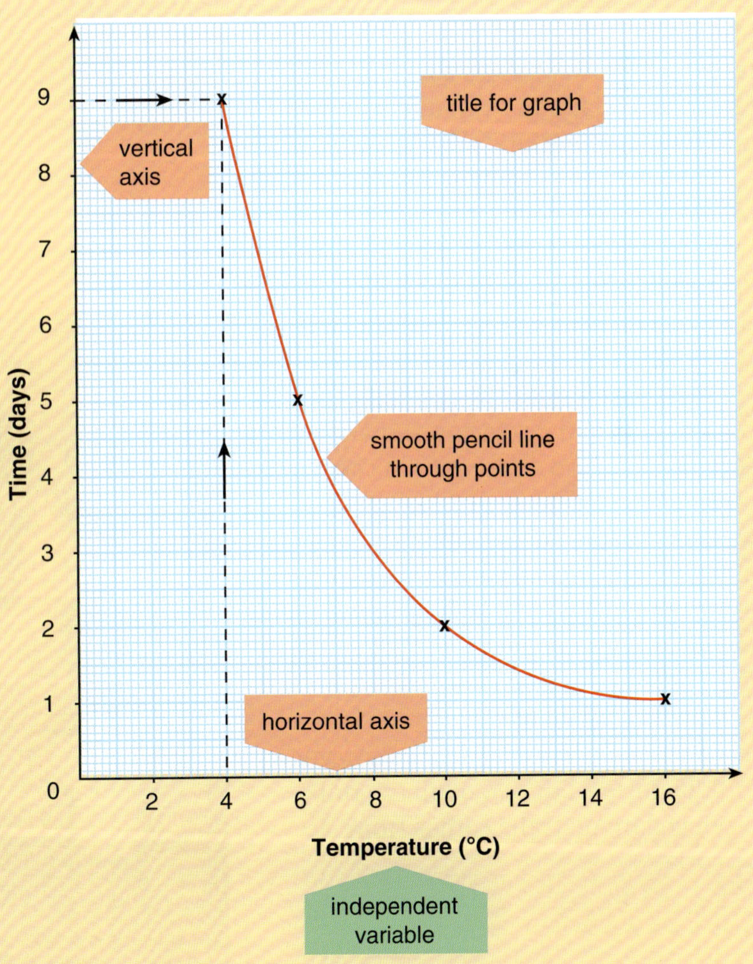

ISBN 978 1 4202 3735 1

ACTIVITY

Use the graph on the previous page to answer these questions.

1 How long does milk last when stored at 8 °C?

2 A carton of milk lasted 1½ days. At what temperature was it probably stored?

3 Describe in your own words what the shape of the graph tells you about the relationship (link) between the temperature of the milk and how long it lasts.

4 Complete this hypothesis. *The lower the temperature ...*

SCIENCE AS A HUMAN ENDEAVOUR

Making milk safe

Elaine Perriman is a food technologist. She works in the laboratory of a country milk factory that makes a range of full fat, reduced fat and skim milks, cream and flavoured milks.

She routinely samples the pasteurised milk from the factory and tests for the presence of disease-causing bacteria. In this way she can tell that the pasteurisation process is working correctly.

She also samples the raw milk that comes in from dairy farms to make sure there are no antibiotics in the milk. When farmers treat sick cows with antibiotics, the antibiotics pass into the milk. Some people are allergic to certain antibiotics, so it is Elaine's responsibility to make sure that the raw milk does not contain antibiotics.

Milk is a very important food in most people's lives. Most states have milk factories, but 61% of all the milk produced in Australia comes from Victoria.

ISBN 978 1 4202 3735 1

SKILLBUILDER

Experimenting

Food poisoning

The Browns were on holidays when Mr Brown and the two children, Ryan and Lia, became ill. Mrs Brown took them to the hospital where they saw Dr Singh.

Mr Brown, who had a recurring stomach ulcer, complained of pains in the stomach. Ryan had been sick in the car and was badly sunburnt because he hadn't put on any sunscreen that morning. Lia said she was dizzy and had a headache. Mrs Brown was worried because Lia had been stung by a bee earlier in the day.

Dr Singh listened to the Browns' problems and made notes. He also asked a lot of questions, including what they had eaten that day. Mrs Brown didn't think that had anything to do with the fact that all three of them were sick, but she told the doctor they had fish at a café on the beach. Mrs Brown didn't have any—just a salad.

Dr Singh thought their symptoms were like those of several other patients he had seen. He suspected food poisoning from the fish they had eaten for lunch. He ordered several tests on the Browns and on samples of fish from the café. When he got the tests back it was fairly obvious that they did have food poisoning.

Copy and complete the table below to show how Dr Singh used science skills to solve the problem.

Science skills	What Dr Singh did
1 Identify the problem	
2 Make observations	
3 Make a hypothesis	
4 Test the hypothesis	
5 Make a conclusion	

Rice flow

Lachlan saw his little sister pouring rice through a funnel. He wondered whether the size of the hole in the funnel makes any difference to how fast the rice flows out. So he designed an experiment with a paper cone, as shown below. He used scissors to cut a bit off the bottom of the cone to make different-sized holes. For each cone he used the same amount of rice, and measured the time for it to flow through. Here are his results.

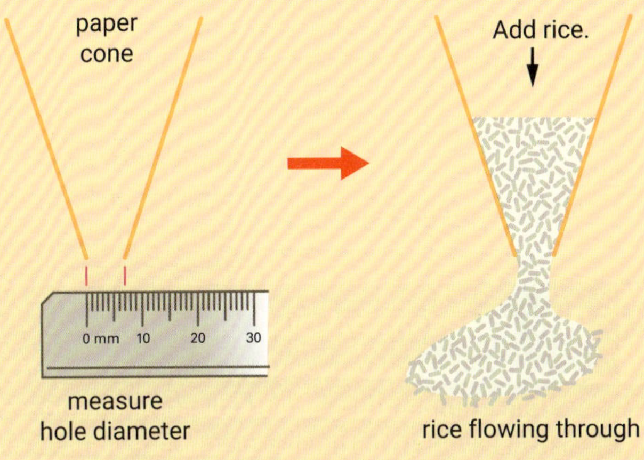

Diameter of hole (mm)	Time taken (seconds)		
	Trial 1	Trial 2	Trial 3
4	38	41	39
6	28	24	26
8	18	16	16
10	11	9	11
12	5	4	5

1 Which variable did Lachlan purposely change?
2 Which variable did he measure?
3 Which was the independent variable and which was the dependent variable?
4 List two variables that Lachlan controlled.
5 Use graph paper to draw a line graph of Lachlan's results.
6 Write a generalisation that Lachlan could use as a conclusion for his experiment.

ISBN 978 1 4202 3735 1

INVESTIGATION 1.1 Dissolving time

Aim

To write and test a hypothesis about how temperature affects the time it takes an antacid tablet to dissolve in water.

Materials

- beaker, e.g. 250 mL
- thermometer
- stopwatch or watch with a second hand
- 4 antacid tablets, e.g. Alka-Seltzer
- hot water (from hot tap)
- ice water
- sheet of graph paper

Note: Clear aspirin tablets can be used instead of Alka-Seltzer.

Risk assessment and planning

1 Write down your hypothesis about how you think temperature affects dissolving time. (Base your hypothesis on your previous experience of making hot and cold drinks or doing the washing up.)

2 Prepare a data table like the one below in which to record your results.

	Temperature (°C)	Time to dissolve (seconds)
Ice water		
Room temperature		
Warm water		
Hot water		

Write down all the variables that could affect the dissolving time. Which variables will you need to keep the same?

Make a list of any safety issues.

Method

1 Fill the beaker with water from the tap. Use the thermometer to measure the temperature of the water.

Record this temperature in your data table.

2 Drop in an antacid tablet. Do not stir. Time how long it takes for the tablet to dissolve; that is, how long before it disappears completely.

Record this time in your data table.

3 Repeat steps 1 and 2 for the other temperatures. Remember to control the variables you listed in the Risk assessment and planning.

Record your results.

4 Plot a graph with water temperature on the horizontal axis and dissolving time on the vertical axis. Draw a smooth curve through the four crosses. This line shows how the dissolving time depends on the temperature of the water.

5 Write a report of the investigation using the usual headings.

Discussion

1 Which is the independent variable, and which is the dependent variable in this experiment?

2 What does your graph tell you about the relationship between temperature and dissolving time?

3 Do your results support (agree with) your hypothesis from the Risk assessment and planning? If not, write a better hypothesis.

ISBN 978 1 4202 3735 1

1 Rebecca, Alistair and Ian compared the hardness of three different types of wood. They did this by measuring how far a dart went into the wood, as shown below. Was this a fair test? If not, explain how the test could be improved.

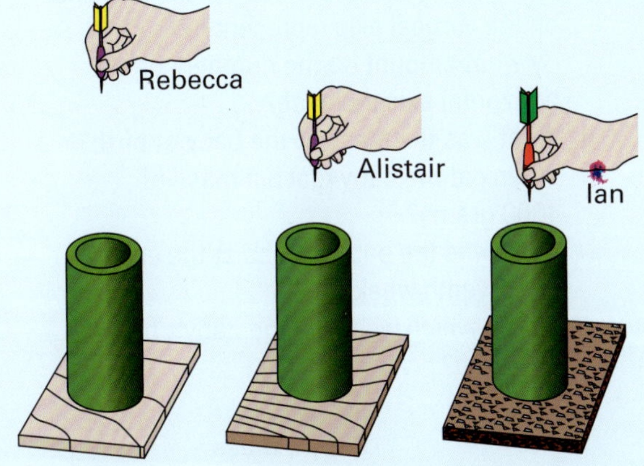

2 What are the variables that affect how long it takes you to get to school?

3 You have three different powders. You want to find out which one dissolves most rapidly in water.
- a Which variables will you need to control in your test?
- b Which variable will you purposely change?
- c What will you measure?

4 Which of the following are inferences and which are hypotheses?
- a This piece of iron must be a magnet.
- b All things fall towards the Earth because of gravity.
- c Plants grow more in summer than in winter.
- d I think the wet road caused this accident.

Justify your answers (explain why they are inferences or hypotheses).

5 Dan used a decibel meter to measure the noise given off by a car travelling at different speeds.
- a Design a data table for Dan's results.
- b Which measurement is the independent variable? Which is the dependent variable?

6 Paul's parents measured his height every year, starting when he was two. They recorded these measurements on a graph.
- a How old was Paul when he was 100 cm tall?
- b Predict how tall he will be when he is eight.
- c Can you predict how tall he will be when he is 20? Explain.

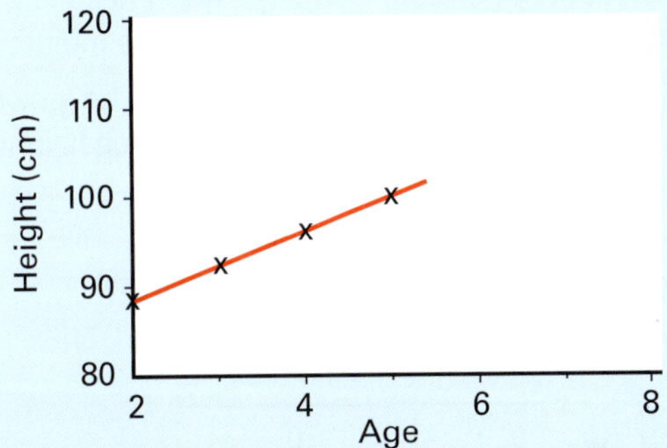

7 Rebecca and Megan want to test whose bike has better brakes. Design a fair test for them. Remember, when designing fair tests you:
- change something
- measure something
- keep everything else the same.

8 Ace planted 2 bean seeds in each of 4 pots of soil. Every 3 days he added water to the pots as shown below.

Pot 1	no water
Pot 2	10 mL of water
Pot 3	20 mL of water
Pot 4	40 mL of water

- a Write a hypothesis for Ace's experiment.
- b Why did he plant 2 bean seeds in each pot and not just one?

ISBN 978 1 4202 3735 1

CHALLENGE

1 The following questions refer to Investigation 1.1 Dissolving time on page 14.

a Suppose you wrote *Antacid tablets dissolve faster in hot water* for your hypothesis. What would you need to do to test this hypothesis?

b Use the graph you drew to predict how long a tablet would take to dissolve in water at 35 °C.

c What temperature would the water need to be for a tablet to dissolve in exactly one minute?

2 A group of students was investigating the growth of seedlings. They measured the average height of the seedlings every day.

a Draw a graph of their data.

b Is the graph a straight line or a curve?

Time (days)	Height (cm)
0	0
1	1.0
2	2.1
3	2.6
4	3.8
5	5.0
6	5.8

3 Mark and Dylan used a datalogger and temperature probe to find out how quickly the temperature of ice changed as it was heated. They obtained the data list below on their calculator screen.

a Draw a graph to display their results.

b Use the graph to find out approximately how long it took the melted ice to reach a temperature of 70 °C.

c What was the approximate temperature of the heated ice after three minutes?

Time (min)	Temperature (°C)
0	0
2	5
4	30
6	75
8	93
10	98

4 Use the graph below to answer the following questions.

a What is the graph about?

b Which is the independent variable?

c Which is the dependent variable?

d By what amount do the numbers on the vertical axis increase?

e How much mass does each small grid line on the vertical axis represent?

f By what amount do the numbers on the horizontal axis increase?

g What was the mass of the baby at birth?

h When did the baby reach a mass of 4000 grams?

i What was the baby's mass at the end of the seventh week?

j During which week did the baby's mass decrease?

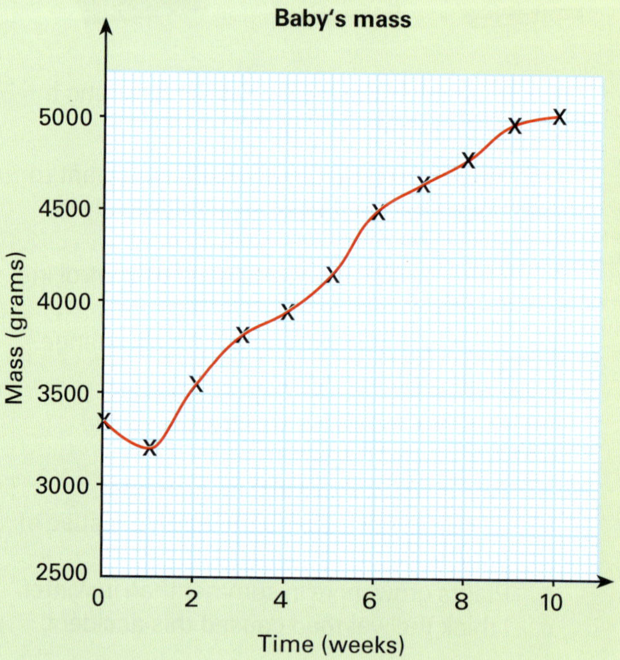

Inquiry

Use the ideas from Investigation 1.1 on page 14 to design an experiment to test the effect of stirring on dissolving time.

ISBN 978 1 4202 3735 1

1.3 Solving problems

Josh is playing a computer game in which he has to find the buried treasure. He is using his science skills to solve this problem.

Read carefully through the six steps on this page. Note that at Step 6 you should be prepared to change your hypothesis if necessary. You cannot ignore some data or change it to fit in with what you think should happen. Also, not all problems are easy to solve. And you may have to do many experiments.

In Investigation 1.2 you can try to solve a problem yourself.

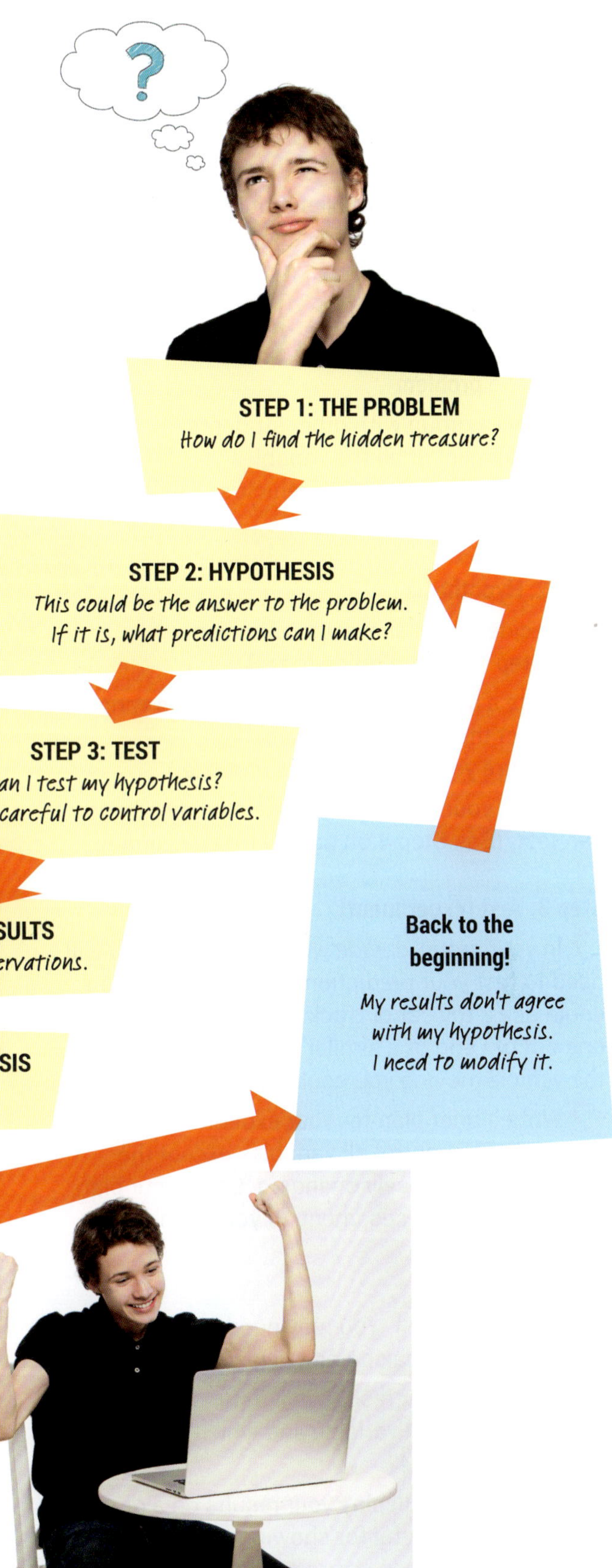

ISBN 978 1 4202 3735 1

INVESTIGATION 1.2 Stopping distance

Aim

To investigate the variables that affect the distance it takes a moving vehicle to stop (stopping distance).

Method

Step 1: The problem

Form a group with other students and make a list of all the variables you think may affect a moving vehicle's stopping distance.

Step 2: Hypothesis

Decide which one of the variables from Step 1 you are going to test.

Write a hypothesis that says how this variable will affect the stopping distance. (Make sure your hypothesis is testable.)

Using your hypothesis, write a prediction you can test. (See Step 4 on page 9 for an example.)

Step 3: Test (experiment)

In your group, decide what equipment you will need to test your prediction. For the vehicle you could use a toy car or truck. Or you could build one out of Lego or a similar building kit. To get the vehicle moving you could run it down a ramp.

Write a brief plan for your experiment. Remember to control all variables except the one you are purposely changing. Show your plan, including a risk assessment, to your teacher.

Step 4: Results

Do your experiment. You may need to do some trial runs before making any measurements.

Record all your results in a data table. Which is the independent variable and which is the dependent variable?

You may want to display your results on a graph. (This would be useful for showing to the rest of the class.)

Repeating the measurement

If you repeat a measurement of the stopping distance, you will probably get a slightly different value. This is because there are some variables you cannot control, e.g. whether the vehicle runs straight or not. For this reason it is a good idea to do each measurement three times and calculate the average.

The more measurements you make, the more reliable the average will be, but three measurements are usually enough.

Step 5: Check hypothesis

Do your results support your hypothesis? That is, was your prediction in Step 2 correct?

Write a conclusion, giving an answer to the question you investigated.

Which variables did other groups investigate? What did they find? How do their results compare with yours?

Step 6: Think again

How accurate do you think your results are? Can you think of ways to improve your experiment?

Write a report of your experiment using the usual headings.

ISBN 978 1 4202 3735 1

Science at work

It is fun to solve everyday problems by experimenting. Choose one or more of the problems below or think of your own problem. In designing your experiment, use the six steps in investigating you used in Investigation 1.2, starting with a hypothesis you can test.

PROBLEM A

What sorts of liquids flow through the funnel most easily? Liquids you could try are water, glycerine, cooking oil, sugary water ...

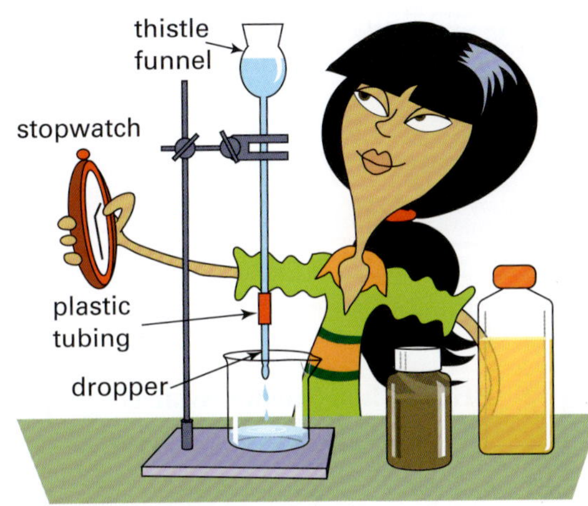

PROBLEM B

Does the shape of a boat's hull affect its speed?

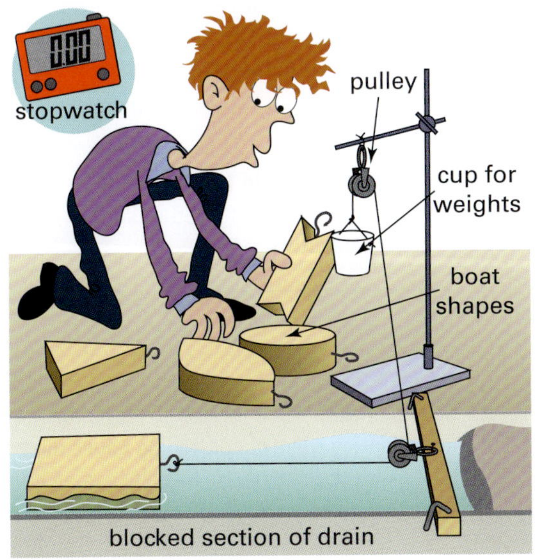

PROBLEM C

Which colour cloth is the coolest in summer? Which is the warmest in winter?

PROBLEM D

Florists say that a vase of flowers will last longer if the stems of the flowers are crushed and if you add a little sugar to the water. Do these variables really affect the life of the flowers?

ISBN 978 1 4202 3735 1

ACTIVITY

Doing a project

Any of the problems on the previous page would make a good student research project. Here are the steps you need to follow in doing a project.

1 Choose a topic

Pick something you are interested in. There are project ideas in some of the Challenges and Explore activities in this book, and many of the experiments can be extended into projects. Check the websites on this page to see what other students have done. Make sure your ideas are feasible. Are there experiments you can do on this topic? Can you get the equipment and materials you need? Can you finish it in the time available? Talk with other people about your ideas.

2 Plan your project

Write a brief outline of what you plan to do and discuss it with your teacher before you start.

3 Do it

Use the skills you have learnt in this chapter to carry out your project. Put your notes straight into a special project logbook so that they are not lost. It is important to record your failures as well as your successes. After each experiment ask yourself *What would happen if ...* ? then try it. Repeat your experiments to make sure you always get the same results, and be prepared to change your ideas in light of your results, as you may not always get the answers you expect.

4 Prepare a report

This may be a written summary, a poster display, a short talk using overhead transparencies or an interactive presentation. The websites below give information if you want to enter your project in a science contest.

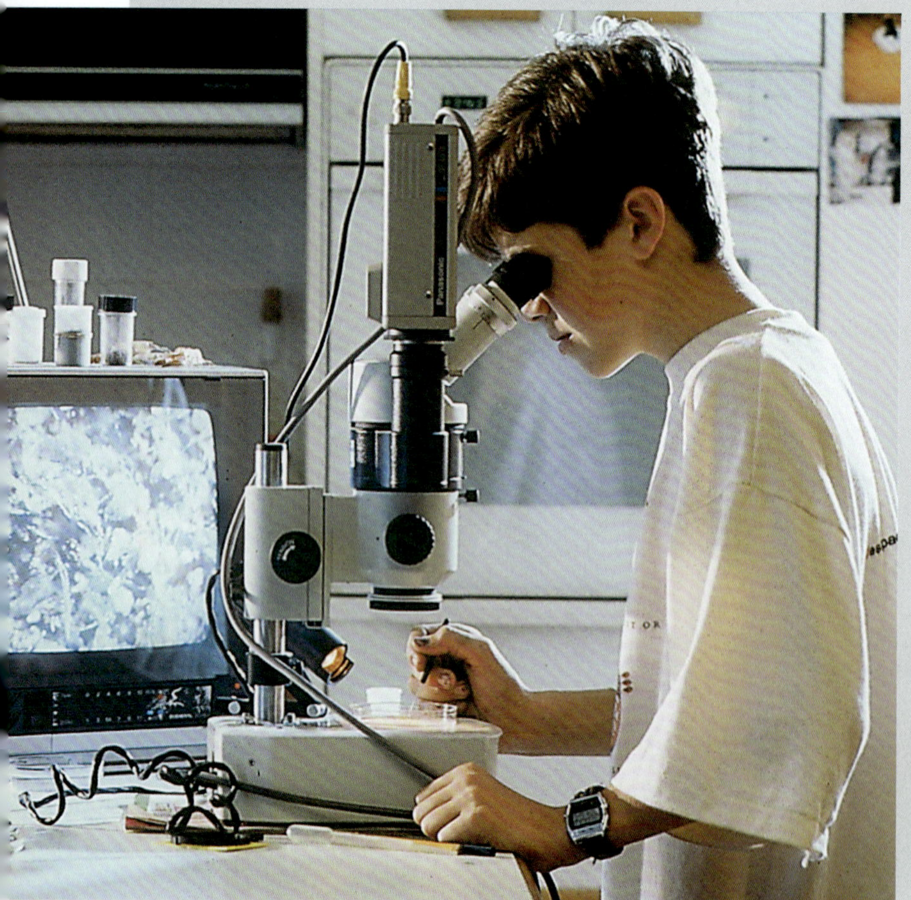

EXPLORE ONLINE

Check out these websites for ideas.

BHP Billiton Science Awards

The site includes what you can win, entry requirements and details of last year's winners.

Science Talent Search (Victoria)

CREST

CREST stands for Creativity in Science and Technology. The site contains many ideas for projects and videos of past projects.

CSIRO Double Helix

This site has links to other sites and ideas for projects.

Figure 1.5 Michael Morris won a BHP Billiton Science Award when he was in Year 8 for investigating ways of controlling house dust mites with tea tree oil. He has since gone on to win other awards for his science projects.

ISBN 978 1 4202 3735 1

SCIENCE AS A HUMAN ENDEAVOUR

Scientists are ordinary people who solve problems using the skills you have learnt in this chapter.

Over the years scientists have made many important discoveries that affect our daily lives. Five of these are described on the following pages. Select at least one of these and answer the questions about it.

Post-it self-stick notes

Some discoveries in science are made by accident, or serendipity.

Art Fry was in church one Sunday in 1974. He sang in the choir but he had a problem. The bits of paper he had put in his book to mark the hymns kept falling out. Suddenly he had an inspiration. Several years ago the 3M company where he worked had made a glue that was thrown out because it wasn't sticky enough. Perhaps it would be sticky enough to make sticky bookmarks for his choir book.

Fry went back to his laboratory at 3M and tried the old glue. It worked, but he spent a year and a half modifying and testing it. When he took it to the advertising department they weren't very keen on his idea for sticky note pads. However, they put them on the market, and soon people around the world were buying Post-it self-stick notes.

Questions

1 Use a dictionary to find the meaning of the word *serendipity*.

2 Did Art Fry work scientifically to make self-stick notes? Explain.

3 Suggest uses for Post-it self-stick notes.

4 For information on Art Fry search for 'Art Fry sticky notes invention'.

Dung beetles

Cattle were introduced to Australia over 200 years ago. We now have a problem of too much cattle dung. It covers grazing land and flies breed in it. George Bornemissza, who came to Australia from Hungary, started studying the problem in 1951. He found that Australia has dung beetles that can break down the dung of native animals such as kangaroos. However, very few of these beetles can break down cattle dung. He therefore suggested bringing dung beetles from other parts of the world to Australia.

The first of these beetles were released in 1967, and today dung beetles are well established in some areas. However, they have not spread far enough, and flies are still a problem throughout Australia. Scientists from CSIRO, Australia's largest scientific research organisation, are therefore still working on the problem.

Questions

1 Why is cattle dung such a problem?

2 Why was it necessary to introduce dung beetles to Australia when there were some here already?

3 How can the spread of dung beetles throughout Australia reduce the number of flies?

4 Suggest a plan to spread dung beetles more evenly across Australia.

5 What precautions must be taken when a foreign animal or plant is planned to be introduced to this country?

ISBN 978 1 4202 3735 1

Twin lambs

Dr Helen Newton Turner was experimenting with the breeding of sheep. In 1951 someone sent her some ewes that produced twins much more often than usual. She knew that twin lambs were rare, and wondered whether she could use the ewes to breed whole flocks of sheep that produce twins more often. She therefore set up a series of experiments with ewes that had produced twins and rams that had been twins. At the same time she did similar experiments with single-bearing ewes mated with single-born rams.

Her results showed that 'twinned' parents produced three times the number of sets of twin lambs as the 'single' parents. She then worked with a farmer near Cooma (NSW) and by 1972 his merino flock was producing 210 lambs each year for every 100 ewes!

The sheep industry has benefited enormously from Dr Turner's work.

Questions

1 Why do sheep farmers like twin lambs?

2 What is meant by 'twinned' parents and 'single' parents for sheep?

3 How did Dr Turner control the variables in her sheep breeding experiments?

Medicines from frogs

Dr Michael Tyler from the University of Adelaide has been studying frogs for over 50 years. The secretions produced by the skin of frogs contain many different chemicals. Some of these are toxic, but others have been found to be useful as medicines. For example, scientists have recently isolated a pain-killer 200 times more powerful than morphine.

Dr Tyler wanted to find a way of extracting the secretions from the frogs without harming them. One day he was having acupuncture for a headache. The acupuncturist inserted needles in his skin and passed a small electric current through him, causing his skin muscles to twitch slightly. This caused him to wonder whether frogs would release their secretions when their skin muscles were twitched using a small electric current.

Back in his laboratory Dr Tyler found that his idea worked, without harming the frog, and without using needles. From a single 'milking' he could obtain up to 100 milligrams of secretions. He found that these secretions contain as many as 70 different chemicals. The secretions kill many different bacteria, fungi and viruses, and his recent work has been to find out which secretions kill which organisms.

Questions

1 Dr Tyler discovered a new laboratory technique. What is it?

2 He repeated his tests several times. Why do you think he did this?

3 Suggest how he could find out which of the 70 chemicals in the secretions kills a particular virus.

4 Suppose he identifies the virus-killing chemical. What do you think he should do next?

Figure 1.6 Dr Tyler with one of his frogs

ISBN 978 1 4202 3735 1

Rabbit plague

Thomas Austin liked to go shooting at the weekends. So in 1859 he imported 24 rabbits from England to his property near Geelong in Victoria. A female rabbit can produce 40–50 young in one year, and with few natural enemies there was soon a plague of rabbits. In one year Tom Austin shot over 14 000, and within 20 years or so the rabbits had spread to almost all parts of Australia. Fences didn't seem to keep them out. They ate every blade of grass and stripped the bushes they could reach, turning once green areas into deserts. Many of Australia's wallabies and native rodents became extinct or endangered.

In 1919 a Brazilian scientist said he knew about a virus called myxoma which infected rabbits and gave them a disease called myxomatosis (MIX-o-mat-toe-sis). However the Australian government ignored his advice because people were making lots of money selling rabbit meat and using the fur to make hats. By the 1940s the rabbit plague was out of control and Jean MacNamara, an Australian expert on viruses, eventually convinced CSIRO to try the myxoma virus. At first CSIRO scientists couldn't get the virus to spread, but they found that it spread more quickly in wet weather, because the disease is carried from one rabbit to another by mosquitoes, which breed when it is wet.

Myxomatosis killed up to 80% of the rabbits in most of Australia and as a result beef and wool production increased. However, as the years passed the virus became less effective. So in 1991 the CSIRO found another weapon against the rabbits—the calicivirus (cal-LEE-sea-virus), which doesn't need mosquitoes to spread it. It was released in 1996 and one year later about 100 million of Australia's 300 million rabbits had died. For the first time in living memory there was lush vegetation across the Nullarbor Plain and in the Simpson Desert.

Questions

1 Why was the rabbit plague such a disaster for sheep and cattle farmers, and for native plants and animals?

2 The rabbit plague resulted in severe soil erosion, with soil washed away during heavy rain. Why did this happen?

3 Suggest why the Australian government was reluctant to introduce the myxoma virus into Australia.

Figure 1.7 The aerial view of a large rabbit warren shows the destruction rabbits can cause to the land.

Figure 1.8 This 1938 photo from South Australia shows the rabbit population explosion.

ISBN 978 1 4202 3735 1

1 When you are doing an experiment, what is the usual order for the following?
 a check hypothesis
 b hypothesis
 c predict
 d results
 e test
 f think again

2 Which of the following are true, and which are false?
 a An experiment is a test containing a series of steps used to solve problems.
 b Hypotheses are always correct.
 c Scientists don't know the answers to some questions.
 d In an experiment all variables must be kept the same.
 e You should ignore data that does not agree with your hypothesis.
 f A good hypothesis allows you to make predictions.

3 a When you write a report of an experiment, what should the section headed 'Aim' tell the reader?
 b What should the conclusion of a report tell you?
 c Under which heading would you describe how you carried out the experiment?

4 Sometimes you have to modify a hypothesis. When would you need to do this?

5 While cooking on the barbecue Tammy was annoyed by all the insects that were attracted to the light. Then she remembered reading that insects are less attracted to yellow light. Use the steps on page 17 to design an experiment to test Tammy's idea. Discuss your design with others.

6 An oil company claims you get more kilometres per litre from their petrol. They say this is because of an additive called Z. How could you test this claim?

7 A group of students wrapped a drink can (can A) in four layers of aluminium foil each 0.25 mm thick. They wrapped an identical can (can B) in a single layer of foil 1 mm thick. They filled both cans with hot water and recorded the temperature of each can every 2 minutes for 10 minutes. (The data table is shown below.)
 a What hypothesis were the students testing?
 b Look at the students' data and decide whether it supports their hypothesis.
 c Write a conclusion for the experiment.

Times (minutes)	Can A (°C)	Can B (°C)
0	90	90
2	87	87
4	85	84
6	84	82
8	84	80
10	83	78

ISBN 978 1 4202 3735 1

EXPLORE ONLINE

Refer to the websites in the Explore online links to complete the following activities.

1 What does CSIRO stand for? If there is a branch of CSIRO in your city or town, see if you can find out what scientists do there. Your teacher may be able to arrange a visit or a scientist may visit your school to talk with you. Visit the **CSIRO** website.

2 Join a science club such as **CSIRO's Double Helix Science Club**. Visit their website. Perhaps your teacher could help you set up a club at school.

3 During the next few weeks check newspapers and magazines and collect articles about new discoveries in science and technology. You can also check out the latest science news at these websites:

Science Daily

New Scientist

Cosmos magazine

CNN Technology

4 Use a library to find out about the life and work of one particular scientist. Once you have collected your information, prepare a three minute talk to the class about your scientist. You may like to use the plan below.

Name of scientist:

Dates born (and died):

Country of birth:

Details of work:

Any other interesting information:

The **Encyclopedia of Australian Science** site has information on more than 3000 Australian scientists.

Figure 1.9 Dr Mark Talbot, a CSIRO scientist, won the *New Scientist* Eureka Prize for Science Photography in 2014 for his amazing images of seeds using a scanning electron microscope.

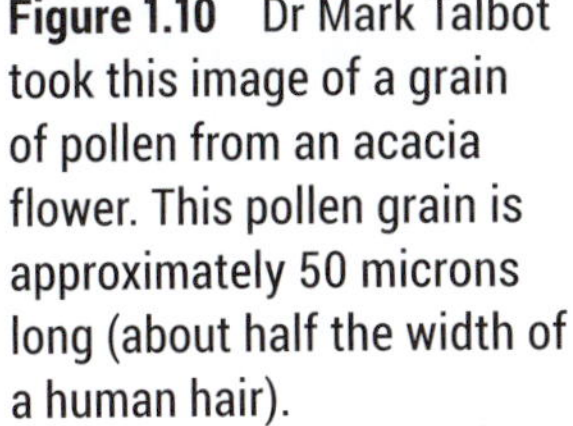

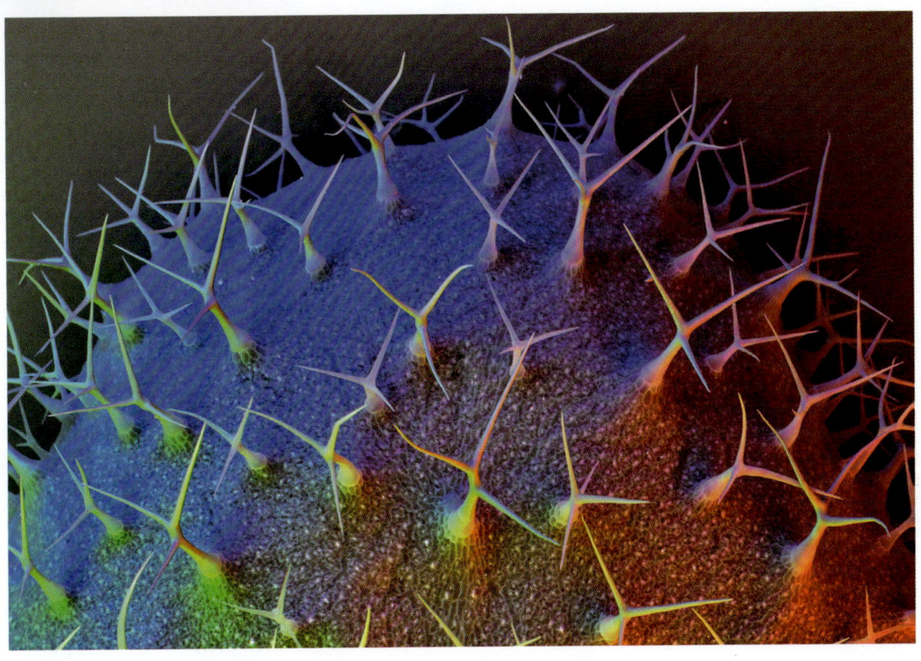

Figure 1.10 Dr Mark Talbot took this image of a grain of pollen from an acacia flower. This pollen grain is approximately 50 microns long (about half the width of a human hair).

MAIN IDEAS

Copy and complete these statements to make a summary of this chapter. The missing words are on the right.

1 _____ is a way of finding answers to questions by doing experiments.

2 Solving _____ by doing experiments involves using skills such as observing, inferring, predicting and _____.

3 A _____ is something which can change the results of an experiment.

4 In an experiment you purposely change one variable and keep all the rest the _____. This process is called _____ variables.

5 A _____ is a generalisation that explains a set of observations or gives a possible answer to a question.

6 Hypotheses can be tested by doing _____. If necessary they can be modified to explain further observations.

7 A _____ is a way of displaying data. It can also be used to show the _____ between two variables.

same
generalising
variable
hypothesis
relationship
problems
experiments
science
graph
controlling

CH•1 REVIEW

1 What name is given to a generalisation that a scientist can test?
A experiment
B hypothesis
C inference
D observation

2 Lim is checking the burning of a candle. He finds that after 2 hours, one-quarter of the candle has burnt. Predict how long it will take the whole candle to burn.
A 1 hour
B 2 hours
C 4 hours
D 8 hours

3 Sally and Bonita both bought the same kind of rubber ball. Sally said: *My ball will bounce better than yours*. Bonita answered: *I'd like to see you prove that*. What should they do to find out which ball bounces better?
A Drop both balls from the same height and see which ball bounces higher.
B Hit the balls against a wall and see how far each bounces off the wall.
C Throw the balls against the floor and see how high they bounce.
D See which ball can be squeezed the most.

4 Tamika tested a number of substances to see whether or not they conduct electricity (allow an electric current to pass through them). She also noted whether the substances were metals or non-metals. Her results are shown below.

Substance	Metal or non-metal?	Does it conduct electricity?
sulfur	non-metal	✗
zinc	metal	✓
copper	metal	✓
iodine	non-metal	✗
lead	metal	✓
phosphorus	non-metal	✗
steel	metal	✓

ISBN 978 1 4202 3735 1

a Use Tamika's results to write down two specific observations about steel.

b Write a hypothesis about metals and non-metals and electricity.

Tamika tested two more substances:

Substance	Metal or non-metal?	Does it conduct electricity?
carbon	non-metal	✓
tin	metal	✓

c Do these results support your hypothesis? If not, modify it.

5 The amount of salt that will dissolve in 100 mL of water is called its *solubility*. This solubility was measured at different temperatures and the results graphed.

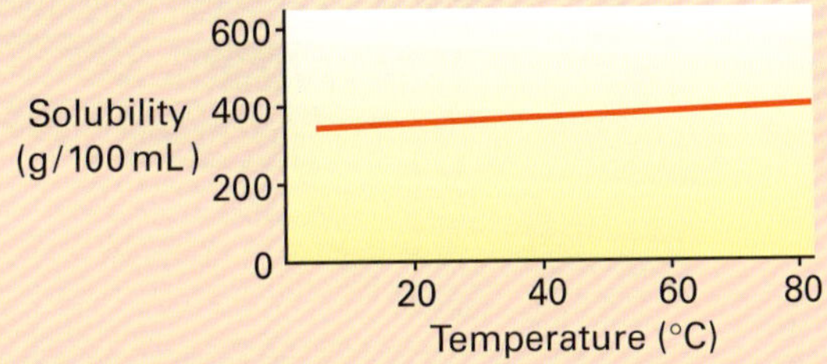

Which of the following statements best describes this graph?

A As the temperature changes the solubility stays the same.

B As the temperature increases the solubility decreases.

C As the temperature increases the solubility increases slowly.

D As the temperature increases so does the solubility, slowly at first, then more quickly.

6 The graph above right is from a TV news weather report. It shows the amount of UVB radiation received on a particular day.

a At what time did the UVB radiation reach its peak?

b During what times of the day was the UVB reading in the very high range?

c In which range was the UVB reading at 10.30 am?

d How could you explain the dip in the graph around 12 noon?

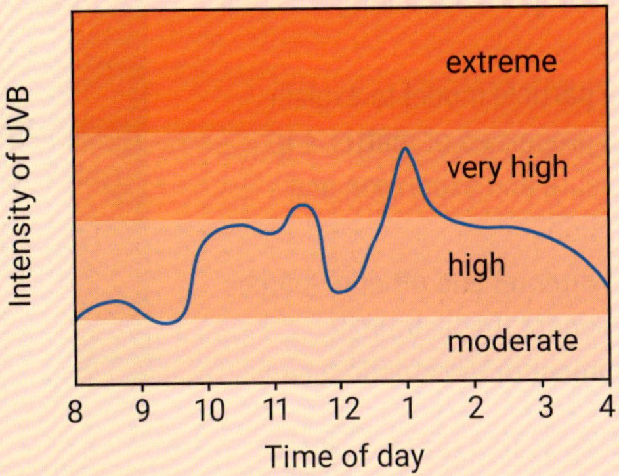

7 You see this advertisement on TV.

You decide to do an experiment to see if Sudso is in fact better than other washing powders.

a Write a brief plan for your experiment.

b Which variables will you need to control?

c Which variable will you purposely change?

d Which variable will you measure?

Check your answers on page 293.

ISBN 978 1 4202 3735 1

IN THIS CHAPTER

Science Understanding

- model the arrangement and motion of particles in solids, liquids and gases
- give examples of how energy is either taken in or given out during a change of state
- use the particle model to explain various properties of matter, e.g. air pressure and expansion and contraction
- use the internet and other resources to investigate the historical development of the atomic theory

Science Inquiry Skills

- measure, record and graph the temperature as ice is melted and water is boiled
- write inferences to explain various properties of matter in terms of the particle theory

CH•2 Solids, liquids and gases

ISBN 978 1 4202 3735 1

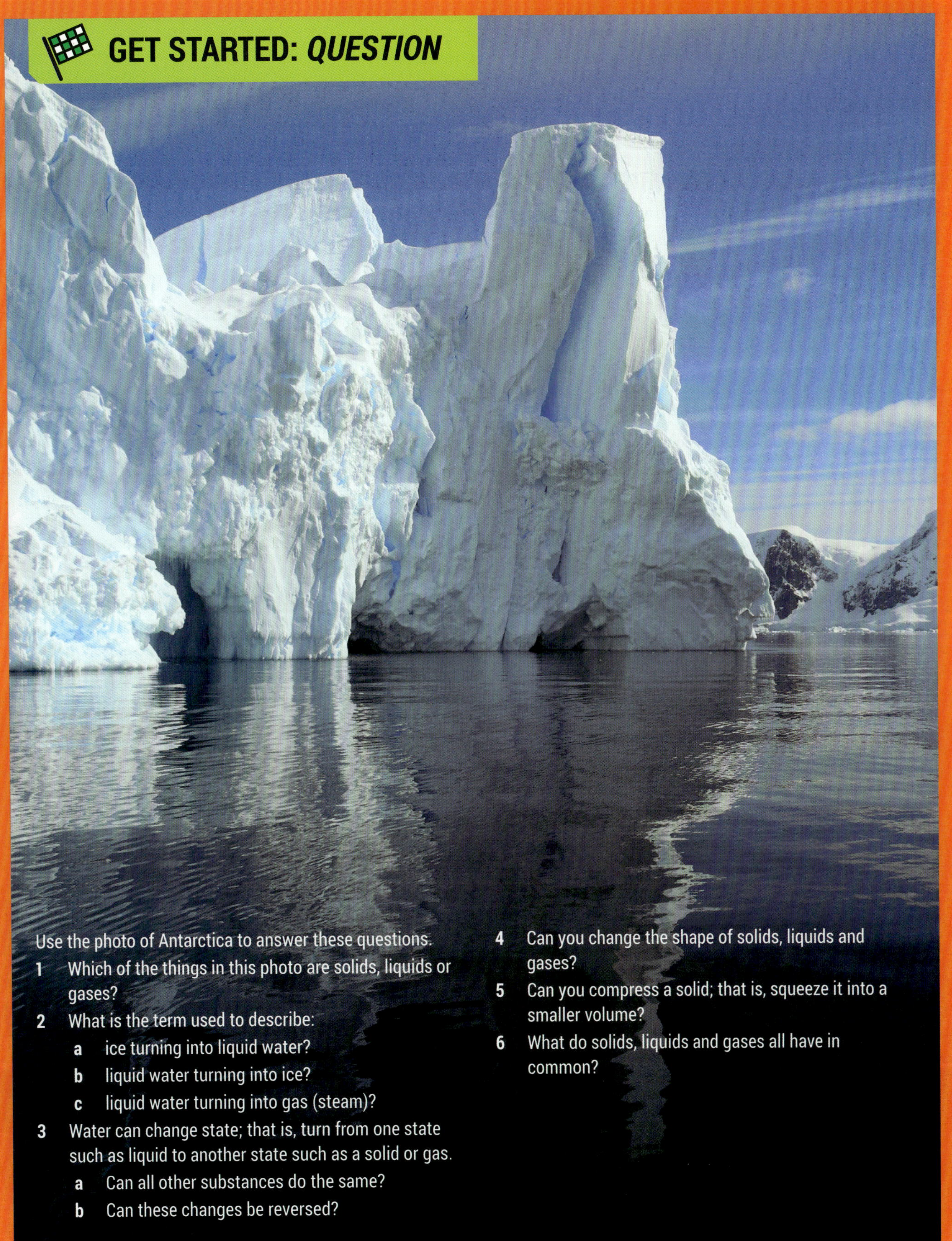

GET STARTED: *QUESTION*

Use the photo of Antarctica to answer these questions.

1 Which of the things in this photo are solids, liquids or gases?

2 What is the term used to describe:
   a ice turning into liquid water?
   b liquid water turning into ice?
   c liquid water turning into gas (steam)?

3 Water can change state; that is, turn from one state such as liquid to another state such as a solid or gas.
   a Can all other substances do the same?
   b Can these changes be reversed?

4 Can you change the shape of solids, liquids and gases?

5 Can you compress a solid; that is, squeeze it into a smaller volume?

6 What do solids, liquids and gases all have in common?

ISBN 978 1 4202 3735 1

2.1 Properties of matter

What is matter?

Everything around you is made up of **matter**—the desk, your shirt, the water in a swimming pool, the hair on your head, even the air you breathe. Most matter can be classified into one of three main groups: solids, liquids and gases. These are usually called the three **states of matter**.

Solids, liquids and gases have two important properties—they all have *mass* and they all take up space. To find the mass, you use a balance. To find the amount of space occupied by something, you measure its *volume*. So all matter has mass and occupies space.

Solids

Solids include such things as steel girders, this book, and most of the objects you can see. They all have mass and occupy space. The shape of most solids cannot easily be changed, and nor can their volume. Powders are also solids but their shape can be changed.

Liquids

Water, juice and oil are all examples of matter in liquid form, and they all have mass. The volume of a quantity of liquid does not change, but its shape can. For example, pour some juice from a carton into a glass. The volume of the juice doesn't change, but its shape does. And if the juice is spilt, it has another shape.

Gases

The air around us is a gas. In fact, it is a mixture of gases, mainly nitrogen and oxygen. Other common gases are helium and carbon dioxide. All these gases have mass and occupy space. Gases do not have a fixed shape or volume. A gas fills its container, no matter what the shape or size of the container. For example, helium gas fills a metal gas cylinder. The gas can be let out through the tap to fill balloons of various shapes and sizes. If the balloon bursts, the gas will escape and spread out into the air. Gases can also be compressed (squeezed into a smaller volume, like the helium in the cylinder). You cannot do this with liquids and solids.

Figure 2.1 The volume of a quantity of liquid does not change, but its shape may. A liquid will take the shape of its container.

Figure 2.2 Gases do not have a fixed shape or volume.

ISBN 978 1 4202 3735 1

ACTIVITY

1 Crumple a tissue, and fit it tightly into the bottom of a glass. Push the glass, mouth down, into a large container of water until most of the glass is under water.

What do you observe?

Pull the glass out of the water and check whether the tissue is wet.

Write an inference to explain your observations.

2 Use a balloon and an electronic balance to test whether air has mass. Design your test carefully!

3 Place your finger over the end of a syringe containing air. Try to push the plunger in.

Can air be compressed?

Draw some water into the syringe.

Can water be compressed?

4 To summarise what you know about solids, liquids and gases, copy the table below. Complete it by putting a ✔ or a ✘ in each box.

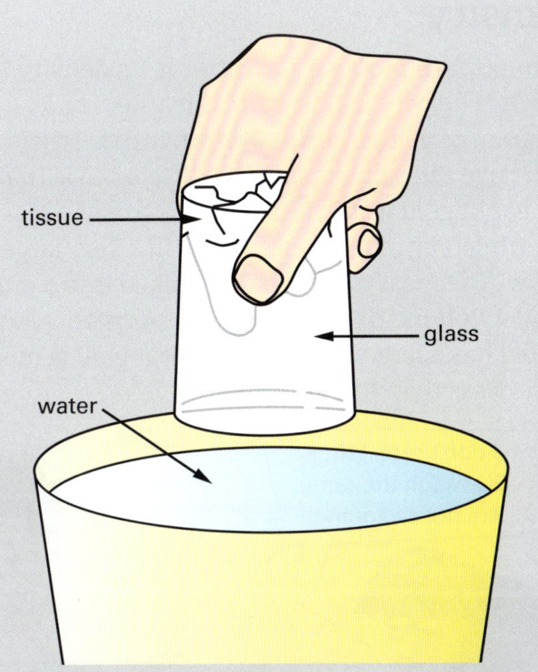

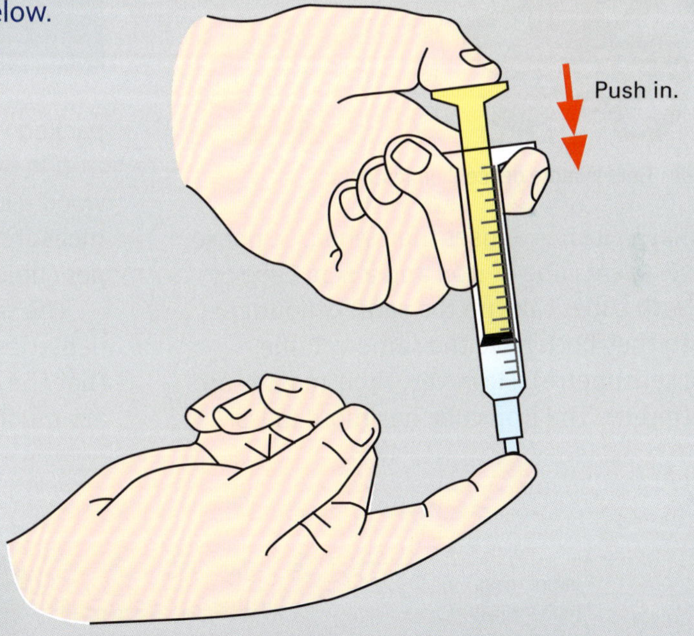

State of matter	Properties of matter				
	Have mass	Occupy space	Fixed shape	Fixed volume	Can be compressed
solids	✓				
liquids	✓				
gases	✓				

ISBN 978 1 4202 3735 1

Density

An important property of matter is its **density**. Which is heavier, a kilogram of feathers or a kilogram of gold? The answer is neither—they both have the same mass. The difference is that a 1 kg bar of gold would be about the size of a Mars bar, while 1 kg of feathers would fill a very large pillow. The mass of the gold is packed into a much smaller volume than the feathers. A small volume of gold has a large mass. We say that gold is much more *dense* than feathers.

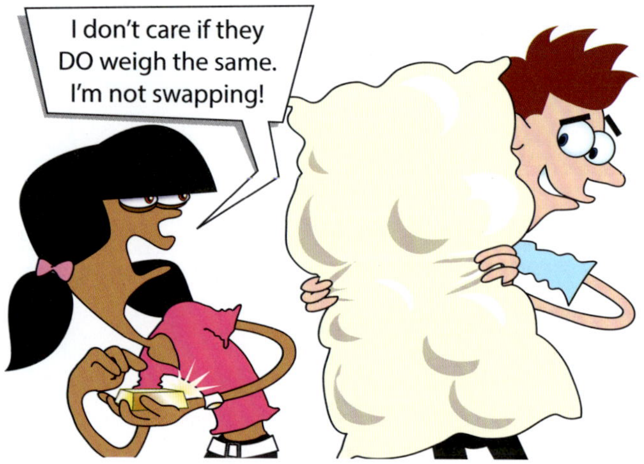

Figure 2.3 Gold is more dense than feathers.

Table of densities (g/cm^3)		
helium gas	0.00018	FLOAT IN WATER
air	0.0013	
carbon dioxide gas	0.002	
polystyrene foam	0.1	
cork	0.2	
pine wood	0.4	
petrol	0.7	
polythene plastic	0.9	
ice	0.9	
water	1.0	
sea water	1.03	SINK IN WATER
aluminium	2.7	
granite	2.7	
iron	7.8	
nickel	8.9	
lead	11.3	
gold	19.3	
osmium	22.5	

Similarly, iron is denser than wood. Suppose you have a 1 cm cube of iron and a 1 cm cube of wood. Both cubes take up the same amount of space, so they both have the same volume (1 cubic centimetre). However, their masses are very different. The iron cube has more mass packed into 1 cubic centimetre. The density of iron is therefore greater than the density of wood. Density is how much mass is packed into a measured volume. It is usually measured in grams per cubic centimetre (g/cm^3).

The table at the top of the page shows the densities of some common substances. Notice that the density of water is 1 g/cm^3, and that gases are much less dense than solids and liquids.

Figure 2.4 Two objects of the same volume with different densities. The more dense object will have the greater mass.

ISBN 978 1 4202 3735 1

Measuring density

To find the density of something, you must first measure its mass and volume. You then divide the mass by the volume to find the density.

$$\text{density (g/cm}^3\text{)} = \frac{\text{mass (g)}}{\text{volume (cm}^3\text{)}} \quad \text{or} \quad d = \frac{m}{V}$$

Example 1: Find the density of a block of aluminium with a mass of 13.5 grams and a volume of 5 cm^3.

Solution: Since we know the mass (*m*) and volume (*V*) we can use the formula to calculate the density (*d*).

$$d = \frac{m}{V} = \frac{13.5\text{ g}}{5\text{ cm}^3} = 2.7\text{ g/cm}^3$$

Example 2: A block of wood is 5 cm × 3 cm × 1 cm in size. Calculate the density if the mass is 9 grams.

Solution: First we need to calculate the volume using the dimensions.

$$V = \text{length} \times \text{width} \times \text{height} = 5 \times 3 \times 1 = 15\text{ cm}^3$$

Now we can calculate the density.

$$d = \frac{m}{V} = \frac{9}{15} = 0.6\text{ g/cm}^3$$

Measuring the volume of a regular solid such as a cube is easy, but how would you measure the volume of an irregularly shaped object such as your body? The secret is to drop the object into water, and measure the volume of water it displaces (pushes out). This method was discovered by Archimedes in Greece in about 250 BCE. In Investigation 2.1 you can use this method to find the density of a small object.

Figure 2.5 One way to measure your volume

Floating and sinking

Anything will float in water if its density is equal to or less than the density of water, that is 1 g/cm^3. For example, a piece of pine wood (density 0.4 g/cm^3) floats in water, but a piece of granite (density 2.7 g/cm^3) sinks. Fruit and vegetables sometimes float and sometimes sink. For those that float, the lower their density the more they stick out above the water. You can try this at home with a bowl of water.

Humans, like most animals, float in water, but only just. This is because we are mostly water. However, we have a layer of fat under our skin, and this has a lower density than water. There are also air spaces, such as lungs, inside our bodies. Sharks are unusual in that they are denser than water. If they don't keep swimming they sink to the bottom.

Figure 2.6 Anything will float in water if its density is equal to or less than the density of water.

ISBN 978 1 4202 3735 1

Measuring density

Aim

To measure the density of two different objects.

Materials

- measuring cylinder, 100 mL
- balance
- piece of wire
- 2 small objects—one that floats (e.g. wooden cube) and one that sinks (e.g. marble)

Risk assessment and planning

Read the six steps carefully and draw up a data table like the one below.

Make a list of any safety issues.

Method

1 Using the balance, find the mass of each object. Check with your teacher if you have forgotten how to do this.

Record the masses in the data table below.

2 About half fill the measuring cylinder with water. It is best if you fill it to a set mark, say 60 mL. Make sure the *bottom* of the meniscus (the curved water surface) is exactly on the mark.

Record this initial volume (V_1) in the data table.

3 Holding the cylinder at an angle, carefully slide in the first object. If it floats, you will have to hold it under the water with a piece of wire, as shown.

Record the water level in the cylinder with the object completely under water (V_2).

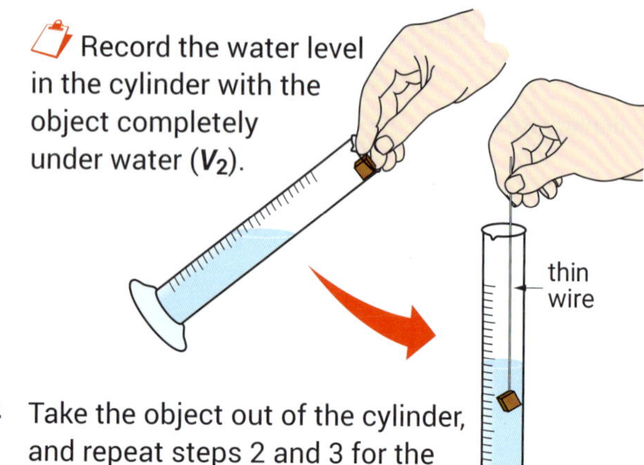

4 Take the object out of the cylinder, and repeat steps 2 and 3 for the other object.

Record the water level for the second object.

5 Calculate the volume of each object by subtracting the initial volume of water (V_1) from the final volume (V_2).

Record your results in the data table. (Note: 1 millilitre = 1 cubic centimetre.)

$$\text{volume of object} = V_2 - V_1$$

6 Calculate the density of each object using the formula:

$$\text{density} = \frac{\text{mass of object (g)}}{\text{volume of object (cm}^3\text{)}}$$

Give your answer to the nearest 0.1 gram per cubic centimetre.

Discussion

1 Compare your results with those found by other students. If they are different, suggest possible reasons.

2 Which object is more dense?

3 Suggest another way of finding the volumes of the objects. Try it, and check your results.

Object	Mass (g)	Initial volume of water, V_1 (mL)	Final volume of water, V_2 (mL)	Volume of object $V_2 - V_1$ (cm^3)	Density (g/cm^3)

ISBN 978 1 4202 3735 1

SCIENCE AS A HUMAN ENDEAVOUR

Figure 2.7 Gold is used in its natural state in jewellery

Using materials

All the materials around us are taken from or made from the Earth's natural resources. For example, we use cotton, wood and rubber from plants and wool, leather and silk from animals. We breathe the air and extract various gases from it, such as oxygen, nitrogen and argon. We use the rocks of the Earth and extract metals such as iron, copper and gold, and other useful materials such as coal, oil and limestone. We eat seafood from the oceans and extract salt from sea water.

Some of these materials we use in their *natural* state. For example, a gold nugget can be made into jewellery and wool can be woven into clothing. Often we process these materials to improve or alter their properties. For example, we may treat the wool to make it shrink-resistant, and we grind up corn to make flour, which we use to make bread. These are *processed* materials.

Over the years, however, we have made many totally new materials. For example, 2000 years ago the Chinese discovered how to make paper from wood. In recent times, we have made an incredible range of materials such as concrete, glass, plastics, paints and pesticides. These materials do not occur naturally and are said to be *synthetic*.

Synthetic materials

The properties of a material determine what it can be used for. Wool is used for winter clothes because it keeps your body heat in. Aircraft are made of aluminium metal because it is light. Copper is used to make electric wires because it is a good conductor of electricity, and because it can be shaped to form wire. Drills are sometimes diamond-tipped, because diamond is much harder than most other substances.

Synthetic materials are continually being developed with special properties to do particular jobs. Here are four examples:

1 Since 1996 Australia's banknotes have been made from polypropylene plastic. These last longer than paper notes, stay cleaner and are very difficult to counterfeit. They can also be recycled to make compost bins, plumbing fittings and other useful household and industrial products.

ISBN 978 1 4202 3735 1

2 In 1999 CSIRO developed a new sunscreen called Sunsorb. It is similar to zinc cream, but because the powder it is made from is so fine, it is virtually invisible.

3 If you break open a disposable nappy, you will find a white powder called WaterSorb. It forms a gel when water is added to it. It can soak up a large volume of urine, keeping the baby's bottom dry. It is also used to prevent pot plants drying out.

4 Polystyrene packing beads are being replaced by beads made from wheat or corn starch. This makes sense because wheat and corn are renewable, unlike polystyrene, which is made from oil and is non-renewable. These beads cause less damage to the environment since they form a suspension in water and are biodegradable.

Figure 2.8 Corn starch packing beads are biodegradable.

ACTIVITY

For this activity you will need some peanut-shaped starch packing beads.

How many starch packing beads do you think will 'disappear' in two teaspoons of water?

To test this, add one starch bead at a time, stirring well to form a suspension. Observe how the beads change and how the water changes.

Could you get the beads back again? How?

EXPLORE ONLINE

Check out some of the websites below.

The secret of the disposable nappy

This site has an experiment to test the superabsorbent powder in a disposable nappy.

Reserve Bank of Australia

This site has information on the famous Australians on our banknotes, how the notes are made and recycled, and how to detect counterfeit notes.

Try searching under the names of some of the newer synthetic materials, e.g. Kevlar, Mylar, Nomex, silica gel, Teflon, Tyvek. Keep notes on the properties and uses of each material you research.

Figure 2.9 Silica gel is a porous form of silicon dioxide (common sand, which can absorb water.

ISBN 978 1 4202 3735 1

1 In which state would a substance be if it had:
 a no fixed volume?
 b a fixed volume and shape?
 c a fixed volume but took the shape of its container?

2 Each of the cartoons below illustrates at least one property of matter. Which shows that a:
 a solid has a fixed shape?
 b liquid can be made to have any shape?
 c gas can be compressed?
 d gas does not have a fixed volume?

3 a How many kilograms are there in 2000 g, 100 000 g, 1530 g?
 b How many grams are there in 2 kg, ½ kg, 6.7 kg?

4 Look at the data table below.

	Mass (g)	Volume (cm^3)
object A	39	6
object B	54	20
object C	6	5

 a Which object has the greatest mass?
 b Which object has the greatest density?

5 Look at the diagrams below. Suppose you keep your finger over the end of the syringe, starting in position A. You push in the plunger to B, then pull it back to C.
In which position is the air in the syringe most dense? Explain your choice.

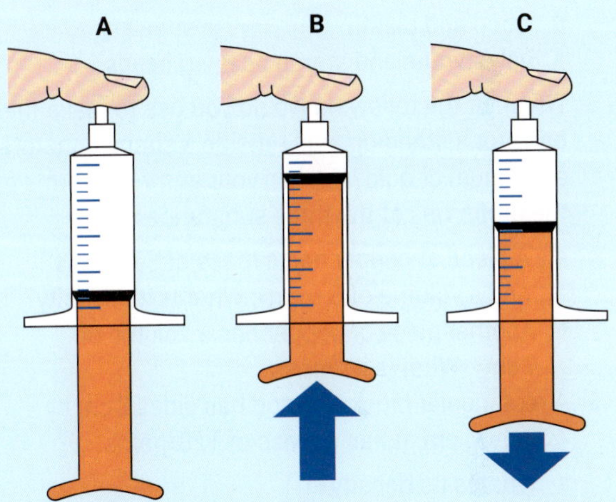

6 It is easier to float in sea water than in fresh water. Use your knowledge of density and the table of densities on page 32 to explain this.

7 A balloon filled with helium rises when you let it go. A balloon filled with carbon dioxide sinks. Explain the difference.

8 In each of these pairs, which is the object, and which is the material it is made from? Describe the properties of each substance that make the object useful. Record your answers in a table with three columns.
 a window / glass
 b styrofoam / coffee cup
 c plastic / ruler
 d aircraft / aluminium
 e banknote / polypropylene plastic

ISBN 978 1 4202 3735 1

9 Classify the following materials as natural, processed or synthetic:

concrete	milk	petrol
flour	natural gas	soft drink
marble	nylon	superphosphate
marijuana	oxygen	uranium

10 Make a table listing the properties of the four synthetic materials described on pages 35–6.

11 What is the difference between a renewable material and a non-renewable one? Give examples.

CHALLENGE

1 Many people incorrectly say that lead is heavier than steel. What should they really say?

2 Which properties allow you to distinguish between the substances in each of the following pairs?
- a steel and aluminium
- b lemonade and water
- c salt and sugar
- d wood and plastic
- e polystyrene and starch packing beads

3 Which of the following would you use to make the base for a stand-up sign outside a shop—concrete, aluminium or gold? Explain your answer in terms of the properties of the three substances.

4 a A piece of copper has a mass of 50 grams and a volume of 5.6 cm^3. What is its density?
b Another piece of copper has a volume of 7 cm^3. What is its mass?

5 A rectangular block of wood has sides 8 cm by 4 cm by 5 cm. It has a mass of 120 grams.
- a What is its density?
- b Would this block of wood float in water?

6 What is the mass of air in a room measuring 10 m × 5 m × 3 m if the density of air is 1.3 kg/m^3?

7 Suggest some uses for a plastic that dissolves in water.

8 The balloons in the photo are made of a material called Mylar. They are filled with helium gas and stay inflated for months. Suggest which properties of Mylar make it suitable for use in these special balloons.

EXPLORE

1 One-third fill a glass vial with glycerine. Carefully pour an equal volume of coloured water down the inside of the vial so that it flows gently onto the glycerine, as shown. Drop a small piece of perspex into the vial. Observe what happens, and try to explain it in terms of density.

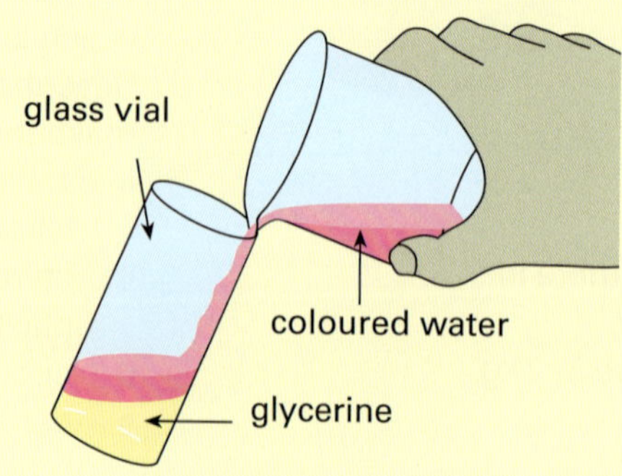

2 Does a fresh hen's egg sink or float in water? Try it. Now add salt to the water, while stirring carefully, and observe what happens. *Explain* your observation.

A rotten egg floats in fresh water. Suggest why.

ISBN 978 1 4202 3735 1

2.2 Solid–liquid–gas

Changing states

The three different states of matter can be changed from one to another by adding or removing heat. These changes are called **changes of state**.

If you heat a solid, it will form a liquid. For example, ice melts to produce liquid water. Metals such as iron and gold also melt if you heat them enough.

Heating also causes **evaporation** of liquids to produce gases. For example, when water is heated, it evaporates to form water vapour, which is a gas. The hotter the water gets, the more quickly it evaporates. When bubbles of water vapour appear in the water it is said to be *boiling*. The water vapour forms more quickly, and is now called steam. This occurs at 100 °C, the boiling point of water. Water can evaporate at any temperature, but boiling occurs only at the boiling point.

Cooling causes gases to condense and form liquids. For example, steam is invisible, but when it meets cooler air, it forms a cloud. This is because the steam condenses to form tiny droplets of water. A similar thing happens in the bathroom when you have a hot shower. Some of the hot water evaporates and changes into water vapour. Because the air in the bathroom is cooler, the water vapour condenses to form tiny drops of water that float in the air and 'fog up' the mirror. Similarly, as water from the Earth's surface evaporates, it forms water vapour. As this water vapour rises, it becomes cooler and may condense to form clouds and perhaps rain.

Cooling also causes water to freeze or *solidify*. This occurs naturally when snow and hail form. We use the same process to make ice blocks and ice-cream. Molten metal can be poured into moulds to solidify into various shapes (see Figure 2.10).

Some solids do not change to a liquid when they are heated. Instead they turn straight into a gas in a process called *sublimation*. For example, 'dry ice' is solid carbon dioxide. When it sits on the bench it soon warms up and changes directly into gaseous carbon dioxide, which is invisible.

Energy in and out

Another way to look at changes of state is to think of the three states of matter as rungs on an energy ladder. To change state by climbing up the ladder, energy must be added to the matter—it must be heated. To change state by going down the ladder, energy must be taken from the matter—it must be cooled.

Figure 2.10 Solid gold melts at about 1000 °C. The liquid gold can then be poured into moulds.

Figure 2.11 Dry ice sublimes, turning from a solid directly to a gas.

ISBN 978 1 4202 3735 1

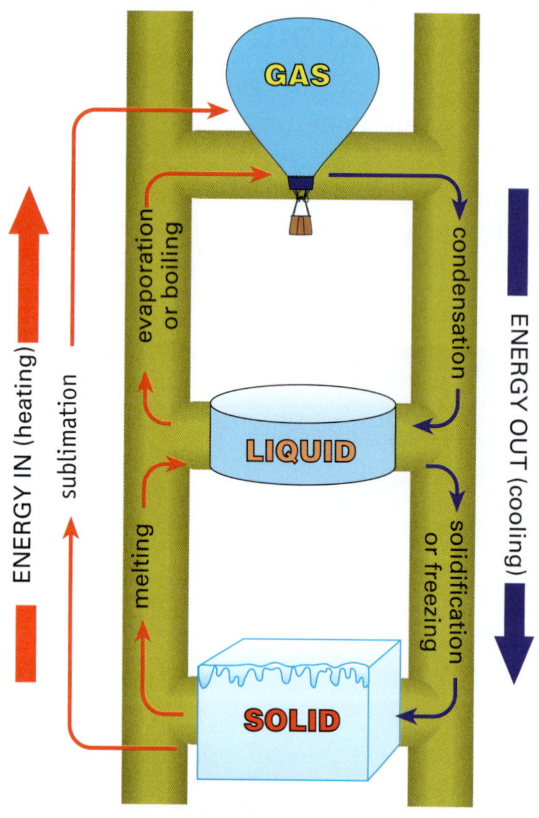

Figure 2.12 The energy ladder

The particle theory

More than 2000 years ago in ancient Greece, a philosopher called Democritus suggested this hypothesis: *All matter, living and non-living, is made of tiny particles too small to be seen*. His idea was that if you kept cutting something into smaller and smaller pieces, you would eventually come to the smallest possible particles—the building blocks of matter. He used the word *atomos* (which in Greek means 'cannot be divided') to describe these tiniest particles. This is where the word 'atom' comes from. You will learn more about atoms in Chapter 8.

Since then scientists have done many tests with matter, and the results have always agreed with Democritus' hypothesis. A hypothesis that is supported by many experimental results is called a **theory**. So the hypothesis that matter is made up of tiny particles too small to see is now called the **particle theory** of matter.

This particle theory can be used to explain the properties of solids, liquids and gases.

The particle theory of matter

1. All matter is made up of tiny particles too small to see.
2. There are spaces between the particles.
3. There are attractive forces between particles. The weaker these forces are, the further apart the particles are.
4. The particles are always moving.
5. At high temperatures the particles move faster than they do at low temperatures.

Figure 2.13 Ice makes you feel cold because it is absorbing your heat energy, which is used to make the ice melt.

Figure 2.14 All matter in the universe is made up of particles.

ISBN 978 1 4202 3735 1

SOLIDS

The particles in a solid (e.g. steel) are packed tightly in a fixed pattern. There are strong forces called **chemical bonds** holding them together, so they cannot leave their positions. The only movements they make are tiny vibrations to and fro.

We cannot see these invisible particles, but we can use a **model**. For example, we can represent the particles by the students in your class. When everyone is sitting down, very close together with shoulders touching, the class is a model for a solid.

The word 'model' has a special meaning in science. It is a way of representing something that is too small to be seen, or too large or complicated to be studied easily. A model is not the real thing. It is only a representation that helps you understand or explain something.

GASES

The particles in a gas (e.g. air) are far apart, and they move about very quickly. There are still attractive forces between them but they are very weak. The particles collide with each other and the walls of the container, and bounce off in all directions.

When the lesson is over students go in many different directions. Some may stay in the room, while others go to different parts of the school. When this happens the class is a model for gas.

HEAT

HEAT

LIQUIDS

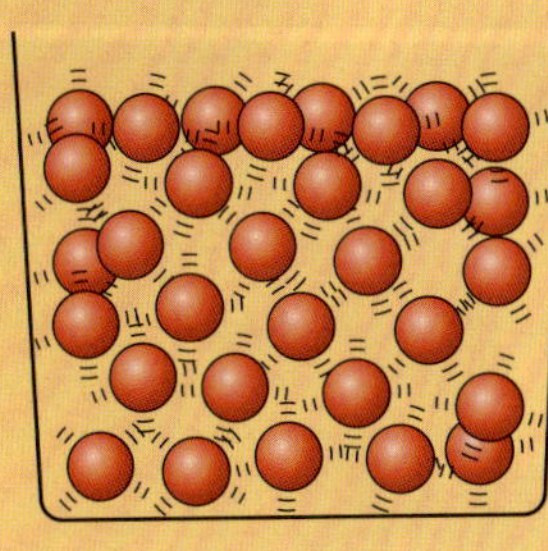

The particles in a liquid (e.g. water) can move about and slide past each other. They are still close together but are not in a fixed pattern. The forces (bonds) that hold them together are weaker than those in a solid.

When the students are moving about busily doing practical work, the class is a model for a liquid.

ISBN 978 1 4202 3735 1

ACTIVITY

1 Make a model for matter by putting some ball bearings in a flat dish or box.
   - What do the ball bearings represent? What does the dish or box represent?
   - Draw the arrangement of the ball bearings.
   - What state of matter does this represent?

2 Shake the dish gently so that the ball bearings move about.
   - Describe the new arrangement of ball bearings.
   - What state of matter does this represent?

3 Shake the dish vigorously.
   - Describe the new arrangement. What state of matter does it represent?

Your teacher may demonstrate this model using a dish on an overhead projector.

Teacher note: It is possible to buy special magnetic marbles for this activity.

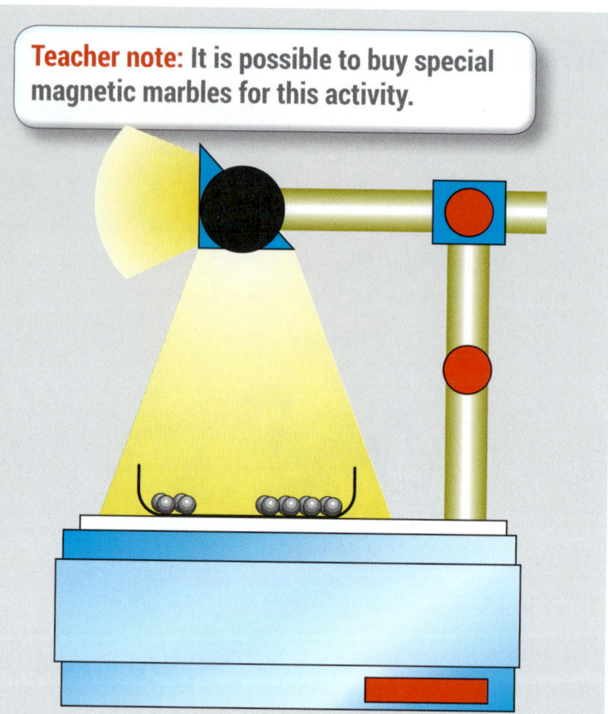

SCIENCE AS A HUMAN ENDEAVOUR

A fourth state of matter

We are familiar with the three states of matter we find on Earth—solids, liquids and gases. However, there is a fourth state of matter called **plasma**, which makes up 99% of the universe. Plasma consists of charged particles that are even further apart than the particles in a gas. You don't see much plasma on Earth because it requires very high temperatures. However, the sun is made of plasma, as are all the stars.

Lightning is a type of plasma that occurs naturally on Earth. You may have seen a plasma sphere in a science centre. The glass sphere contains a gas at a very low pressure and when very high voltage passes through it, it glows and looks like 'bottled lightning'. Plasmas also occur in neon signs and fluorescent bulbs.

Because the particles in a plasma are charged, they are affected by a magnetic field. Loops of plasma erupt from the surface of the sun and follow the curved magnetic field of the sun.

Scientists are experimenting with plasmas as hot as 100 million degrees. They are trying to make a fusion reaction that produces energy as the sun does. They use powerful electromagnets to create a 'magnetic bottle' to contain the super-hot plasma.

EXPLORE ONLINE

To find out more about plasmas, go to **Amazing plasmas**.

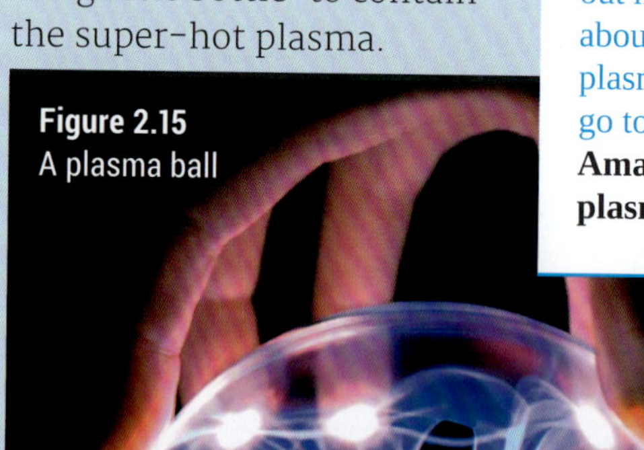

Figure 2.15
A plasma ball

ISBN 978 1 4202 3735 1

Explaining melting

We can use the particle theory to explain changes of state. When a solid is heated, its particles gain more energy and vibrate more. This makes the solid expand—get bigger. At the melting point, the particles vibrate so much that they break away from their positions. When this happens, the solid becomes a liquid.

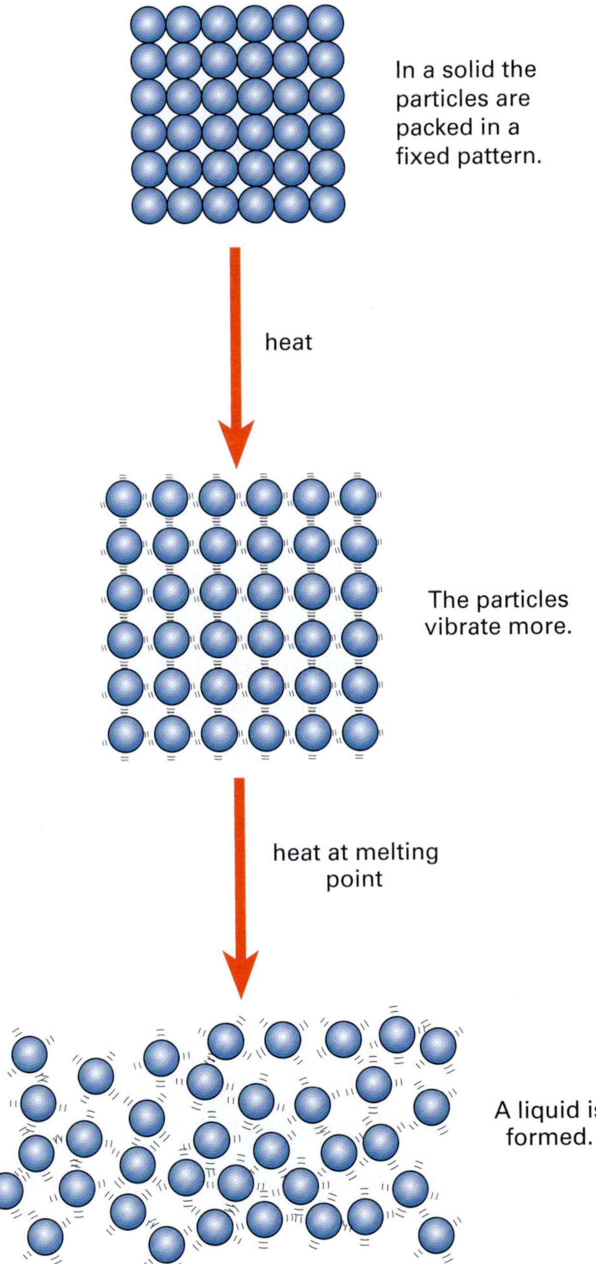

Figure 2.16 The particle theory can be used to explain the melting of a solid.

Explaining boiling

When a liquid is heated, its particles have more energy and move faster. They bump into each other more energetically and bounce further apart. This makes the liquid expand. At the boiling point, the particles have enough energy to break the bonds holding them together. They break away from the liquid and form a gas.

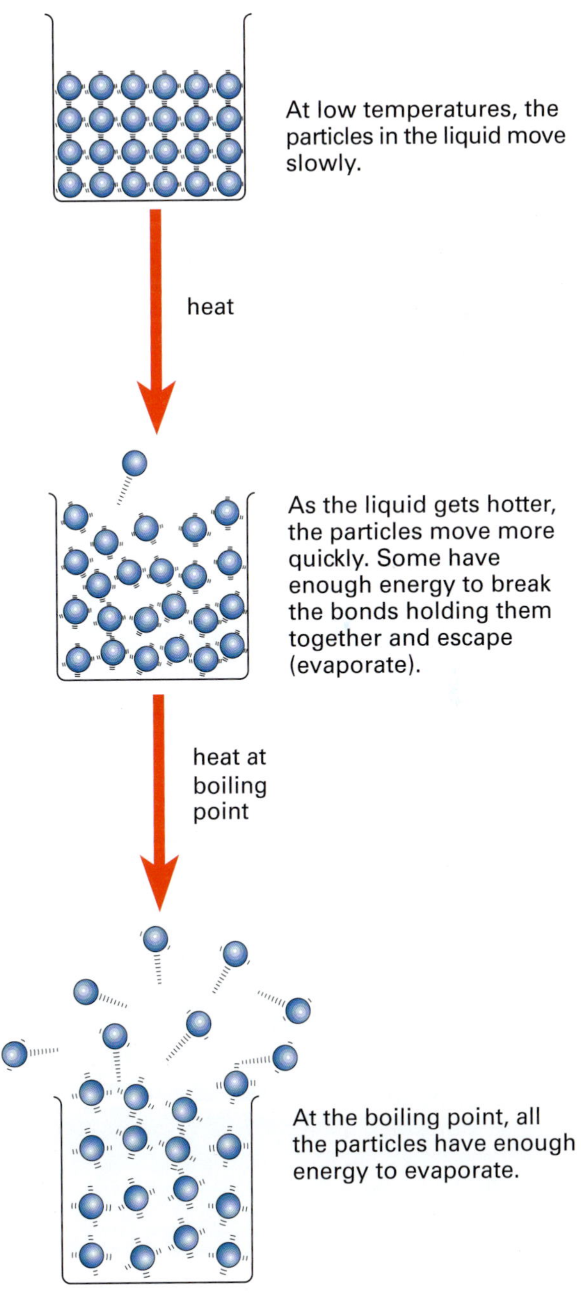

Figure 2.17 The particle theory can be used to explain the boiling and evaporation of a liquid.

ISBN 978 1 4202 3735 1

ACTIVITY

Set up the apparatus shown on the right and observe carefully what happens.

Where does evaporation occur?

Where does condensation occur?

Where does water exist in a:

a solid state?
b liquid state?
c gaseous state?

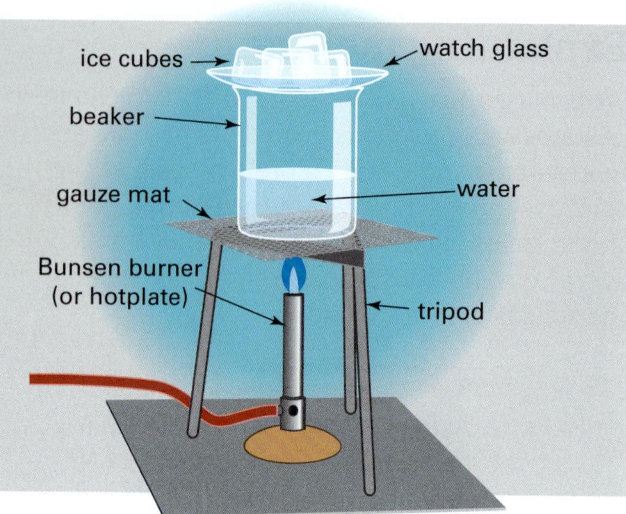

INVESTIGATION 2.2 Melting and boiling

Aim

To measure and graph the temperature as ice melts to water and then boils.

Materials

- small beaker, e.g. 250 mL
- crushed ice
- thermometer (−10 to 110 °C) or datalogger and temperature probe
- burner, tripod, gauze and heatproof mat
- stopwatch
- stirring rod
- retort stand and clamp
- stopper with hole to hold thermometer in clamp
- graph paper

Risk assessment and planning

Read through the Method.

- What safety precautions will be necessary?
- Which is the independent variable and which is the dependent variable? How do you know which is which?

Design a data table with columns for the two variables to be measured.

Method

1 Set up the apparatus as shown.

2 Half-fill the beaker with crushed ice, and measure its temperature. (Remember to wait until the reading is steady.)

Record the temperature of the ice in your data table.

3 Light the burner and adjust it to a medium flame. Put it under the beaker and immediately start timing.

4 Measure the temperature every minute. Use the stirring rod to stir gently before each reading. Continue your measurements until the water has been boiling for 3 or 4 minutes.

Record the data in the data table.

5 Graph your results or print them from the datalogger.

Discussion

1 What caused the ice to melt?

2 What did you notice about the temperature as the ice melted?

3 What did you notice about the temperature as the water boiled?

ISBN 978 1 4202 3735 1

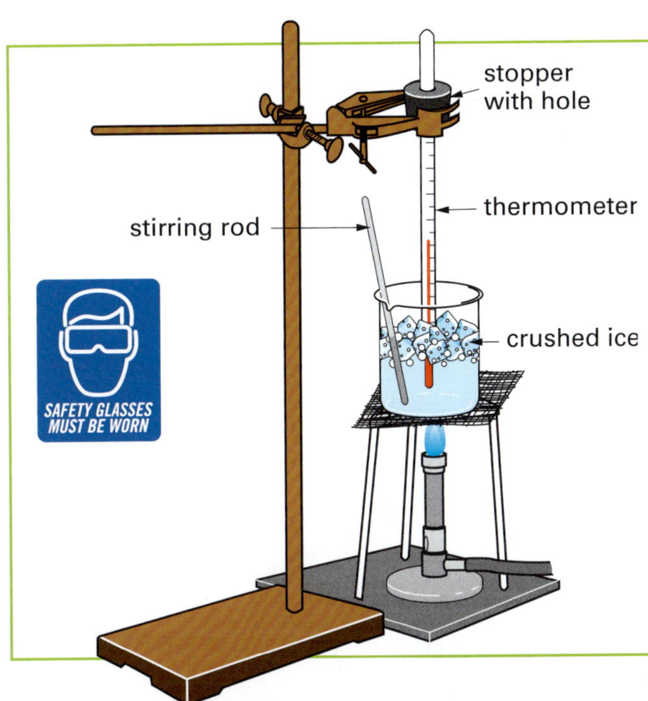

4 Your graph has two flat sections joined by a slope. What does it mean where the graph is flat? On the graph, mark when the ice is melting. Also mark when the water is boiling.

5 Use your graph to find the temperature of the water 10 minutes after you started heating.

6 Predict the temperature of the water 10 minutes after it started to boil.

7 The temperature did not increase while the ice was melting and while the water was boiling—even though there was a constant supply of energy from the burner. Use the particle model to explain where that energy was going.

SCIENCE AS A HUMAN ENDEAVOUR

From idea to theory

Figure 2.18 Democritus (de-MOK-rit-us) was a Greek philosopher who lived from 460 to 357 BCE. He had the idea that everything is made of atoms.

Figure 2.19 John Dalton (1766–1844) turned Democritus's idea into a scientific theory. There is more about Dalton on page XXX.

Use the internet and other resources to research the following questions about Democritus and Dalton. You could search under 'atomic theory history'. Do this individually or in a group. Because there are so many questions, you may need to divide the task between different individuals or groups.

1 Why did Democritus use the Greek word *atomos* to describe invisible particles of matter? (See page 185.)
2 Were Democritus's particles all the same? Explain.
3 How did Democritus explain the difference between a solid and a liquid?
4 Who was Leucippus?
5 Did the ancient Greeks do scientific experiments? Explain.
6 We know very little about Democritus and his ideas. Suggest a reason for this.
7 Aristotle was a famous Greek philosopher. Did he agree with Democritus's ideas? Explain.
8 Were the Greeks the only ancient people to come up with the idea of atoms?
9 In the Middle Ages the church associated atomic thinking with Godlessness. Suggest a reason for this.
10 Suggest why Democritus's idea was not developed any further for 2000 years.
11 When did people start doing proper scientific experiments?
12 What experiments were done by Dalton and other scientists around 1800?

ISBN 978 1 4202 3735 1

13 What was Dalton's atomic theory?

Once you have finished your research, share your findings with the class. Then use what you have found out to discuss these final questions.

14 Why was Dalton able to convert Democritus' idea into a scientific theory? Why did this process take more than 2000 years?

15 Do you think the atomic theory has changed since the time of Dalton? Explain.

1 Copy the diagram below. Put one word in each box and on each arrow to summarise what you know about changes of state.

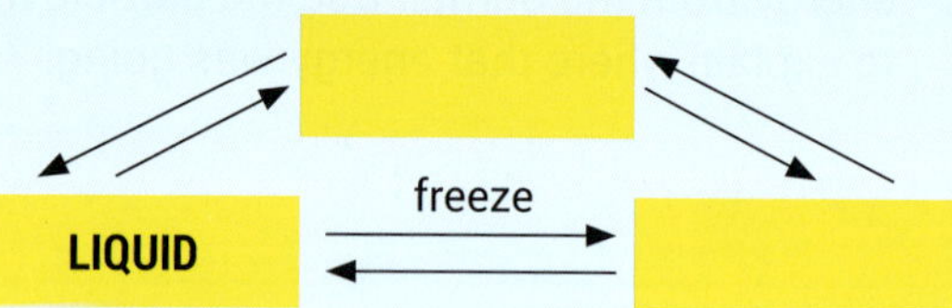

2 Complete these sentences.
- a The melting point of ice is the temperature when it changes from a _____ to a _____.
- b The melting point of ice is _____°C.
- c The boiling point of water is the temperature when it changes from a _____ to a _____.
- d The boiling point of water is _____°C.

3 Choose from the words solid, liquid or gas to say what type of substance will be formed when a:
- a gas condenses
- b liquid freezes
- c solid melts
- d liquid boils.

Write your answers in complete sentences.

4 Name the change of state that occurs when:
- a dew forms on the grass
- b a bottle of perfume is opened and can be smelt on the other side of the room
- c a puddle of water on the road disappears when the sun shines
- d moth balls placed in a suitcase of clothes are gone after a few months
- e lava flows from a volcano and slowly forms a rock called basalt.

5 The table below lists five changes of state. For each change decide whether heating or cooling is needed. Copy the table and tick the correct columns.

Changes of state	Heating	Cooling
a solid to liquid		
b liquid to gas		
c gas to liquid		
d liquid to solid		
e solid to gas		

6 Which of the following statements are true and which are false? Rewrite the false ones to make them correct.
- a Melting occurs when a solid changes to a liquid.
- b All matter consists of particles.
- c To change a liquid to a gas, you have to cool it.
- d Solids have a definite shape because their particles are free to move around.
- e Water can evaporate at any temperature.
- f If water boils for a long time, its temperature rises above 100 °C.
- g Condense is the opposite of evaporate.
- h The particles of a gas are so far apart that they do not attract each other at all.
- i The particles of a solid do not move.

7
- a In which state do the particles move fastest?
- b In which state are the particles closest together?
- c In which state are the particles close together but not arranged in a regular pattern?
- d In which state of matter are the bonds between particles greatest?

ISBN 978 1 4202 3735 1

1 Luigi wears glasses. He finds it hard to see when he enters a hot steamy bathroom. Use what you have learnt in this chapter to explain this.

2 If gases and liquids are both made of particles, why are their properties so different? Explain your answer in terms of particles and bonds.

3 Answer these questions in complete sentences.
   a How can you make the particles in a solid move faster?
   b What are the particles doing if a liquid is evaporating?
   c What can happen to a gas if its particles slow down?

4 When you cook food in a saucepan with the lid on, you might notice water on the inside of the lid. Why is this?

5 Use your knowledge of the particle theory to explain each change in the diagrams below.

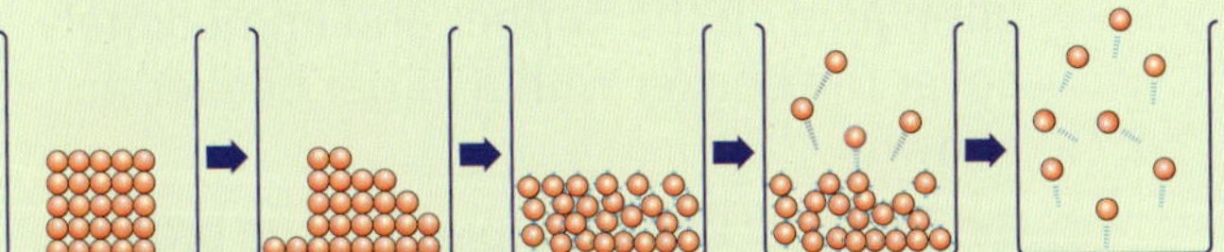

6 The graph below shows the change in temperature over time as wax is heated.

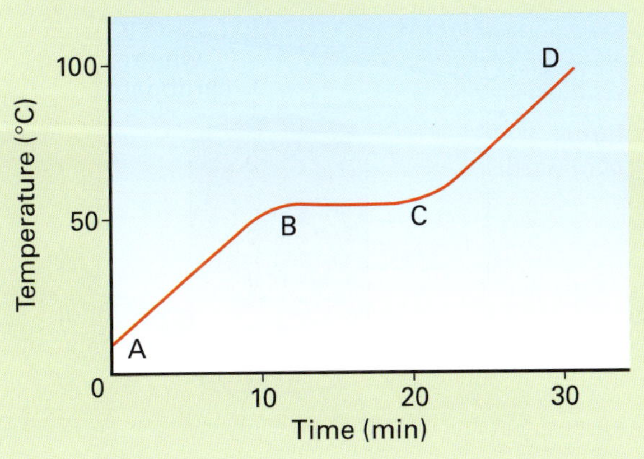

   a Which part of the graph shows that a change of state is taking place?
   b What is the melting point of the wax?
   c What is the state of the wax during the first 10 minutes of heating?
   d What is the state of the wax between C and D?
   e In which part of the graph are the bonds between the wax particles greatest?

7 Hexane is used as an industrial solvent. It has a melting point of −94 °C and a boiling point of 69 °C.
   a Is hexane a solid, a liquid or a gas at room temperature (20 °C)?
   b If hexane is heated to 90 °C, would you expect it to be a solid, a liquid or a gas?

8 Which would evaporate more quickly: water in a flat tray, or water in an open bottle? Explain your answer in terms of the particle theory.

9 Why do clothes dry faster on a windy day than on a calm day?

10 Angie wanted to keep her yoghurt cool, so she put it in a jar with some ice and screwed the lid on tightly. When she went to put it in her locker 15 minutes later, she saw that the outside of the jar was quite wet.

Everyone had a go at explaining what had happened (see the cartoon).
   a Whose inference do you agree with? Why?
   b Can you suggest a better inference?

11 Dry ice is sometimes used to create fog and mist on stage. If carbon dioxide is invisible, how can you see the dry ice fog?

12 On a hot day you perspire (sweat). As this perspiration evaporates it cools you. Use the particle theory to explain how evaporation produces cooling.

ISBN 978 1 4202 3735 1

2.3 Using the particle theory

We have used the particle theory to explain solids, liquids and gases, and how they can change from one state to another. In this section we will use it to explain some other properties of matter.

Diffusion

If someone opens a bottle of perfume in the middle of the classroom, you soon smell it in other parts of the room. The fragrance spreads through the air in all directions. This gradual mixing of substances is called **diffusion**.

Let's explain how perfume diffuses. When the lid is on, the gas particles remain inside the bottle. When the lid is taken off, the liquid perfume evaporates easily. As there are only weak forces between the particles, the particles can spread out, moving away from the crowded bottle to places where there are fewer particles of perfume. Eventually the particles spread evenly throughout the air in the room.

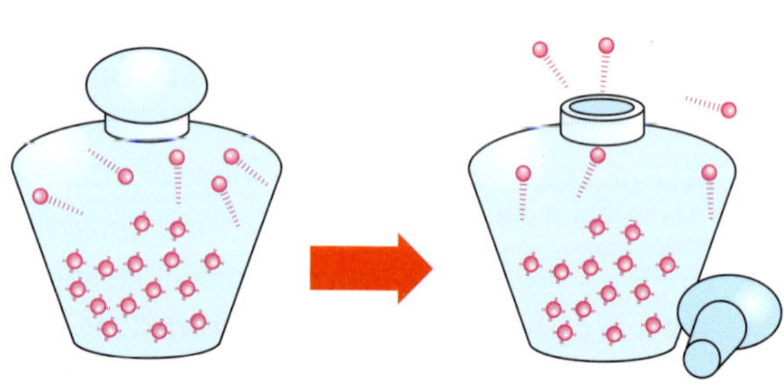

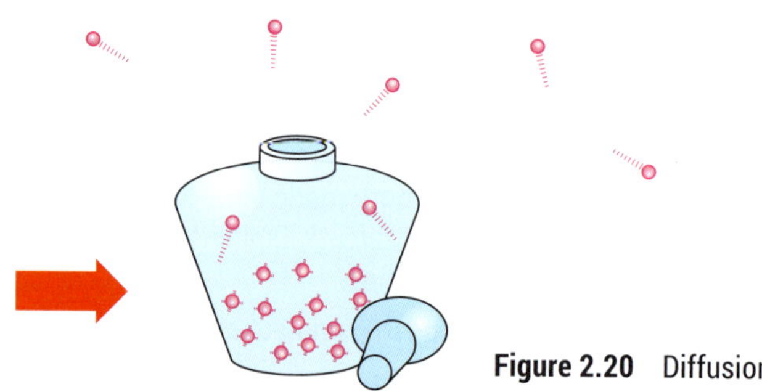

Figure 2.20 Diffusion

ACTIVITY

A Put a beaker on a sheet of white paper and half fill it with water. Let it stand for a while to let the water become perfectly still. Use a pair of tweezers to drop a single crystal of potassium permanganate (Condy's crystals) down a drinking straw as shown. Then leave the dish undisturbed overnight.

Explain what you observed in terms of the theory of particles.

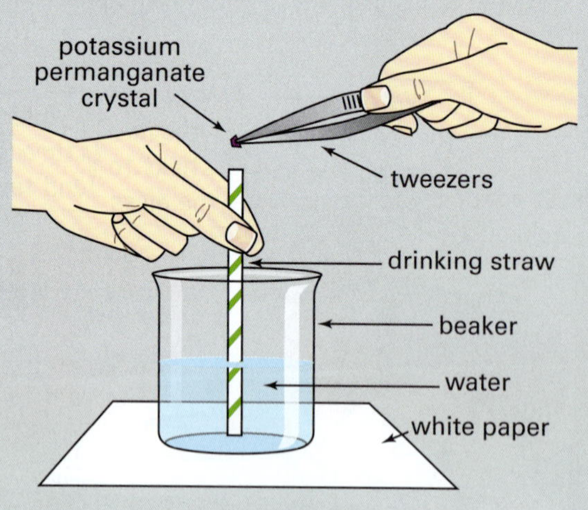

B *This activity involves the poisonous gas nitrogen dioxide and can only be done as a teacher demonstration in a fume cupboard.*

Place a few pieces of copper in a small beaker. Pour a few drops of concentrated nitric acid on the copper and immediately cover the beaker with a larger beaker, as shown.

Observe what happens to the brown gas. (It helps to put a piece of white cardboard or paper behind the beakers.)

ISBN 978 1 4202 3735 1

As you saw in the activity on the previous page, when a crystal of potassium permanganate is placed in water, the water slowly turns purple. Both the crystal and the water are made of particles. As water is in the liquid state, its particles are moving and bump into the particles in the crystal. This causes some particles to leave the crystal and move into the spaces among the water particles, as shown below. This is the process of dissolving. As the particles continue to move, they diffuse throughout the water and the purple colour spreads evenly.

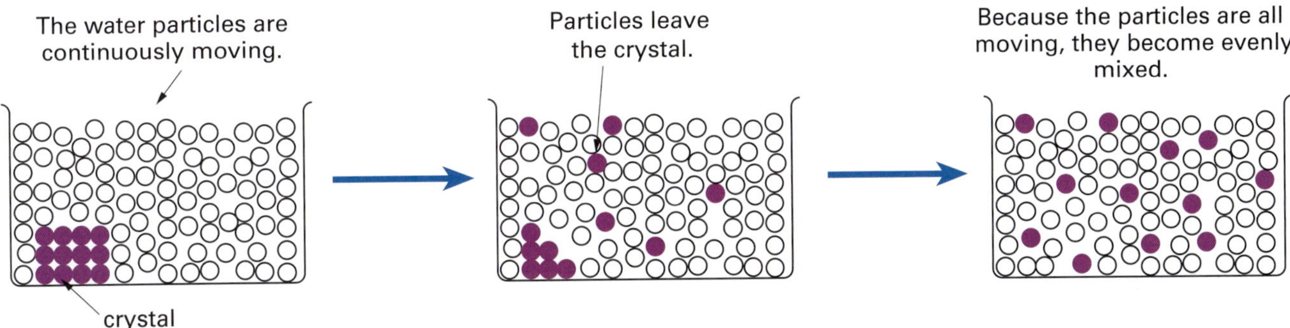

Figure 2.21 Diffusion of potassium permanganate in water

Your teacher may demonstrate the following activities. For each activity, *predict* what you think will happen, then *observe* what happens and finally *explain* what happened.

A Using a ball and ring apparatus (or other metal shapes), put the ball through the ring. Then heat the ball strongly and try to put it through the ring again.

What do you predict will happen if you heat the ring and try again? Try it.

B Fill a flask with coloured water and fit a stopper with a piece of glass tubing through it. The coloured water in the tube should reach just above the stopper. Mark this level with a marking pen. Put the flask in a container of hot water for a few minutes. Now put it in a container of *cold* water.

How could you use this apparatus to measure temperature?

C Put a balloon over the mouth of a flask. Heat the flask gently using a Bunsen burner.

Write a generalisation to explain the results of *all three* activities.

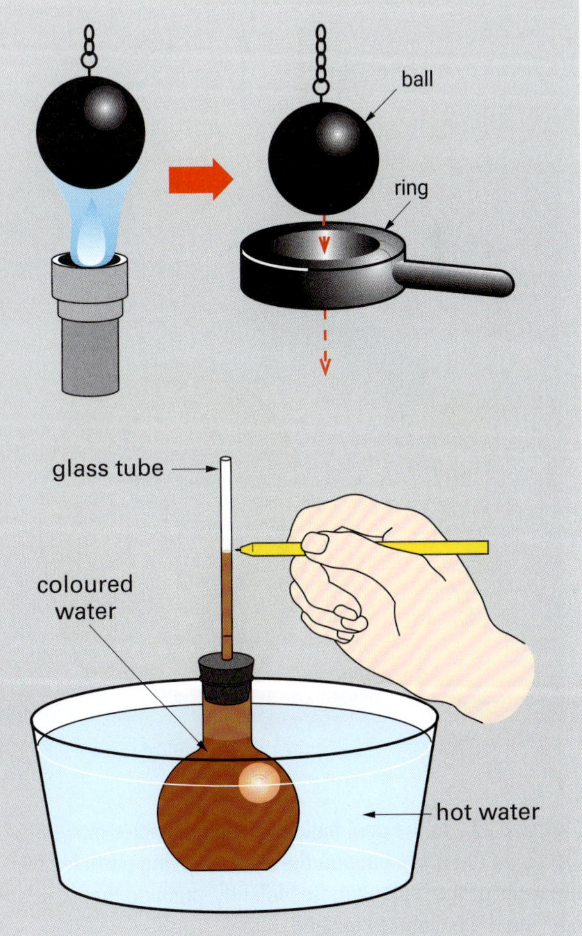

ISBN 978 1 4202 3735 1

Expansion and contraction

As you saw in the activity, solids, liquids and gases all expand (get larger) when they are heated. That is, they occupy more space. Similarly, when they are cooled they contract (get smaller) and occupy less space.

In solids the particles vibrate in fixed positions. As the solid is heated, the particles absorb energy, vibrate more violently and start to bump into each other. This causes them to move further apart so that they have more room for their violent vibrations. As a result, the solid as a whole expands. When the solid is cooled, the particles lose energy. The particles slow down again and occupy less space (contract).

Expansion and contraction of liquids and gases can be explained in a similar way.

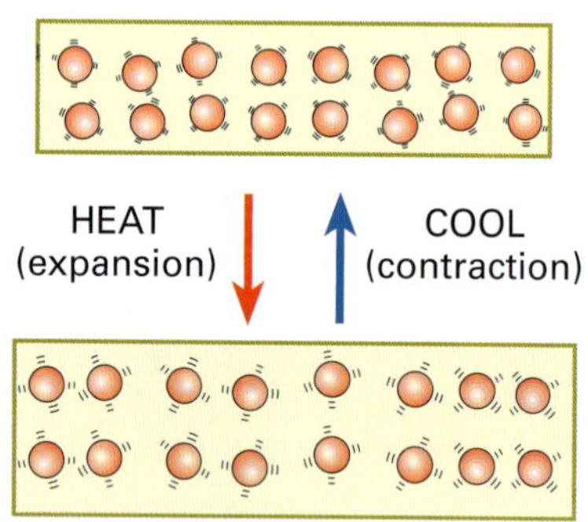

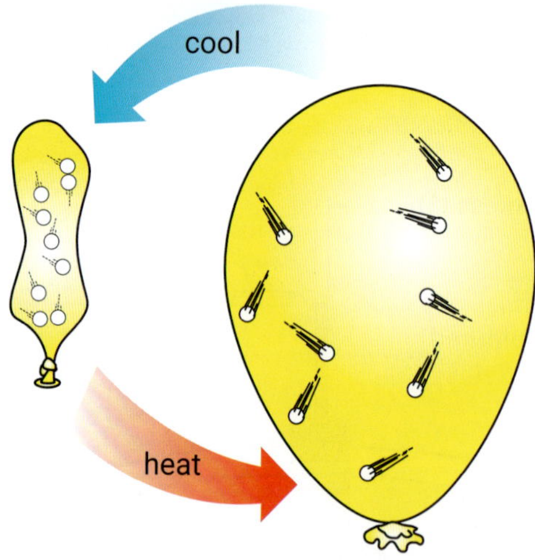

Figure 2.22 In the cool balloon the air particles move slowly. In the warm balloon the faster-moving particles hit the walls of the balloon more violently, pushing them out and causing the balloon to expand.

Air pressure

When Rhys pulls into the service station, he discovers his car has a flat tyre. What keeps the tyre inflated is *air pressure*. The invisible particles of air are only tiny but they move very rapidly—about the speed of a rifle bullet. These tiny bullets bombarding the walls of the tyre cause the air pressure. What has happened to the flat tyre is that some of the air has escaped and there are not enough particles to give the pressure needed to keep the tyre inflated. Rhys gets the tyre fixed and pumps it up with compressed air. Now the air pressure is back to normal.

Figure 2.23 A flat tyre contains few air particles and the air pressure is low. An inflated tyre contains many air particles and the pressure is high.

It is a hot day and Rhys drives non-stop for two hours to get to the beach. When he checks the tyre pressure, he finds it has gone up. There can't be any more air particles in the tyre. What has happened is that friction between the tyre and the road has caused the air inside the tyre to heat up. This means the particles have more energy and are moving faster and hitting the walls of the tyre harder. Hence the air pressure is higher. When the tyre cools down, the particles will lose energy and slow down, and the pressure will return to normal.

You have seen how the particle theory can be used to explain changes of state, diffusion, expansion and contraction, and air pressure. You can now try to explain some other properties of matter for yourself.

ISBN 978 1 4202 3735 1

For each observation below write an inference to answer the question in terms of the particle theory. Draw a model to explain what is happening to the invisible particles.

1 Observation: *Lead is four times denser than aluminium.*

Question: How can you explain this?

2 Observation: *When you add a teaspoon of sugar to a glass of water and stir, the sugar disappears.*

Question: Explain what happens.

3 Observation: *You can pour water from one container to another, but honey is much harder to pour, especially when it has been in the fridge.*

Question: How can you explain this?

4 Observation: Crystals have a definite shape, *with straight edges and sharp corners. For example, salt crystals are cubes, and quartz crystals (below) are like pointed columns.*

Question: How can you explain these different shapes?

5 Observation: *Hold your finger over the end of a bicycle pump and push in the plunger. Let go of the plunger and it moves back to where it came from.*

Question: What pushed the plunger back?

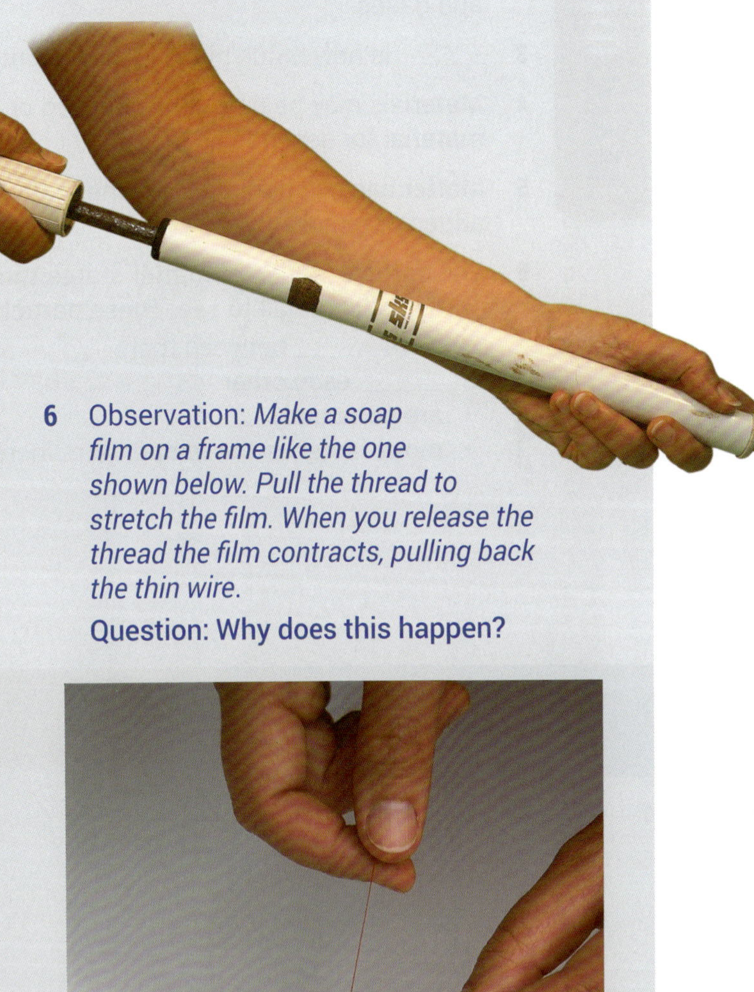

6 Observation: *Make a soap film on a frame like the one shown below. Pull the thread to stretch the film. When you release the thread the film contracts, pulling back the thin wire.*

Question: Why does this happen?

ISBN 978 1 4202 3735 1

MAIN IDEAS

Copy and complete these statements to make a summary of this chapter. The missing words are on the right.

1 Matter has ______ and takes up space (its ______).

2 There are three common ______ of matter on Earth: solids, liquids and gases.

3 ______ is how much mass is packed into a measured volume.

4 Materials may be natural, processed or ______. What you use a material for depends on its ______.

5 Matter can be changed from one state to another when ______ is added or removed.

6 The particle ______ of matter states that all matter is made of particles too small to see. These particles:
- have ______ between them
- ______ each other
- are constantly ______
- move faster as the temperature increases.

density
attract
spaces
heat energy
moving
states
volume
synthetic
mass
properties
theory

CH•2 REVIEW

1 The statement *All matter is made of particles* is:
A an observation.
B an inference.
C a prediction.
D a generalisation.

2 Copy the diagrams below and label them by replacing the letters (A–G) with one of these words: condensation, evaporation, gas, liquid, melting, solid, solidification.

3 A substance has no fixed shape. From this information it would be correct to say that the substance is a:
A gas.
B solid or liquid.
C liquid or gas.
D solid or gas.

4 Give three examples of how the use of a substance depends on its properties.

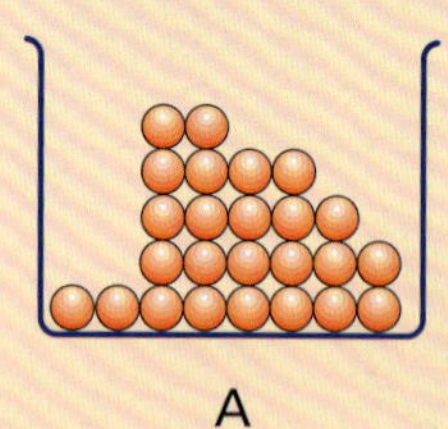

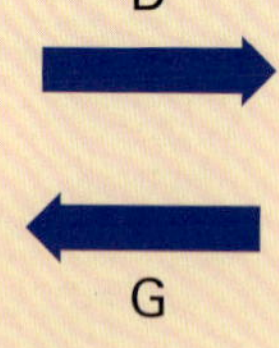

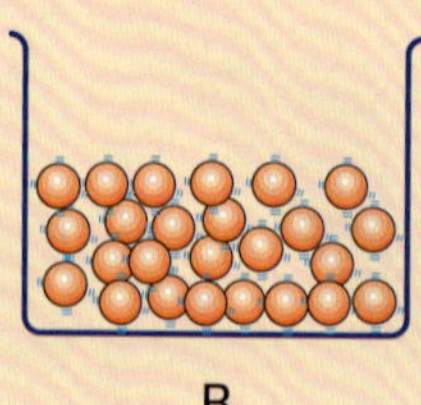

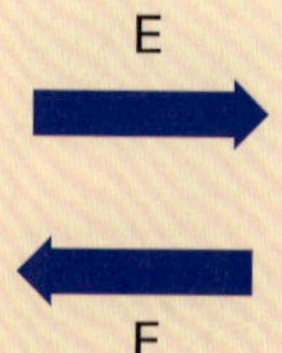

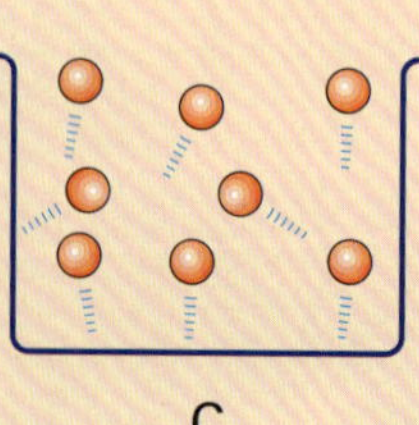

ISBN 978 1 4202 3735 1

5 Mercury is a liquid with a density of 14 g/cm^3. Use the table below to find something that would sink in water but float in mercury.

Table of densities (g/cm^3)	
aluminium	2.7
lead	11.3
platinum	21.5
polystyrene foam	0.1
petrol	0.7
water	1.0

6 Archimedes was asked to find out if the crown belonging to the king of Syracuse (in ancient Greece) was made of pure gold. The king thought some silver might have been added to reduce the amount of gold needed. Archimedes decided to use his knowledge of density to solve the problem. He found the volume and mass of the crown. He also found the mass of an equal volume of pure gold. Here are his results.

volume of crown = 100 cm^3

mass of crown = 1500 g

mass of 100 cm^3 of pure gold = 1930 g

- **a** What is the density of pure gold in g/cm^3?
- **b** What was the density of the crown?
- **c** Was the crown made of pure gold?

7 Write a paragraph describing the advantages and disadvantages of plastic, paper and cotton canvas for making supermarket bags. In your answer use these words: processed, non-renewable, renewable and synthetic.

Questions 8 and 9 refer to the following list of some of the possible properties of particles.

Rate of movement
- **A** vibrating or moving very slowly
- **B** moving around freely but slowly
- **C** moving freely and rapidly

Spaces between particles
- **D** very close together, almost touching
- **E** fairly close together
- **F** wide spaces between them

Forces between particles
- **G** very strong bonds
- **H** held together to some extent but free to move around
- **I** very weak bonds

8
- **a** Aluminium is a solid. Which *three* properties listed above would probably be true of its particles?
- **b** Ozone is a gas. Which *three* properties would probably be true of its particles?
- **c** Petrol is a liquid. Which *three* properties would probably be true of its particles?
- **d** Steel cannot easily be compressed. Which description in *Spaces between particles* would best explain this?
- **e** Diamond is a very hard substance. Which *two* properties above would best explain this?
- **f** Property E in the above list can sometimes be changed to property F by:
  - **A** heating the substance.
  - **B** cooling the substance.
  - **C** putting it in another container.
  - **D** compressing the substance.

9 Use the particle theory to answer each of the following questions.
- **a** Why do gases have much lower densities than solids and liquids?
- **b** Why do gases condense to form liquids when cooled?

Check your answers on page 294.

ISBN 978 1 4202 3735 1

IN THIS CHAPTER

Science Understanding

- recognise that potential energy is stored energy, such as gravitational, chemical and elastic energy
- recognise that kinetic energy is the energy possessed by moving bodies
- describe how energy can be *transferred* from one object to another and *transformed* from one form to another
- use energy chains to show changes between different forms of energy
- investigate energy transformation in everyday devices, including where energy is wasted as heat
- participate in an inquiry to decide whether a nuclear power station should be built
- describe some energy sources

Science Inquiry Skills

- use a model to understand energy

CH•3 Introducing energy

ISBN 978 1 4202 3735 1

GET STARTED: *IMAGINE*

You have probably heard the word 'energy' and used it many times. But what does it really mean? In this activity you are going to work in a small group to brainstorm what you know about energy.

1 Sit in a group of four to six people and have a large piece of paper and pen ready. One person can write down everything as you go.

2 What does your group know about energy? Everyone should try to give at least one idea. If you are having trouble getting started, try one of these:
   - > Think of a sentence with the word 'energy' in it.
   - > Draw something with lots of energy.
   - > List what we use energy for.
   - > List different types of energy.

3 Imagine you are the gymnast on the left.
   - > What is going to happen to your energy as you start to move from this position?
   - > Where does your energy come from? Where does it go?
   - > Will you have more energy at the bottom of the swing or at the top?

4 Imagine you are about to drive this solar car.
   - > Where does the energy to run this car come from?
   - > The energy collected by the solar panels makes the car move. Describe the steps or changes the energy goes through to get the wheels and the car moving.
   - > Can energy be stored? Explain how this car can store energy.
   - > Does this solar car waste any energy? If so, explain how you think that happens.

ISBN 978 1 4202 3735 1

3.1 What is energy?

If you use a torch for a long time, the light gradually gets dimmer and dimmer until the torch no longer shines. We say the batteries are 'flat'—they have run out of energy. In a similar way you can't pedal a bicycle or dig in the garden for too long because your body runs low on energy. If you don't eat food, your body becomes weaker and weaker. This is why we say that food gives us energy.

Everything around us depends on energy. Plants need energy from the sun to make food. Cars depend on the energy stored in petrol. Energy is used in homes, offices and industry to run all sorts of machines. It is used for lighting and heating our homes, and for cooking and storing food.

Obviously energy is very important to us, but you probably found in Getting started that it is difficult to say exactly what it is. It is easier to say what energy can do. If you have a lot of energy, then you can do a lot of work. You do **work** when you use a force to move something. **Energy** is the ability to do work.

Energy and work

The more energy something has, the more work it can do. A gale-force 100 km/h wind has more energy than a gentle 10 km/h breeze. It can therefore do more work; for example, turning windmills or ripping off roofs. Also, a raised sledgehammer has more energy and can do more work than an ordinary hammer, because it is heavier (has more mass).

Anything that does work must have a supply of energy. A motorbike will not keep running unless it is supplied with petrol. Petrol provides energy that the engine uses to do work. When you pedal a bicycle, the energy comes from the muscles in your body, and your muscles get their energy from the food you eat. If you have a higher intake of energy than you need, then the extra energy is stored in your body as fat. However, a diet inadequate in energy will lead to a thin and unhealthy body.

Figure 3.1 Heavier objects can do more work.

Figure 3.2 The bicycle and the motorbike both need energy to move them.

Measuring energy

In talking about how much energy something has, it is important to have a unit for measuring energy. In the same way that the litre is the unit for measuring volume, energy has a unit called the **joule** (J). This unit was named after a British scientist called James Joule. You use one joule of energy to lift a 100-gram mass one metre. Because a joule is only a small amount of energy, it is common to use **kilojoules** (kJ) and megajoules (MJ).

1 kilojoule = 1000 joules
1 megajoule = 1 000 000 joules

The table on page 58 shows you how much energy is involved in various everyday activities.

To find how much energy is stored in food, you can turn it into heat and measure what that heat can do. In Investigation 3.1 you will burn some food to do that. Of course, there are no fires burning inside you. The food combines with oxygen in your cells in the chemical reaction called respiration, and heat energy is released.

ISBN 978 1 4202 3735 1

INVESTIGATION 3.1 Energy from food

SAFETY GLASSES MUST BE WORN

Aim

To find out how much energy is released when a small piece of food burns.

Materials

- small piece of food, e.g. Nutri-Grain or Tiny Teddy
- Bunsen burner
- wire to make holder
- small test tube
- thermometer
- measuring cylinder
- stand and clamp

Teacher note: When selecting foods, remember some students may be allergic to burning peanuts.

Risk assessment and planning

- Read through the investigation, then describe to your partners what you have to do, measure and record.
- What data will you need to record?
- What safety precautions will be necessary?

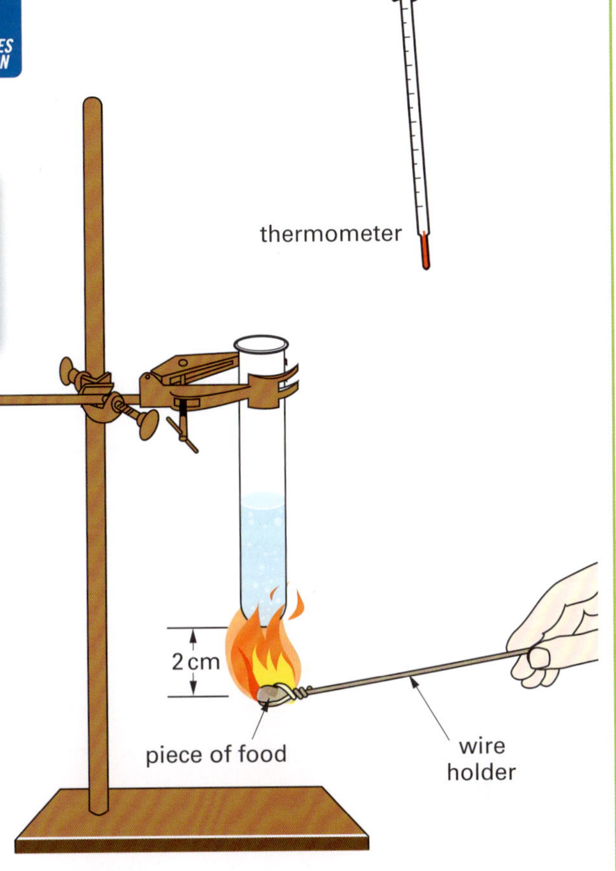

Method

1. Use the measuring cylinder to measure *exactly* 10 mL of water into the small test tube.
2. Clamp the test tube as shown.
3. Use the thermometer to measure the initial temperature of the water.
4. Use the wire to make a holder for the piece of food.
5. Light the Bunsen burner, then put the food in the flame. As soon as it catches fire, hold it about 2 cm under the test tube.
6. When the food stops burning, stir the water *gently* with the thermometer, and measure the final temperature.
7. If you have time, repeat the experiment with other foods such as potato crisps, nuts, bread, rice, spaghetti.

Discussion

1. By how many degrees did the temperature of the water increase?
2. It takes 4.2 joules to raise the temperature of 1 mL of water by 1 °C. So, to calculate the heat energy gained by 10 mL of water, multiply the temperature rise by 42. Your answer will then be in joules.
3. Do you think all the energy from the burning food went into heating up the water in the test tube? Explain.
4. Were there any problems with the investigation? If so, suggest how these problems could be fixed.

ISBN 978 1 4202 3735 1

CHECK

1 For each of the following words, write a sentence to show that you understand its scientific meaning.

force work energy

2 How do you know if something has energy?

3 Why can a cricket ball do more work than a golf ball moving at the same speed?

4 How many joules are there in a:
a kilojoule?
b megajoule?

5 Use the table below to answer these questions.
a How much energy does the average person get from the food they eat in a day?
b How many kilojoules of energy does a burning match produce?
c Which has more energy stored in it—a car battery or 1 litre of petrol?
d Is there enough energy stored in a car battery to boil a kettle of water?

Energy involved in everyday activities (in kilojoules)

Activity	Energy (kJ)
Energy produced by a burning match	10
Energy you gain by eating a chocolate biscuit	300
Energy needed to boil a kettle of water	700
Energy you use in walking 5 km	1000
Electrical energy stored in a car battery	2000
Energy used during one day's hard work	7000
Average energy gained from the food you eat in a day	11 000
Electrical energy used by a family home each day	80 000
Energy stored in 5 litres of petrol	160 000
Energy made by a power station every second	2 000 000

6 a Where does the energy needed to start a car come from?
b If you leave the lights on while your car is parked for a few hours, you may have trouble starting it. Why?

7 In a science lab, Alex and Holly are doing an experiment on the chemical energy stored in foods. Look carefully at the illustration. List at least five things they are doing that are unsafe.

8 Why do you puff and pant after running quickly or exercising?

9 In Get started on page 55, a student said that whenever a change occurs, energy is involved. For example, a kettle boils when you supply heat energy. Give as many examples as you can to illustrate this idea.

ISBN 978 1 4202 3735 1

3.2 Forms of energy

There are many different forms (types) of energy.

Kinetic energy

Any moving object has **kinetic** (kin-ET-ic) **energy**. When you run, you have kinetic energy. A moving train has a large amount of kinetic energy. The kinetic energy of the strong winds in a cyclone or tornado can cause a lot of damage. As a moving object slows down, it loses kinetic energy. When it stops, it has no kinetic energy.

The amount of kinetic energy an object has depends on its speed. The faster the object moves, the more kinetic energy it has. For example, a cricket ball bowled by a fast bowler has more kinetic energy than one bowled by a spin bowler.

Kinetic energy also depends on the mass of the moving object. The larger the mass, the greater its kinetic energy. A cyclist and a bus may be travelling at the same speed, but the bus has much more kinetic energy because it has greater mass.

Gravitational potential energy

Much of the energy around us is *stored energy*. We notice it only when it changes to other forms. It has the *potential* to do work, so stored energy is called **potential energy**. For example, the stored energy that something has when it is high up is called **gravitational potential energy**. This energy is there ready to be used because of the pull of gravity. When you are at the top of a slide, you have gravitational potential energy—you have the *potential* to slide to the bottom. The heavier you are, and the higher the slide, the more potential energy you have. As you slide down, this gravitational energy is changed to kinetic energy. Energy can easily change back and forth between potential and kinetic.

Figure 3.3 The winds in cyclones and tornadoes have huge amounts of kinetic energy.

Figure 3.4 A person at the top of the slide at the Melbourne Show has gravitational potential energy.

ISBN 978 1 4202 3735 1

Elastic potential energy

When you jump on a trampoline, what pushes you into the air? Try to visualise what happens in slow motion. The trampoline consists of a frame with a flexible mat attached by springs. When you land on the mat, it moves down, stretching the springs and storing energy called **elastic potential energy** in them. As the stretched springs return to their original size and shape, they release their stored energy. The mat is pulled back up, and you are thrown into the air.

A wind-up toy stores elastic potential energy. So does a stretched elastic band. The more it is stretched, the more elastic energy the band has, and the more work it can do.

Figure 3.5 The elastic energy stored in the stretched trampoline springs throws you into the air.

Make a motormouse as shown.

Step 1

Thread a rubber band through the cotton reel.

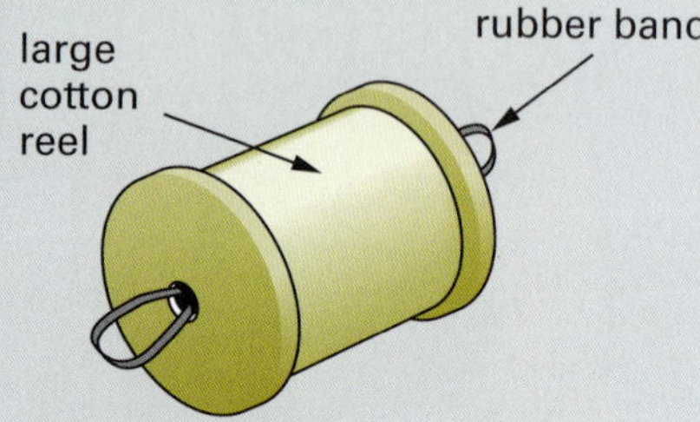

Step 2

Put a piece of broken match through one end of the rubber band. Tape the match to the reel so it will not move.

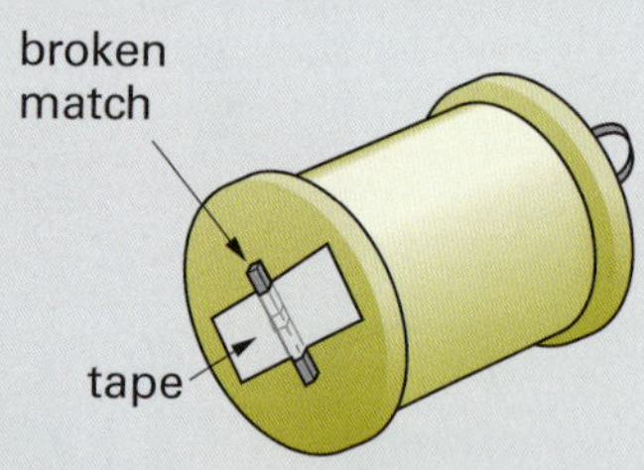

Step 3

At the other end, put a washer over the rubber band, then put a pencil through the rubber band.

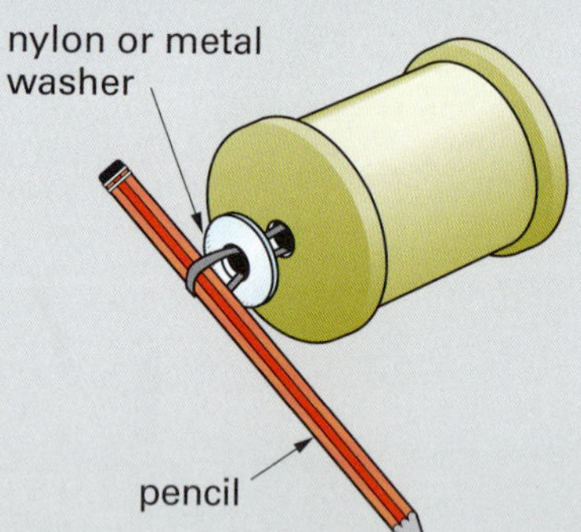

To make it go, simply wind up the pencil until the rubber band is tightly twisted. Then put the motormouse on the floor, with the pencil still in place, and let it go.

What type of energy does the motormouse have when you let it go?

What type of energy did it have before you let it go?

Energy is needed to wind up the motormouse. Where did this energy come from?

Investigate the relationship between the number of turns of the pencil and the distance the motormouse travels.

ISBN 978 1 4202 3735 1

Chemical energy

Energy is stored in chemicals as **chemical energy**. When fuels such as wood and petrol are burned, this stored energy is released as heat and light. Foods also contain chemical energy that can be used by our bodies.

Figure 3.6 Cyclists need to eat during a long race to keep up their energy store.

Nuclear energy

Energy is also stored inside atoms as **nuclear energy**. It can be released from some atoms (e.g. uranium atoms) in nuclear power stations.

Figure 3.7 Nuclear energy stored in hydrogen atoms is the source of the sun's energy.

Sound energy

Sound is a form of kinetic energy caused by vibrating objects. It travels from place to place as sound waves. The louder the sound is, the more energy it has, and the more work it can do by vibrating things such as your eardrums.

Heat energy

Heat is a form of energy that hot objects have. If heat energy is taken away from an object, it becomes cooler. This is what happens in refrigerators and in air-conditioned rooms.

Light energy

Burning chemicals, very hot objects and stars all release light energy. It travels through space in waves (as do radio and TV waves, microwaves and ultraviolet waves). Light energy from the sun, called **solar energy**, is used by plants to make their food.

Electrical energy

Electrical energy is widely used because it is easily transmitted by wires to the place where it is needed. It can be changed into other forms of energy by the many electrical devices that have been invented. It can also be stored in batteries as chemical energy.

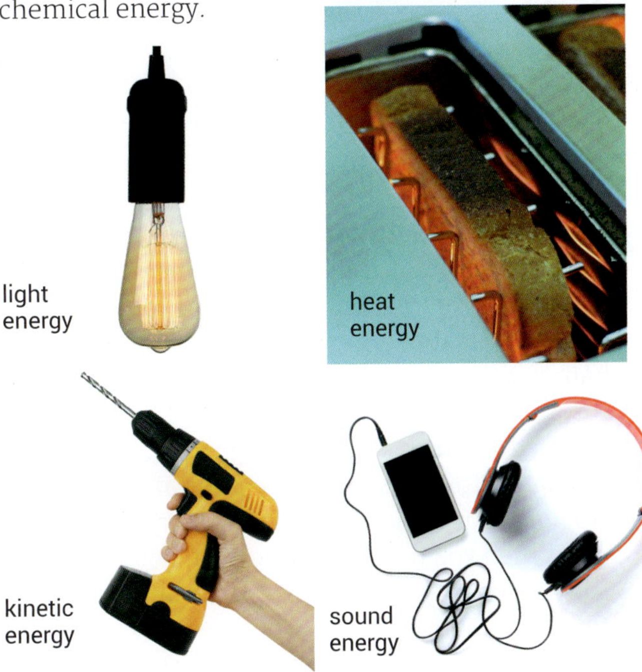

Figure 3.8 Electrical energy is very useful because you can easily convert it into other forms of energy. Four different energy converters are shown here.

ISBN 978 1 4202 3735 1

Energy changes

Energy can be *transferred* from one object to another. In golf, a ball at rest is made to move by a moving golf club. Some of the kinetic energy of the club is transferred to the ball.

Figure 3.9 The golf club transfers kinetic energy to the ball.

Another everyday energy transfer occurs when you heat water on a stove. Heat is transferred from the gas flame or the electrical heating element to the water, causing it to boil.

Energy can also be *converted* or *transformed* from one form into another. For example, if you rub your hands together, they become warm. You have converted the kinetic energy of your moving hands into heat energy. You can describe this change with an arrow, as shown in Figure 3.10.

Sometimes more than one form of energy is produced when an energy change occurs. A candle is designed to convert stored chemical energy into light, but some of the stored energy becomes heat. When you use an electric drill, not all of the electrical energy is converted to the kinetic energy of the drill. Some is lost as sound energy and some as heat energy (the drill becomes hot).

Figure 3.10 Rubbing your hands together converts kinetic energy into heat energy.

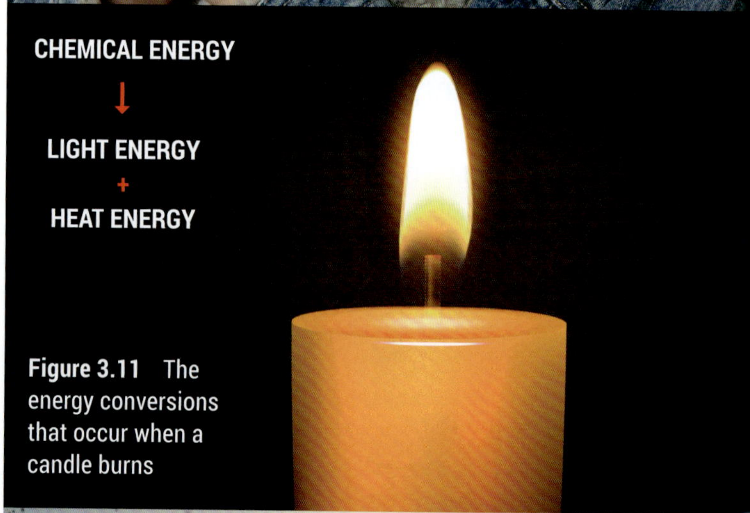

KINETIC ENERGY
↓
HEAT ENERGY

CHEMICAL ENERGY
↓
LIGHT ENERGY
+
HEAT ENERGY

Figure 3.11 The energy conversions that occur when a candle burns

Figure 3.12 Can you describe the energy changes in this bike as it goes downhill? What happens to the energy when the brakes are applied?

ISBN 978 1 4202 3735 1

INVESTIGATION 3.2 Observing energy changes

Aim

To observe the energy changes that occur in a variety of situations.

Risk assessment and planning

Discuss the safety issues for each part of the investigation. Draw up a data table like the one shown below. For each part, you will record the energy conversion(s) that occur, and any energy transfer(s) from one place to another without an energy conversion. (You may need to discuss this with others.)

Part	Observations	Energy conversion(s) that occurred	Energy transfer(s) that occurred

PART A

Materials

- piece of magnesium ribbon 1–2 cm long
- pair of metal tongs
- Bunsen burner
- heatproof mat

Warning: Do not look directly at the burning magnesium. Look to one side. The light is very bright and could damage your eyes.

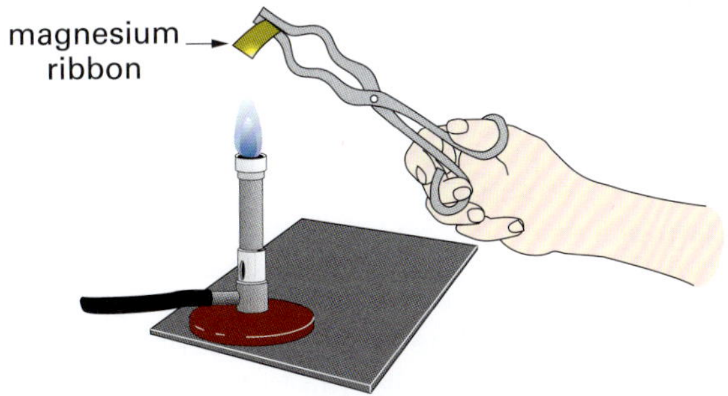

Method

Light the burner. Use the tongs to hold the magnesium in the flame until it starts to burn. Then take it out of the flame and hold it over the heatproof mat.

PART B

Materials

- 6-volt battery
- 3 connecting wires with alligator clips
- heatproof mat
- few strands of steel wool
- switch

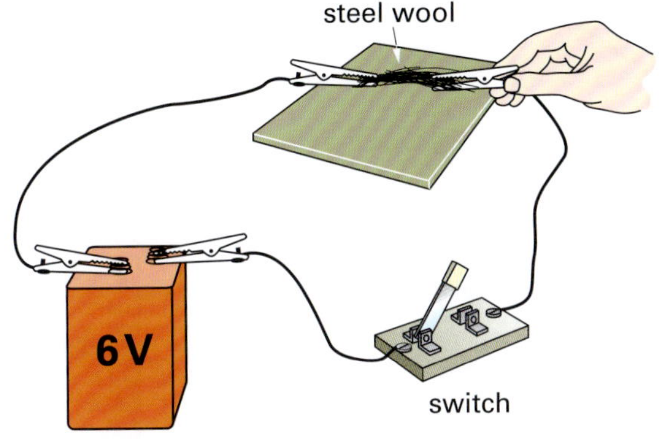

Do not leave the switch on.

Method

Use the wires to connect the battery and switch as shown. Put the steel wool on the heatproof mat. Connect the wires to it. Press down the switch for a few seconds. Observe what happens.

PART C

Materials

- piece of nichrome wire or iron wire about 50 cm long
- 2 pieces of copper wire about 50 cm long
- multimeter
- Bunsen burner
- beaker of crushed ice

Method

Sandpaper the ends of the copper wires, then twist the ends of the three wires together tightly as shown. Connect the ends of the copper wires to the terminals of the multimeter. (The multimeter detects small electric currents.) Put one junction in the crushed ice and heat the other junction until it gets red hot. Observe the multimeter carefully.

What you have made here is called a *thermocouple*. It is used to measure temperatures in ovens and furnaces.

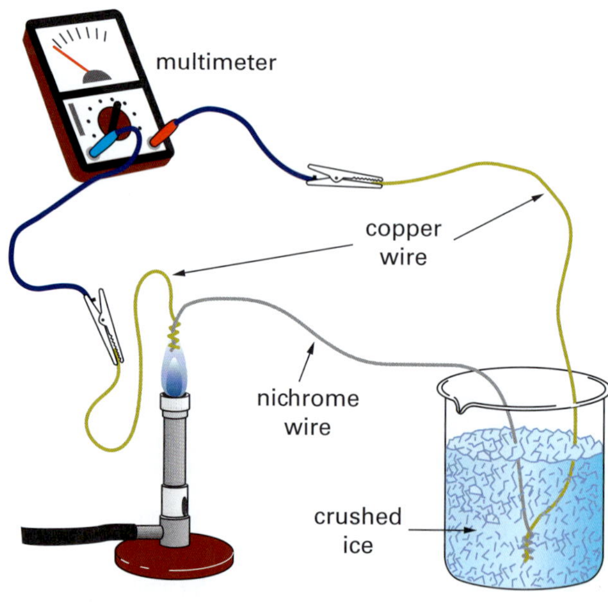

Be careful not to touch the hot wires.

PART D

Materials

- solar cell kit (consisting of several solar cells connected to an electric motor, preferably fitted with a propeller)

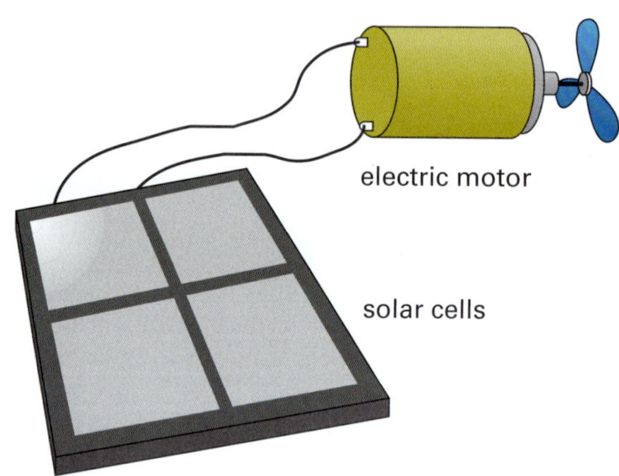

Method

Place the solar cell kit in bright sunshine. What happens if you cover all or some of the solar cells?

PART E

Materials

- beaker of water
- tuning fork

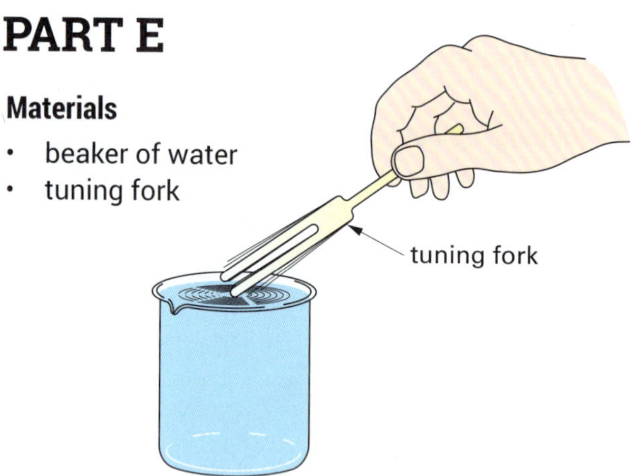

Method

Strike the forked end of the tuning fork gently on the heel of your shoe (not on the bench). Hold the fork near your ear. Strike the fork again, but this time look closely at the prongs.

Strike the fork a third time, and touch the surface of the water in the beaker with the vibrating prongs.

ISBN 978 1 4202 3735 1

1 Copy and complete each of these sentences.
   a A moving object has _____ energy.
   b Energy that is stored is called _____ energy.
   c A boulder rolling downhill is losing _____ _____ energy, but gaining _____ energy.
   d Burning a piece of coal changes _____ potential energy into _____ and _____ energy.
   e Springs can _____ energy, which can be released later.

2 Make two columns, one headed 'Kinetic energy', and the other 'Potential energy'. Place each of the following in the correct column.
   a an archery bow ready to shoot an arrow
   b a running high-jumper just before leaving the ground
   c a jet plane at the point of take-off
   d at the top of your bounce on a trampoline
   e a spring-loaded popgun
   f a child's swing at its highest point
   g a child's swing at its lowest point

3 What are the two types of potential energy?

4 The two rocks below have the same mass. Which one has more potential energy? Why?

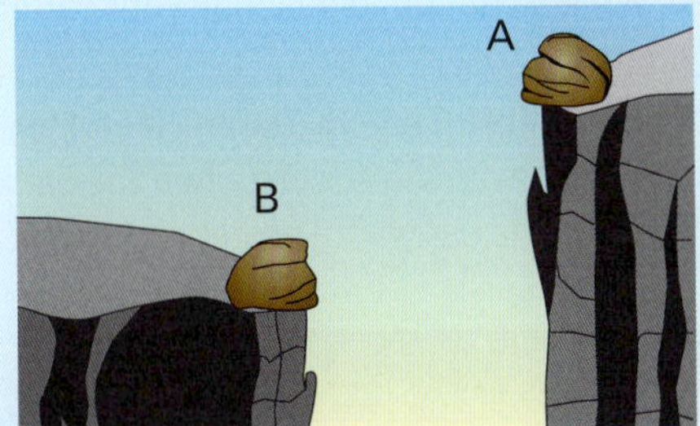

5 What is the difference between an energy transfer and an energy conversion? Give examples.

6 What form(s) of energy do the following have?
   a a diver standing at the top of a tower
   b a bent ruler
   c a block of chocolate
   d a burning log
   e a glowing firefly
   f a lightning flash
   g ocean waves
   h a slice of bread
   i a TV set (turned on)
   j a warm pizza
   k the water in a waterfall
   l a wound-up toy

7 Pair up these lists correctly in your notebook.

Object	**Main energy conversions**
battery	electrical to sound
electric motor	electrical to light and sound
lift going up	chemical to kinetic
solar cell	chemical to heat and light
radio	nuclear to electrical
TV	chemical to electrical to light
torch	light to electrical
car	chemical to electrical
campfire	electrical to kinetic
nuclear power station	electrical to kinetic to gravitational

8 Maria connected a coil of wire to a milli-ammeter, as shown. When she pushed a magnet quickly into the coil, the ammeter showed that there was an electric current flowing. When she stopped moving the magnet, no current flowed. Write a sentence describing what happened in terms of energy changes.

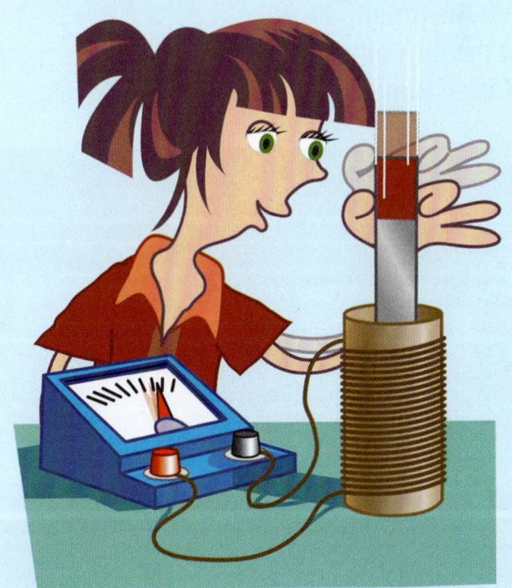

9 Go back to Get started on page 55. How have your ideas about energy changed after working through this chapter?

ISBN 978 1 4202 3735 1

CHALLENGE

1 Copy and complete the table below.

Energy used	Energy converter	Energy produced
________	light globe	________
________	electric fan	________
________	petrol engine	________
kinetic	________	electrical
________	torch cell	________
________	steam engine	________
________	atomic bomb	________
electrical	________	heat
________	slingshot or catapult	kinetic
________	waterwheel	________
kinetic	________	sound

2 What energy changes are being described in each of the following? Use arrows as in Figure 3.10 and Figure 3.11 on page 62.

a The wind blew hard, turning the windmill noisily as it pumped the water from deep underground into the trough.

b At the flick of a switch, the washing machine started turning and churning the clothes.

c '... two, one, zero.' The rocket belched fire and smoke, the ground shook and, with a deafening roar, the rocket left the launch pad.

d The lightning flashed and the thunder crashed. The gum tree was split right down the middle.

3 List at least three different things in which chemical energy is stored.

4 Into what forms of energy does the human body convert the chemical energy in food?

5 If a neon street light converts 300 J of electrical energy into 200 J of heat energy and 90 J of light energy, how much sound energy is produced? (Assume these are the only energy conversions that occur.)

6 Give an example of something that has:

a gravitational energy due to its high position

b elastic energy because it has been stretched

c chemical energy.

In each case, explain how the energy can be used to produce movement.

7 How could you demonstrate that sound is a form of kinetic energy?

8 Draw a cartoon of a jack-in-the-box. Discuss with another student how potential energy is involved, and how this energy changes when the lid is opened. Write a caption to describe your cartoon in energy terms.

9 What is the source of energy for an electric car? What energy conversion occurs when the car is moving? Does the car waste any energy?

Figure 3.13 Tesla electric car

ISBN 978 1 4202 3735 1

EXPLORE

1 Build a mousetrap racer as shown. To make it go, simply wind the string around the axle by turning the rear wheels. Then put it on the floor and release it. What energy changes occur?

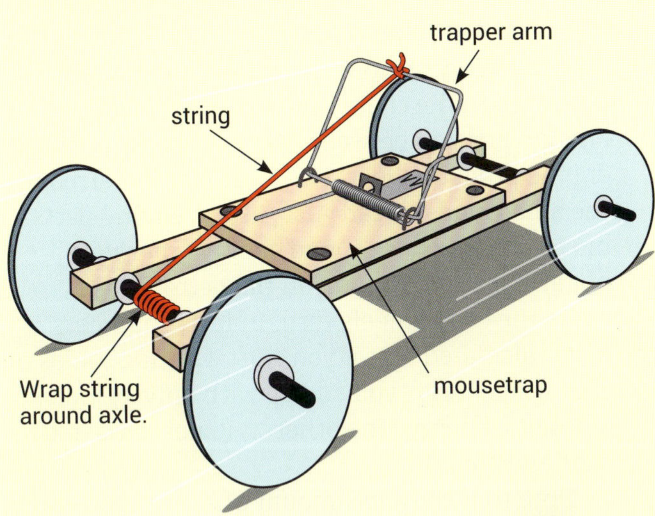

To find out more on mousetrap racers, follow the links to **Mousetrap racers**.

2 Make a working model of a waterwheel as shown below. In this device, the gravitational energy of the water is changed to kinetic energy, which is then transferred to the spinning wheel.

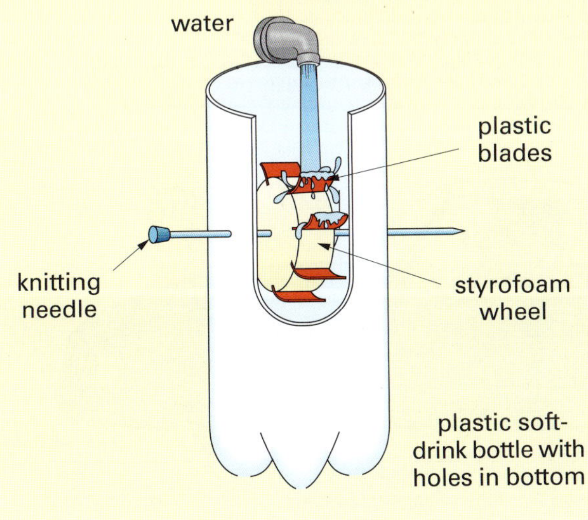

3 To make a windmill you will need two empty soft-drink cans, a wire coathanger, scissors and pliers.

Carefully cut the bottom 4 cm off a soft-drink can (A). Cut the top rim off a second can (B). Cut strips 2 cm wide to within 2 cm of the bottom of can B.

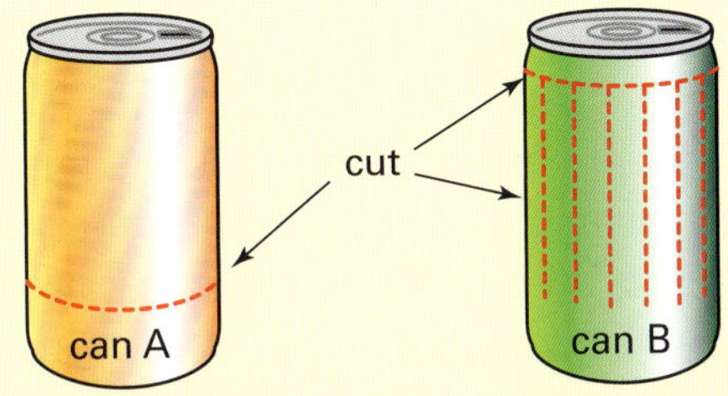

Put the base of can B into can A, then fold back the strips to form vanes, as shown in the photo below. Make a hole in the middle of the base of both cans. Put the coathanger wire through this hole and bend it as shown.

Find some moving air and watch it spin. Can you modify it to make it work better?

ISBN 978 1 4202 3735 1

3.3 Energy comes—energy goes

Wasted energy

When we use energy, it often changes from one form to another. For example, the cyclist in Figure 3.14 is using the chemical energy stored in his muscles to pedal his bike. So chemical energy is converted into useful kinetic energy. But as he pedals he gets hot. So some of his energy is wasted as heat energy. These energy changes can be shown with an energy arrow. The thickness of the arrow shows roughly how much energy is converted into the different types.

Sometimes one energy change follows another. The series of steps is called an **energy chain**.

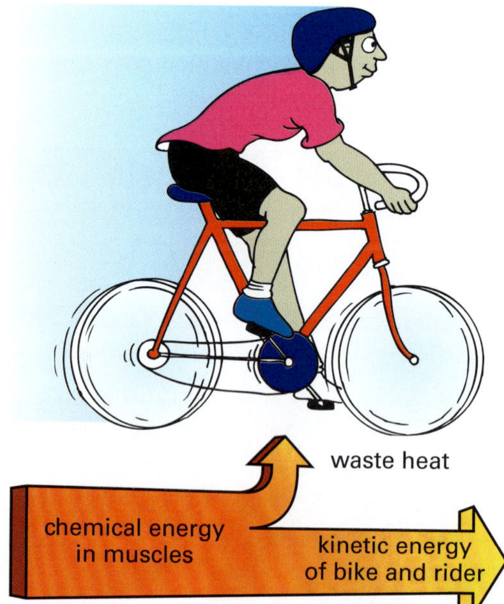

Figure 3.14 An energy arrow for riding a bike

For example, the energy chain for a moving car has three steps:

1 The stored chemical energy of the petrol is converted into heat energy when the petrol is burnt in the car's engine.
2 Some of this heat energy is converted into kinetic energy of the moving engine parts.
3 This kinetic energy is then transferred through the gears to the wheels.

The energy chain is not 100% efficient, since each step in the chain involves some loss of energy. Friction between the moving parts of the engine produces heat. This heat is transferred to the air around the car. Also, as the engine parts move, they produce sound energy. Therefore, not all the stored energy in the petrol is used to make the car's wheels turn. In fact, engineers have calculated that if you start with 100 joules of chemical energy, you end up with only 25 joules of kinetic energy. The other 75 joules is wasted as heat and sound.

Note that the *total* amount of energy you end up with is the same as the amount you started with. The 75 joules of waste heat and sound from the car is not useful, because it cannot be used again. All energy converters waste energy like this—usually as heat. The longer the energy chain, the more energy that is wasted.

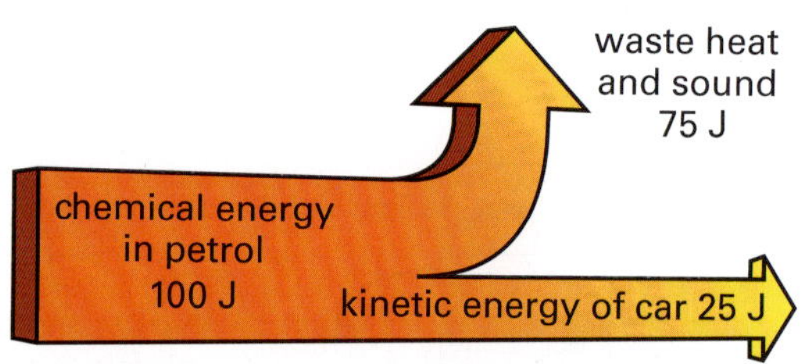

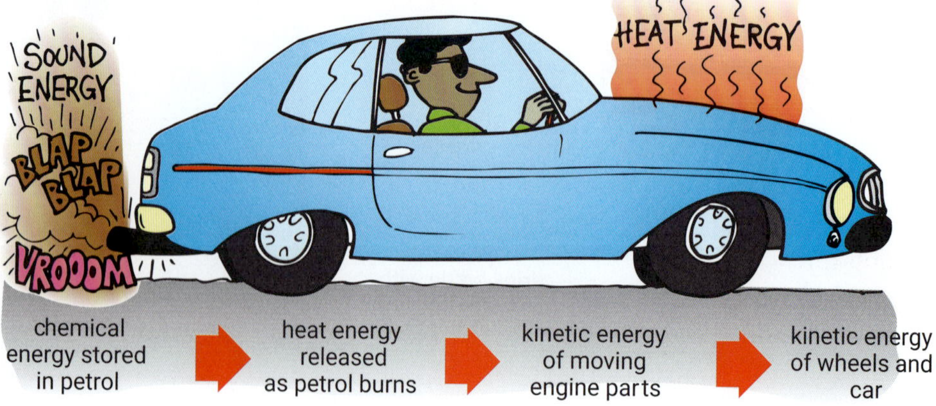

Figure 3.15 Energy chain and energy arrow for a car

ISBN 978 1 4202 3735 1

Energy efficiency

The **efficiency** of an energy converter is the percentage of the input energy that is turned into useful energy:

$$\text{efficiency} = \frac{\text{useful energy}}{\text{input energy}} \times 100$$

For example, the efficiency of a car is about 25%. Because there is always some waste energy, the efficiency of an energy converter is always less than 100%.

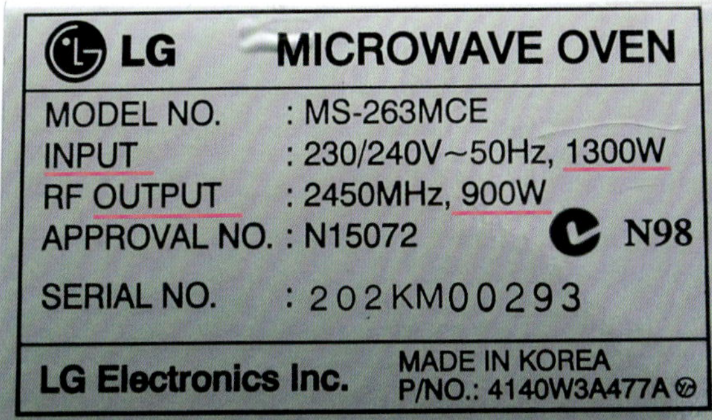

Figure 3.16 This label from a microwave oven shows that for every 1300 watts of electricity (1300 joules per second), the oven produces only 900 watts of heat. It is therefore 69% efficient.

Conservation of energy

You have looked at examples of how energy is converted from one form to another. After thousands of such observations, scientists decided that there is a special rule or *law* that describes energy changes.

> The law of **conservation of energy** says that energy cannot be made or destroyed—it can only be converted from one form to another.

This means that the universe always has the same amount of energy, even though this energy is constantly being converted from one form to another and being transferred from one place to another.

To help you understand the law of conservation of energy, think about a board game such as Monopoly, where money can be used for buying and selling. The money is transferred between players and the bank, but the total amount is always the same. At the end of the game, if all the players add up their cash, the total should be the same as at the beginning, although it will be distributed differently. The same applies to energy. It moves around and changes its form, but the total amount is always the same.

Where does the energy go?

Aim

To find out what happens to the heat energy as a container of hot water cools down.

Materials

- 2 L ice-cream container or similar
- 250 mL beaker or similar
- 2 thermometers
- boiling water
- graph paper
- stopwatch

> **Teacher note:** You could use a data logger with temperature probes.

Risk assessment and planning

Read through steps 1–6 so that you know exactly what you have to do.

You will need to do steps 2–4 quickly.

Make a list of any risks for this activity.

Design and draw up a suitable data table to record your results. You will be measuring the temperature inside and outside a beaker of hot water every minute for at least 15 minutes.

ISBN 978 1 4202 3735 1

Method

1 Use one of the thermometers to measure the temperature of the air (room temperature).

2 Put the beaker in the ice-cream container as shown below. Add 200 mL of hot water to the beaker—be careful not to burn yourself.

3 Pour 1500 mL of cold water into the ice-cream container.

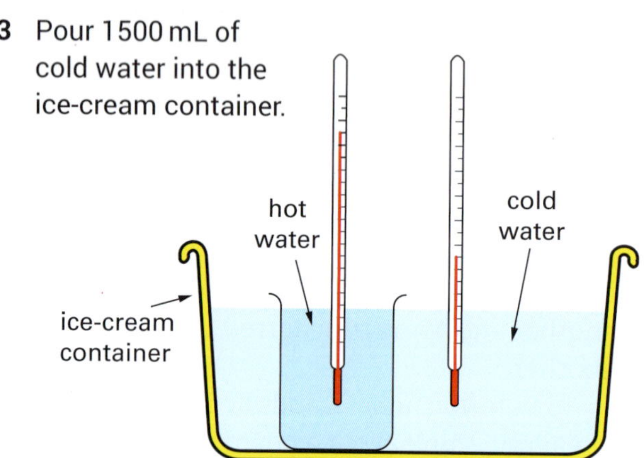

4 Place one thermometer in the beaker and the other in the ice-cream container. Start the stopwatch and measure the temperature inside and outside the beaker.

Record these temperatures in your data table (for time = 0).

5 Measure the inside and outside temperatures every minute, using the thermometers to stir the water gently. (Don't take the thermometers out of the water.)

Keep taking temperatures for 15–20 minutes.

6 Plot both sets of results on a graph of temperature (vertical axis) versus time (horizontal axis). Draw a smooth curve for each set of points. (The curve doesn't have to go through each point—as long as it shows the general trend of the results.)

Label the two curves and draw a third line on your graph to represent room temperature during the experiment.

Discussion

1 Copy and complete the following summary.

As the temperature of the water in the beaker decreased, the temperature in the ice-cream container ____. The water in the beaker ____ energy, while the water in the ice-cream container ____ ____.

2 Which is the independent variable, and which is the dependent variable?

3 Calculate how much heat energy the water in the beaker lost (volume of water in mL × rise in temperature).

4 Calculate how much heat energy the water in the ice-cream container gained.

5 Are the two amounts of heat energy the same? If not, explain why they are different.

6 Describe the transfer of heat energy in this experiment. Do you think that the total amount of energy changed? Explain.

7 On your graph, look at the curve for the water inside the beaker. The curve is steep to start with, then levels out. Suggest a reason for this.

8 Predict what would happen to the temperatures inside and outside the beaker if you continued this experiment for an hour or more.

Where does energy come from?

We use a lot of stored energy every day. We get food from the shops, petrol from the service station, and electricity through power lines. But where does the energy in these things come from in the first place? The sun is the source of all of this energy on Earth.

Figure 3.17 The main source of energy on the Earth is the sun, producing food and supplying the energy we use every day.

ISBN 978 1 4202 3735 1

Green plants use the sun's energy to make food (chemical energy) through the process of photosynthesis. Animals that eat these plants use most of the energy for their body activities and store the rest. This means that all animals that eat plants and/or other consumer animals are using stored energy that originally came from the sun.

Most of our electricity in Victoria comes from power stations that burn coal to produce steam. This steam is used to turn generators that produce electricity. The petrol we use in our cars is produced from oil, and we also use natural gas for heating and cooking. Coal, oil and natural gas are called **fossil fuels** because they were formed from plant and animal remains, which originally gained their energy from the sun.

Even most renewable energy comes from the sun. It can come directly as solar energy, but even wind energy is the result of the sun heating the atmosphere and causing wind.

Renewable or non-renewable?

There is a major problem in using fossil fuels as a source of energy. They are **non-renewable**. They have taken millions of years to form from energy that came originally from the sun. Yet once they have been burnt in our cars or in power stations, they are gone forever. This is why we say they are non-renewable. The process of obtaining energy from fossil fuels is also very inefficient, as shown below. In fact, there is more energy reaching the Earth in 10 days of sunlight than in all the fossil fuels on Earth!

It makes much better sense to use **renewable** energy sources like those you learned about in *ScienceWorld 7*. We now have the technology to capture the sun's energy directly for our use. For example, solar cells are used to provide power supply systems for remote and rural areas. Hydro-electricity and wind power are other renewable energy sources.

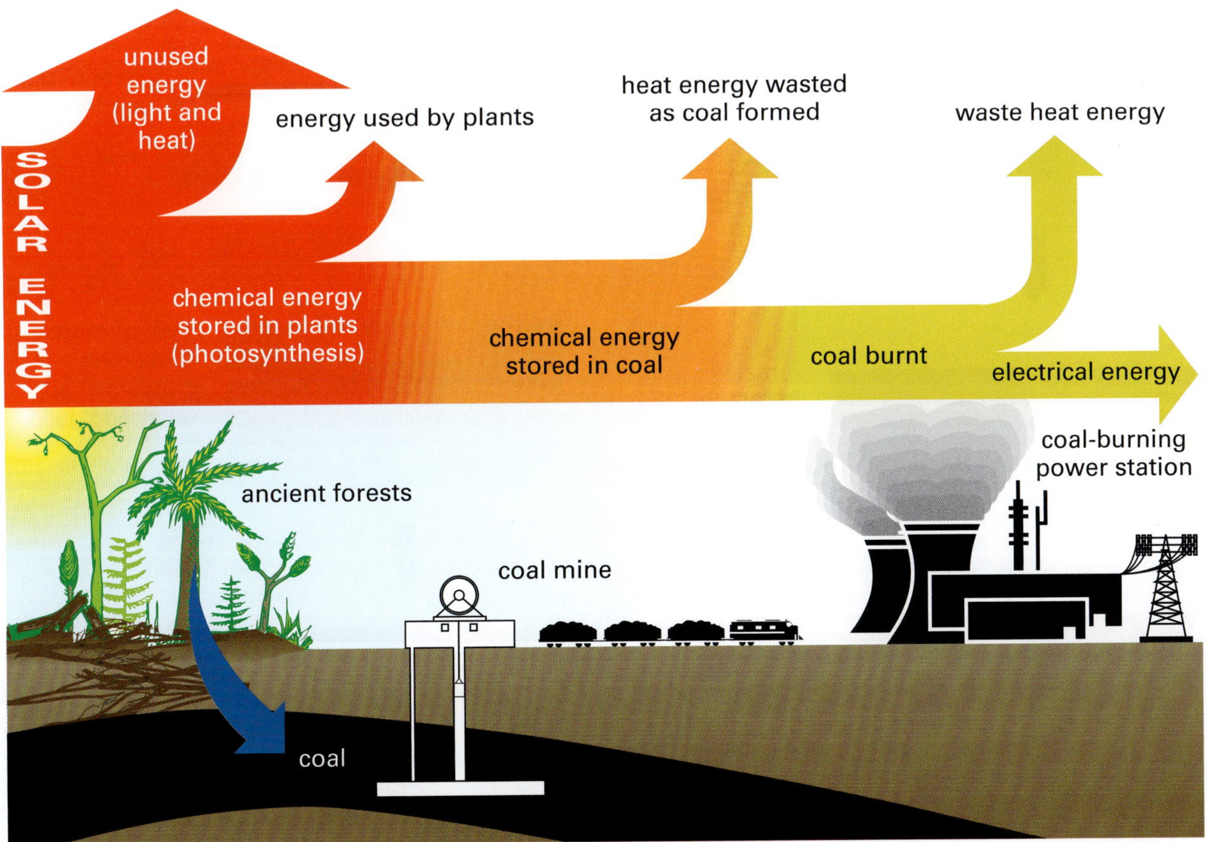

Figure 3.18 An energy arrow showing how the electrical energy we use came initially from solar energy. Notice how much energy is wasted at each step.

ISBN 978 1 4202 3735 1

How coal was formed

When geologists examine the fossilised plants in coal, they find that these remains come from plants that no longer exist on Earth. They infer that these plants probably grew in moist, warm swampy forests about the time dinosaurs roamed the Earth. This suggests that present coal deposits probably formed from ancient plants that existed millions of years ago.

Over a period of time, the climate changed and the plants in these forests died, leaving layers of decaying wood and other plant material. Sediments such as sand and mud were then deposited on top of the old forests, trapping the plant material.

As more and more sediments were deposited, the weight of these layers forced out much of the water and gases from the plant material, making it richer in carbon. Thus began the slow change over millions of years from wood to coal.

How oil was formed

Geologists infer that oil was formed from microscopic plants and animals that died and then settled to the bottom of shallow seas and lakes.

The remains of these marine organisms were quickly covered by sand and mud. After being buried by thick sediments and subjected to heat and pressure over millions of years, biochemical processes formed crude oil, various gases and water. At the same time, the sediments hardened to form rock.

Once formed, the oil and natural gas slowly seeped towards the Earth's surface through porous rocks such as sandstone, which soak up the oil like a sponge. Sometimes the oil and gas were trapped (often under pressure) beneath a layer of non-porous rock such as shale, through which they could not escape. To extract the oil and gas, a pipe has to be drilled down through the rocks above.

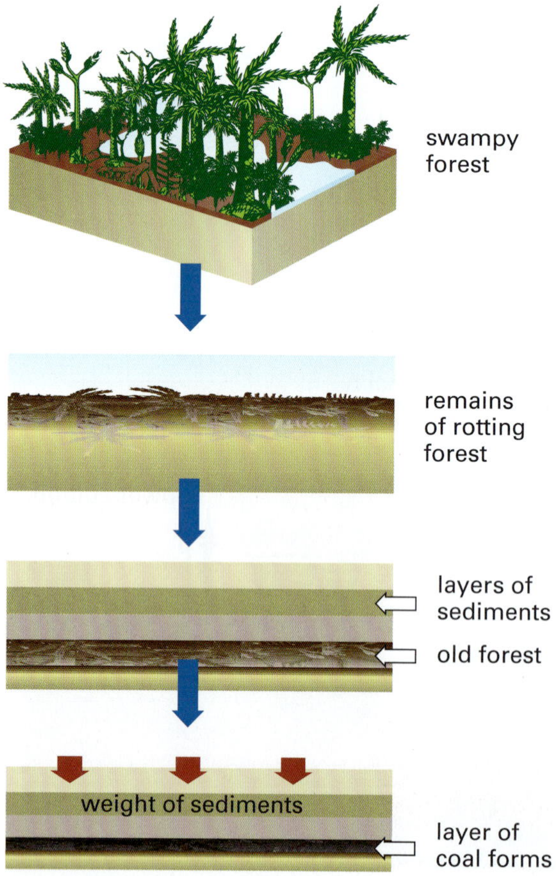

swampy forest

remains of rotting forest

layers of sediments

old forest

weight of sediments

layer of coal forms

Figure 3.19 Formation of coal

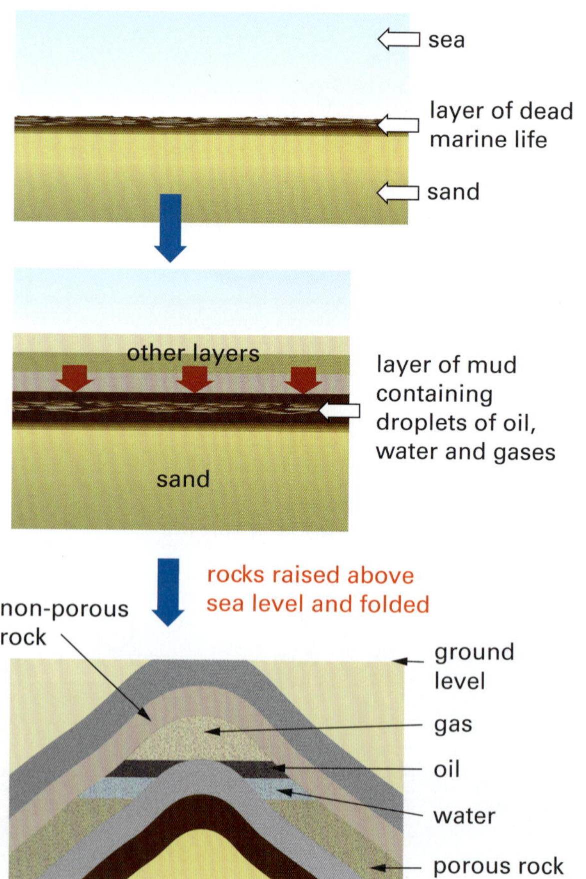

Figure 3.20 Formation of oil

ISBN 978 1 4202 3735 1

SCIENCE AS A HUMAN ENDEAVOUR

Nuclear power station inquiry

Imagine there is a proposal to build Australia's first nuclear power station at Nelson Bay, just north of Newcastle. There are individuals and groups who have many different viewpoints on this proposal. So a public inquiry is to be held in Newcastle to discuss the new power station, and to vote on whether it should be given the go-ahead.

For the inquiry, the class will be divided into seven different groups:

- *For*—three groups are in favour of the nuclear power station.
- *Against*—another three groups are against the power station.
- *Undecided*—the rest of the class is undecided and it is their job to develop a set of questions to ask the speakers before they vote.

Each of the groups is to prepare a 3-minute speech for the inquiry, using the brief notes in the box. You will need to do research to fill out the details of your argument for or against the proposal. You will also need to elect a speaker to present the case prepared by your group.

A chairperson will organise the inquiry and keep order. You will start with a speaker from the 'for' side, then one from the 'against' side, and so on. The undecided group will be given time to ask their questions. Finally, each of the members of the undecided group will vote for or against the power station, based on the arguments presented by the groups.

FOR	AGAINST
Federal government Australia must reduce its greenhouse gas emissions, and nuclear power stations don't produce carbon dioxide as coal-burning power stations do.	**Australian Conservation Foundation (ACF)** Nuclear wastes are radioactive for thousands of years and the nuclear industry does not have a long-term storage plan.
Australian Nuclear Science and Technology Organisation (ANSTO) Coal-burning power stations produce sulfur dioxide gas and ash containing toxic heavy metals. Nuclear power stations don't.	**Australian Council of Trade Unions (ACTU)** Any accident at the power station is likely to release dangerous radiation, and there is a risk of earthquakes like the one that occurred in Japan in 2011.
Economists Australia has huge reserves of uranium, and nuclear power could be produced at a competitive price. The use of nuclear power would also reduce the cost of an emissions trading scheme.	**People for a nuclear-free Australia (PNFA)** Instead of funding a nuclear power station, the government should be encouraging the use of renewable energy sources such as wind and solar.

ISBN 978 1 4202 3735 1

1 Suppose you wind up a toy car and let it go.
  a Where did the energy needed to wind up the toy come from?
  b Where has this energy gone when the toy stops moving?

spring

2 When using a hacksaw to cut a piece of metal, the blade and the metal both become hot. Explain in energy terms why this happens.

3 Classify the following energy sources as renewable or non-renewable: coal, diesel fuel, LPG gas, ocean waves, the sun, uranium, wind, wood.

4 Copy the boxes and complete the two energy chains below.

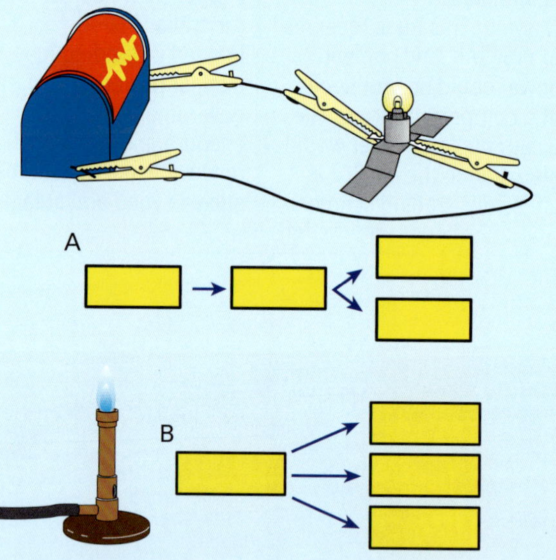

5 Draw an energy chain that shows the energy changes from the sun to the woman.

6 Explain in your own words how the petrol used in cars came originally from energy from the sun.

7 A hot water system is 65% efficient. If it is supplied with 3000 joules of electrical energy, how much heat energy does it produce?

8 To charge a battery you have to supply energy. But you never get as much energy from the battery as you use to charge it. Why is this?

9 The diagram below shows the energy changes in a coal-burning power station.
  a Draw an energy arrow to describe what happens in the power station.
  b How many joules of heat are lost to the environment for each 100 joules of chemical energy stored in the coal?
  c A small amount of energy is lost when the kinetic energy of the turbo-generator is converted to electrical energy. Infer how this energy is lost.
  d What is the efficiency of the turbo-generator?
  e What is the overall efficiency of the power station?

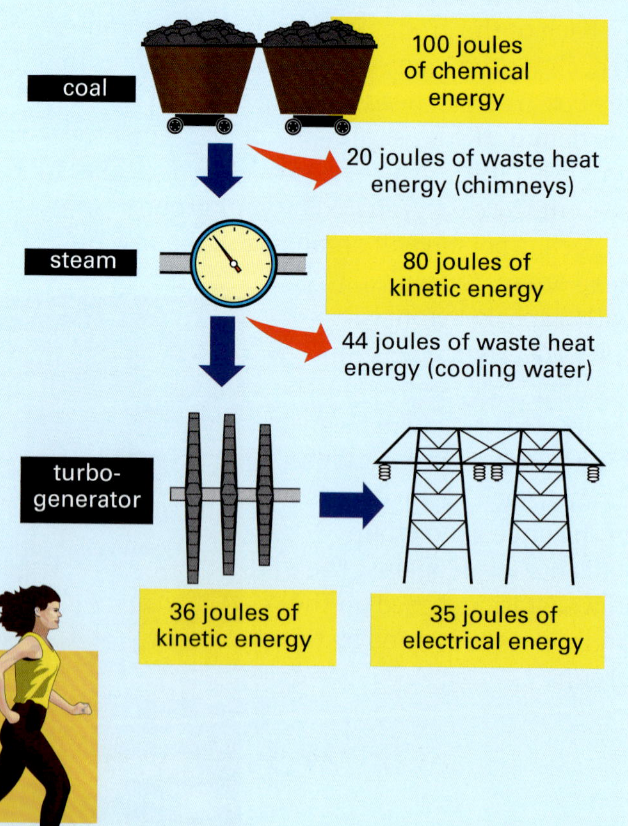

ISBN 978 1 4202 3735 1

1 Here are the efficiencies of five energy converters:

torch battery	90%
solar cell	25%
electric motor	60%
filament light bulb	10%
fluorescent light	20%
LED light bulb	80%

a Draw a bar graph to display this data.

b Draw a table that shows for each of the five energy converters the type of:

- input energy
- output energy
- wasted energy.

c Why is it cheaper to light schools with fluorescent lights than with filament light bulbs?

2 What form of energy does a frictional force usually produce?

3 Peter burnt his finger on a frypan. He immediately put his burnt finger in some crushed ice. Explain in energy terms what happened when he:

a burnt his finger

b put his finger in the ice.

4 Two cars collide head-on. What happens to the kinetic energy that each car had before the crash?

5 Machines that have moving parts can be made to run more efficiently. Use examples to explain how this can be done.

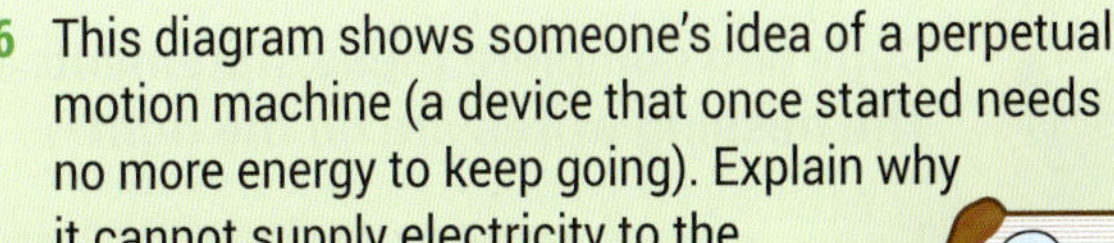

6 This diagram shows someone's idea of a perpetual motion machine (a device that once started needs no more energy to keep going). Explain why it cannot supply electricity to the house.

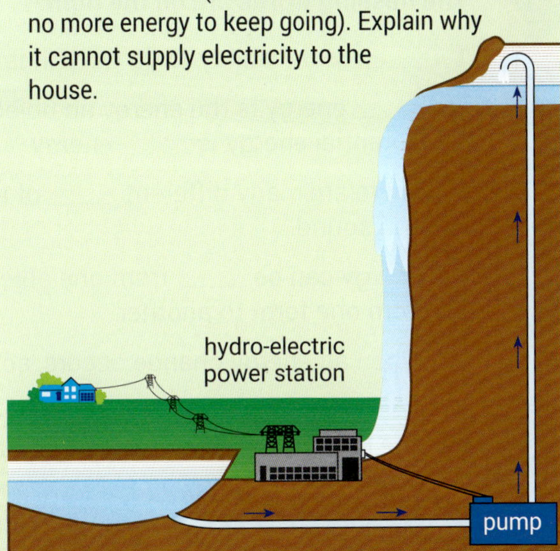

7 State the law of conservation of energy. Illustrate your answer by describing the energy changes that occur when a fireworks rocket takes off and explodes high in the air, emitting coloured balls of light as the remaining pieces fall to the ground.

8 Write a story (approximately a page) about 'The year the sun stopped shining'.

9 Look at the diagram below. Draw an energy chain tracing the energy changes from the sun to the energy user on the left.

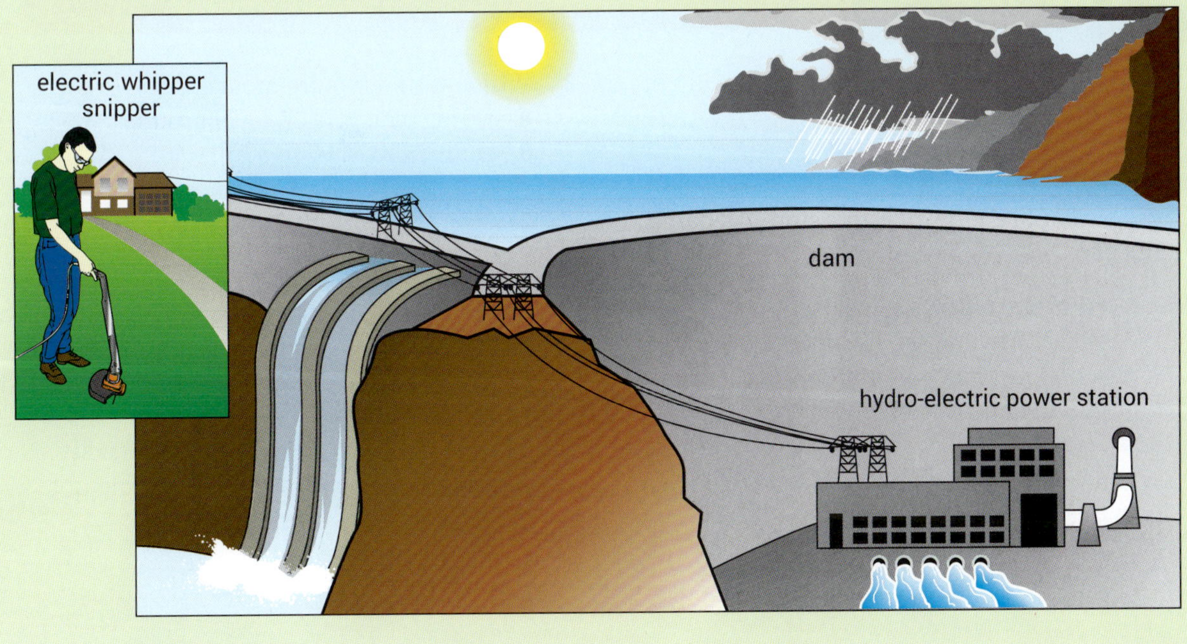

ISBN 978 1 4202 3735 1

MAIN IDEAS

Copy and complete these statements to make a summary of this chapter. The missing words are on the right.

1 _____ is the ability to do work. It is measured in _____ (J).

2 _____ energy is the energy an object has because of its movement. Potential energy is _____ energy.

3 There are many different _____ of energy; for example, light, heat, _____ and sound.

4 Energy can be _____ from one object to another, and it can be _____ from one form to another.

5 When an energy change occurs, some energy is always wasted as _____.

6 The law of _____ of energy says that energy cannot be made or destroyed.

7 Most forms of energy (including fossil fuels) can be traced back to the sun using energy _____.

8 _____ energy sources such as solar energy can be replaced as they are used. Non-renewable sources such as _____ cannot be replaced when they are used.

heat
chains
energy
kinetic
stored
converted
electricity
renewable
forms
coal and oil
transferred
joules
conservation

CH•3 REVIEW

1 The electricity you use in your home is a form of energy that came originally from:
- **A** electricity in thunderstorms.
- **B** coal.
- **C** the potential energy of water stored in dams.
- **D** the sun.

2 Which one of the following is false?
- **A** If an object has energy, it can do work.
- **B** A raised object has potential energy.
- **C** Energy can appear from nowhere and also disappear.
- **D** When you hit something, you are transferring energy.

3 Which would require most energy?
- **A** riding a bicycle on level ground
- **B** riding a bicycle up a hill
- **C** walking
- **D** doing your homework

4 Which of the following involves a transfer of energy from one object to another, rather than a change in the form of the energy?
- **A** Hot tea poured into a cup makes the cup hot.
- **B** A hydro-electric power station uses running water to generate electricity.
- **C** The tyres of a moving car become hot.
- **D** Oil is burnt to heat a room.

ISBN 978 1 4202 3735 1

5 a In which position does the roller-coaster car have the most gravitational potential energy?
  b In which position does it have the most kinetic energy?

6 For every 100 joules of energy used by an electric light bulb, you get only about 5 joules of light energy.
  a What happens to the other 95 joules of energy?
  b What is the efficiency of the light bulb?

7 A rock is held above a concrete path and dropped. Copy and complete the energy chain below, by putting the correct energy forms in the two empty boxes.

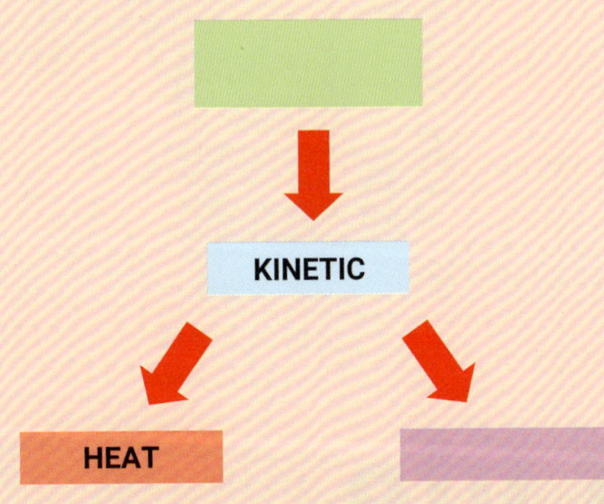

8 David said that electrical energy is made in power stations. Is he correct? Explain using the law of conservation of energy.

9 Bree found this data for Australia's energy use.

Energy source	Percentage of total
coal	41.8
oil	33.8
natural gas	19.6
hydro-electricity	1.1
wood, bagasse and other renewables	3.7

  a Draw a pie chart to display this data.
  b Which fossil fuels are used in Australia?
  c What percentage of Australia's energy use is from renewable sources?
  d Use a dictionary to find out what bagasse is.

10 Write an energy chain to describe the energy changes that occur in a hydro-electric power station (shown below).

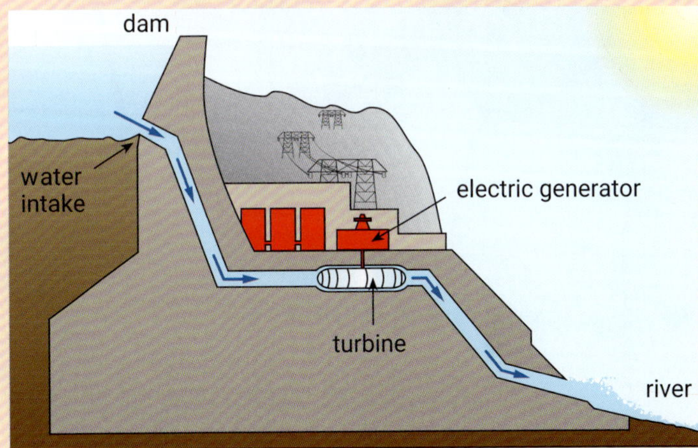

11 A ball bounces because the kinetic energy it has when it hits a surface changes to elastic potential energy as the ball is pushed slightly out of shape. This elastic energy then changes back to kinetic energy as the ball leaves the surface. Design an experiment to compare the efficiency with which different types of ball change their kinetic energy into elastic potential energy when they bounce.

Check your answers on page 295.

ISBN 978 1 4202 3735 1

IN THIS CHAPTER

Science Understanding

- examine a variety of cells using a microscope
- describe the structure of cells and the functions of the parts of cells
- explain how the structure of a cell relates to its function in tissues and organs
- describe how cell respiration provides the energy for many cell functions
- use a model to explain how substances enter and leave cells
- make choices and present arguments about the benefits and ethics of stem cell research

Science Inquiry Skills

- design experiments to investigate cells and their functions

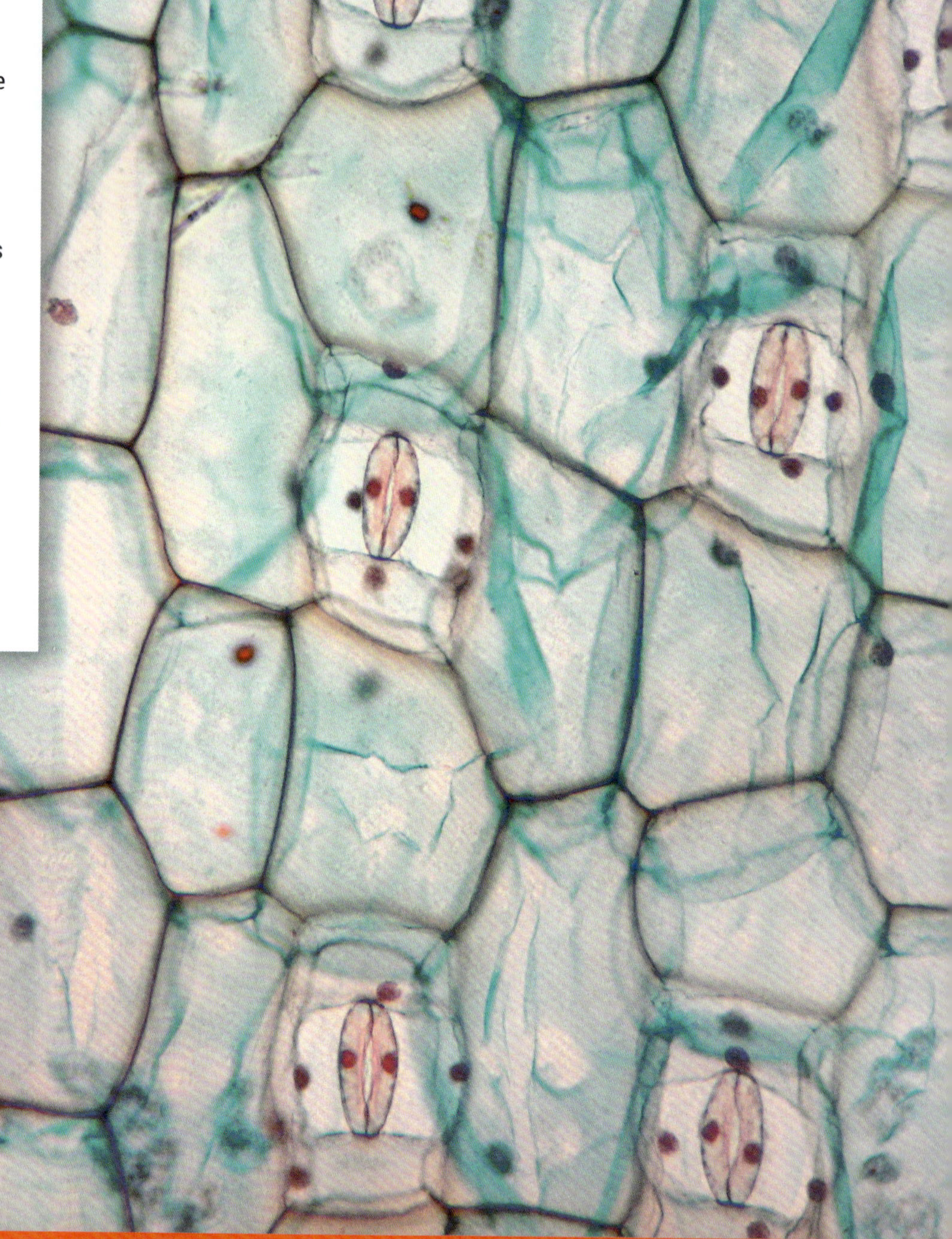

CH•4 Cells of life

ISBN 978 1 4202 3735 1

GET STARTED: *EXPLORE*

- Imagine you are using a magnifying glass to look at a tiny insect on a stick. The magnifying glass has ×2 on it. What does this mean?
- Another magnifying glass has ×4 on it. How is this different from the first one? What will you see if you look at the insect with this magnifying glass?
- The organisms in this photo live in freshwater ponds and creeks. What does the ×100 mean on the photo? Can you think of a way to find out how big these organisms are?
- If you looked at pond water without a magnifying glass, do you think you would be able to see these organisms?

ISBN 978 1 4202 3735 1

4.1 Cells

Cells are the building blocks of all organisms. Your body contains over 3 billion of them. Most cells are very small and can be seen only with a microscope. However, some cells, such as birds' eggs, are large enough to be seen with your eye. The ostrich egg is one of the largest single cells.

Unicellular organisms

Some organisms are *unicellular*. These single cells are complete organisms. The photo below shows microscopic organisms called euglena (you-GLEEN-a), which live in fresh water and contain chlorophyll to make their own food by **photosynthesis** (foe-toe-SIN-thu-sis).

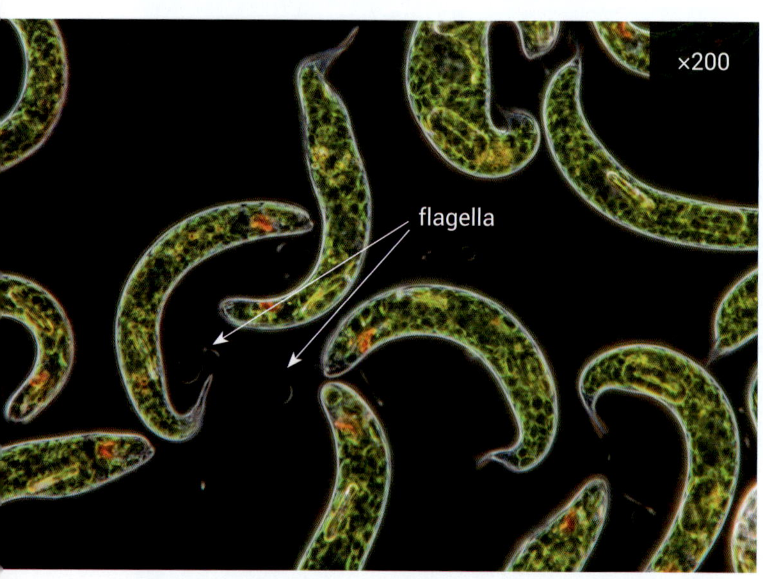

Figure 4.1 Euglena live in freshwater lakes and ponds. Long, whip-like 'hairs' called flagella at one end of the cell help it move through the water.

Multicellular organisms

Multicellular organisms contain many different types of cells and each type of cell is specialised. This means that each type of cell has a different job to do in the organism. For example, in humans, red blood cells carry oxygen, muscle cells contract and relax to move bones and organs, nerve cells conduct nerve messages, and stomach lining cells make substances that help in the digestion of foods.

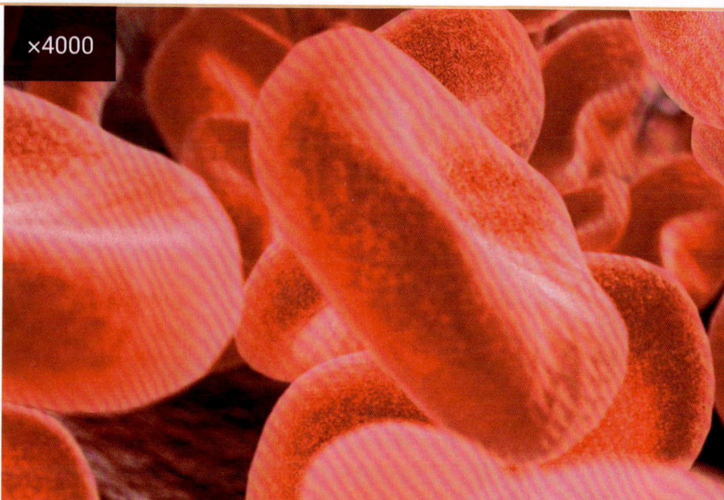

Figure 4.2 Red blood cells are specialised cells that carry oxygen around your body.

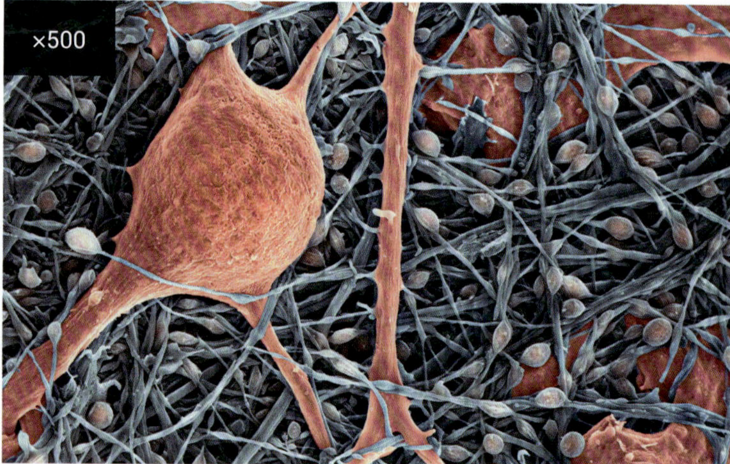

Figure 4.3 Nerve cells have an irregular shape. They carry nerve messages throughout your body.

The photos of the cells on this page are many times larger than the actual size of the cells. Each photo shows the number of times that the cell has been magnified. For example, the ×200 on the euglena photo means that the cells have been magnified 200 times. You can use this information to find the actual sizes of the cells.

Measure an average-sized euglena cell with your ruler. Then divide this by 200 (the magnification) and give your answer in millimetres.

Use this method to find the sizes of the other cells in the photos.

ISBN 978 1 4202 3735 1

SKILLBUILDER

Using a microscope

In this chapter you will be using a microscope to view different types of cells.

Parts of the microscope

Study the diagram below which shows the parts of a microscope. Your microscope may be slightly different from this one. However, the basic parts will be the same. If you are in doubt, ask your teacher for advice.

Setting up a microscope

1. Rotate the objective lenses until the low power lens clicks into position directly above the hole in the stage. (The low power objective lens is usually the shortest one, and has the lowest number stamped on it, e.g. ×4.)
2. Place a hair on a microscope slide and put it on the stage.
3. Looking from the side, turn the coarse focusing knob to move the lens very close to the slide.
4. Now look through the eyepiece lens and move the objective lens away from the slide until the hair is in focus.

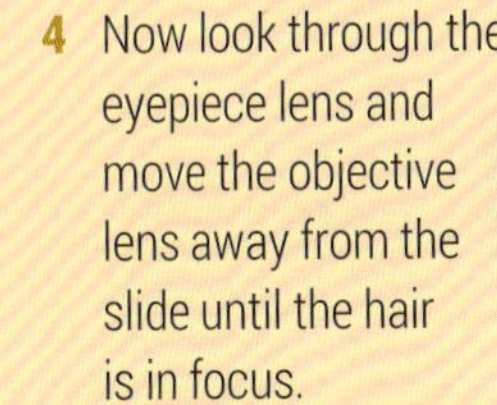

Rotate the higher power objective lens into place. (You might need to use the fine focus knob to make the image clearer.)

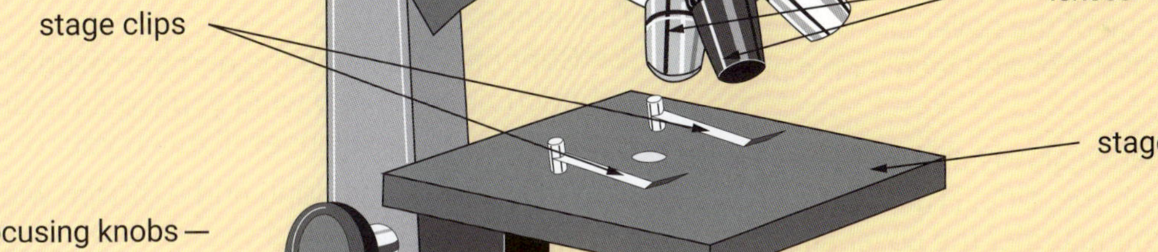

ISBN 978 1 4202 3735 1

What ×10 means

A microscope magnifies things. Each lens of a microscope has its magnifying power marked on it.

Look at the eyepiece lens. You may see the number ×10. This means that this lens magnifies things to 10 times their original size. The objective lenses are marked in the same way.

The total magnifying power of the microscope is found by multiplying the power of the eyepiece lens by the power of the objective lens. If the eyepiece is ×10 and the objective lens is ×10, then the microscope will magnify the object 100 times.

Observing prepared slides

Your teacher will give you a microscope slide containing some cells for you to practise your microscope technique.

Observe the shapes and features of the cells.

Questions

1. A microscope has a ×4 eyepiece and a ×10 objective lens. What is the total magnification of the microscope?
2. When focusing, why do you turn the focusing knob so that the objective lens moves away from the slide?
3. A hair is 0.005 mm wide. How wide would it be if you looked at it with the lenses in Question 1?

Making a wet-mount slide

1. Place a drop of water in the middle of a microscope slide.
2. Cut out a small lower case 'e' from a piece of newspaper and place it on the drop of water on the slide. Cover the 'e' with another drop of water.
3. Place the edge of the cover-slip on the edge of the drop of water, and lean it on a pencil, as shown.
4. Lower the pencil slowly and let the cover-slip fall flat on the slide. (This stops air bubbles forming under the cover-slip.) You should do this a few times to master the skill. Show your slide to your teacher.
5. Place the slide on the stage and observe the letter under low power.

   Record your observations. Is the 'e' the right way up? Move the slide to the left. Which way does the 'e' move when viewed through the lens?

Questions

1. Suppose you place the number '5' under the microscope. Draw what you would expect to see through the lenses. Explain your drawing.
2. A cell is 0.01 mm long and 0.02 mm wide. How big would it be if you viewed it under a microscope with a ×10 eyepiece lens and ×4 objective lens?

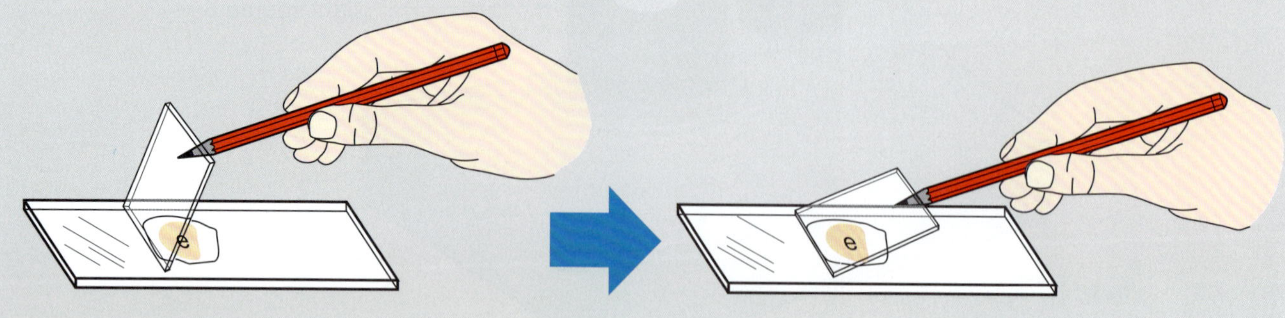

ISBN 978 1 4202 3735 1

Cell structure

The cells of living things vary in shape and function, but they do have features in common.

All cells are surrounded by a thin covering called a **cell membrane**, which acts like a fence controlling the movement of substances into and out of the cell. The cell membrane also helps to hold the cell together and to give it shape.

The round, dark-coloured structure in the cells in the photos below is the **nucleus** (NEW-klee-us). This controls all the cell's activities, and without it the cell eventually dies.

The inside of cells is filled with the jelly-like **cytoplasm** (SIGH-toe-plaz-um). This is where many chemical reactions take place. The cytoplasm also contains many other small bodies and structures called **organelles** (OR-gan-els). These help to keep the cell functioning correctly.

How are plant cells different from animal cells? Plant cells have a **cell wall** on the outside of the cell membrane. This is a thick, tough layer that protects the softer parts inside the cell and also provides stiffness that helps support the plant.

Plant cells also contain large liquid-filled spaces called **vacuoles** (VAK-you-oles) in which water and dissolved substances are stored. Some animal cells have small vacuoles, but most have none at all. Inside the cytoplasm of plant cells there are organelles called **chloroplasts**. These contain the green pigment chlorophyll, which is needed for photosynthesis. Photosynthesis occurs in the chloroplasts.

Animal cells

These cells are from the inside lining of a human cheek. The diagram below will help you interpret the photo.

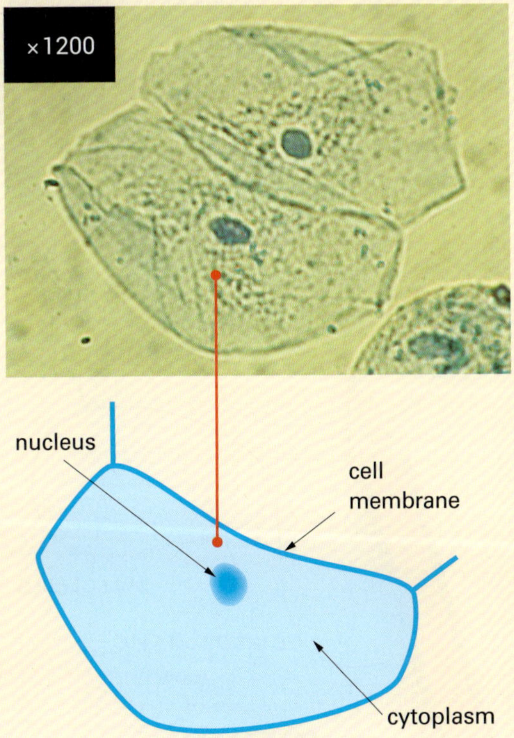

Figure 4.4 Animal cells

Plant cells

These cells are from the leaf of a plant. The diagram below will help you interpret the photo.

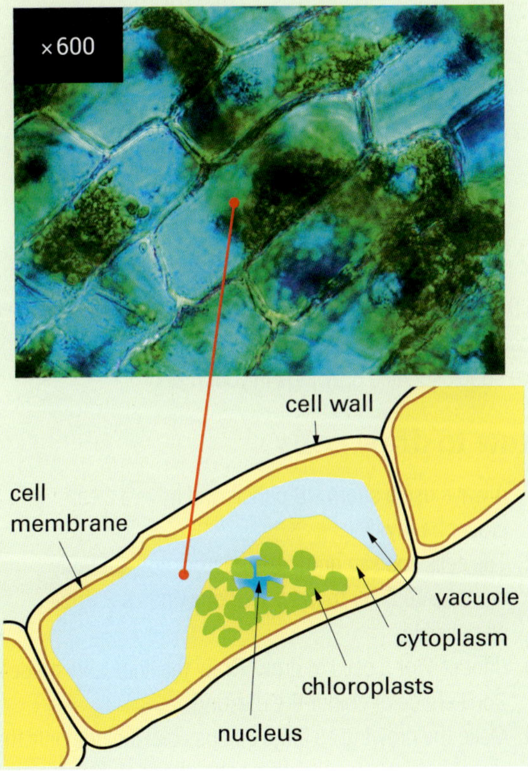

Figure 4.5 Plant cells

ISBN 978 1 4202 3735 1

SCIENCE AS A HUMAN ENDEAVOUR

Yeast

Luke is a baker. He makes different kinds of bread with the help of a unicellular organism called *yeast*.

When making bread, Luke adds the basic ingredients—flour, sugar, water and yeast—and mixes them together to form dough. The dough is then left for a while in a warm place. During this time, the yeast cells grow and multiply rapidly using the sugar as a food.

Yeast cells get the energy needed for growth and reproduction by breaking down the sugar. Carbon dioxide and alcohol are produced as waste products. This process is called *fermentation*.

glucose → carbon dioxide + alcohol

The carbon dioxide gas given off by the yeasts causes the bread to rise and makes the holes in the bread. When the bread is baked, the heat of the oven quickly evaporates the alcohol from the dough.

Figure 4.6 Yeast is used to make bread rise.

SKILLBUILDER

Drawing cells

In the next investigation you will be using the microscope to observe some animal and plant cells. In these observations, you should include drawings in your report.

How to draw cells

1. Always use a sharp HB pencil, and have a clean eraser handy.
2. The cells you see under the microscope are fairly complicated. Try to keep your drawings as simple as possible.
3. Choose 2 or 3 cells to draw. Draw the lines and shapes. Don't shade or colour the drawing.
4. Make the drawing as large as possible. Include only the structures you can identify. Label these structures.

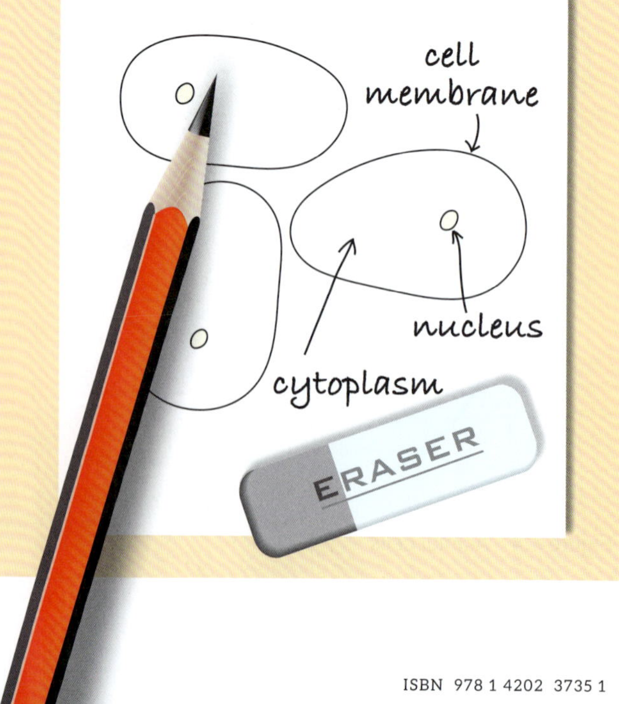

ISBN 978 1 4202 3735 1

INVESTIGATION 4.1 Observing cells

Aim

To use a microscope to observe plant and animal cells.

Materials

- microscope
- 4 microscope slides and cover-slips
- piece of onion
- methylene blue stain
- freshwater plant (e.g. elodea)
- small pieces of apple, mince meat, raw chicken, moss, potato, spirogyra, duckweed etc.

Risk assessment and planning

- Carefully read through Parts A, B and C, and make a list of the materials you will need for each part.
- Ask a partner to describe what they are going to do in Part A. Then you describe what you are going to do in Part B.

PART A Onion skin cells

Method

1 Remove one layer from the onion. Then peel a small piece of the very thin skin from inside the layer.

2 Put a drop of water on a slide then place the piece of onion skin on the drop. Add another drop of water on top of the onion skin. Then add a cover-slip as shown below.

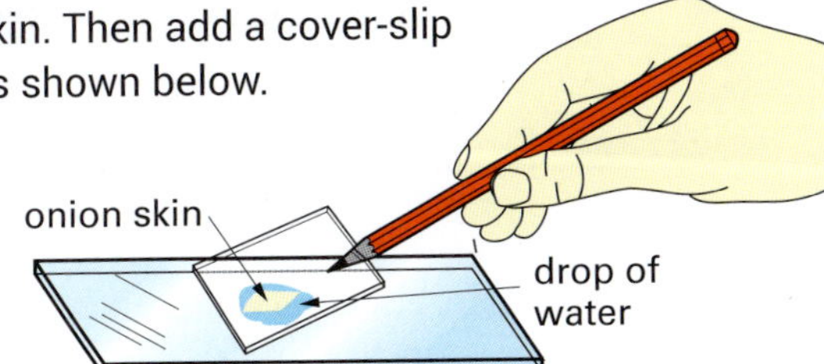

3 Repeat steps 1 and 2 with a second slide, but instead of adding water, add one drop of methylene blue stain. Then place a cover-slip over the onion skin.

4 Observe both slides under low power, then under higher power.

Record the differences between the two slides. In which one are the cells more easily observed? Which parts of the cell can you easily see?

Draw two or three stained cells. Label the cell wall, the nucleus and the cytoplasm.

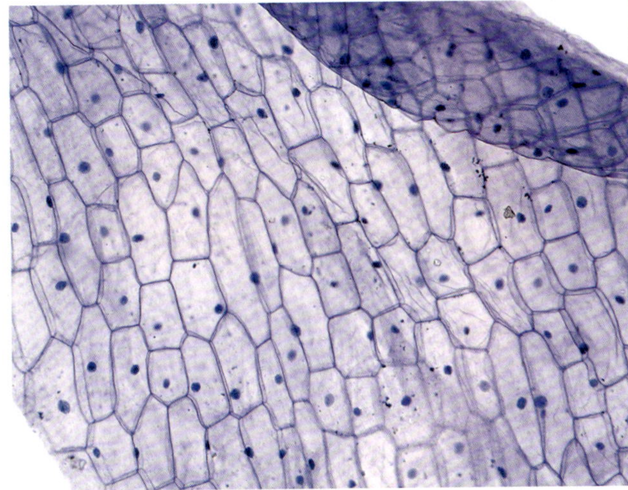

Figure 4.7 Stained onion cells

PART B Looking at chloroplasts

Method

1 Tear a small leaf from the top of the freshwater plant.

2 Prepare a slide as you did for the onion skin, but this time use the leaf. (You can use a drop of water or the methylene blue stain if you wish.)

3 Observe the leaf under low power, then under higher power.

Use the photo of the plant cells on page 83 to help you identify the round chloroplasts, the cell wall, the nucleus and the cytoplasm.

Draw a labelled diagram of what you observe. How do these cells compare with the onion cells from Part A?

ISBN 978 1 4202 3735 1

PART C Other cells

Method

1 For this part you will look at cells in apple, mince meat, chicken, moss, potato, spirogyra, duckweed etc.

2 Place a small amount of material on the end of a toothpick. Scrape it onto a slide.

3 Add a drop of water and a cover-slip. You can add a drop of stain if you wish.

 Observe the cells. Draw and label two or three of the cells.

Discussion

1 Why is a stain used when observing cells?

2 What general shape are the onion cells? Do other types of cells also have a regular shape? Do other cells have the same shape as onion cells?

Video microscope: Your teacher may connect a camera to a microscope to show you different types of cells.

EXTRA FOR EXPERTS

Sizes of cells

You have seen that cells have a variety of shapes depending on their function. Most animal and plant cells are about 0.005 mm to 0.02 mm in diameter. However, the longest cell is a type of nerve cell found in the giant squid and can be up to 7 m in length.

Bacteria

Bacteria are unicellular organisms and have a much simpler cell structure than animal and plant cells. A bacteria cell is usually smaller than other cells, ranging in size from 0.0005 mm to 0.003 mm.

Bacteria are usually classified by their shape. There are rod-shaped ones (bacilli), spherical ones (cocci) and spirals (spirilli). Bacteria have a cell wall, but no nucleus. This is the major difference between a bacterial cell and a plant or animal cell.

Follow the links to **Bacteria**. Use the websites to find out what different bacteria look like, where they're found, what they eat and how they move.

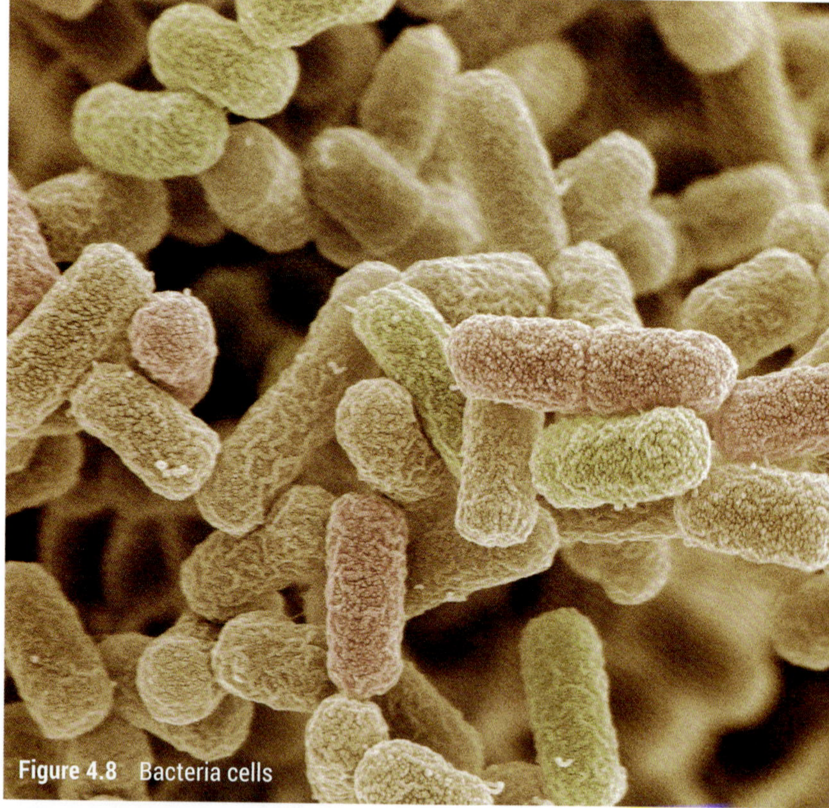

Figure 4.8 Bacteria cells

Questions

1 What is the main difference between an animal cell and a bacterial cell?

2 What is the average diameter of an animal or plant cell? What is the average diameter of a bacterial cell? How much larger is an average animal cell than an average bacterial cell?

ISBN 978 1 4202 3735 1

Cells, tissues and organs

Unicellular organisms such as euglena contain all the structures necessary to exist on their own and be independent from other cells. However, the cells in large, multicellular organisms are generally specialised, and therefore need to work together with other cells for the survival of the organism. For example, a single cheek cell cannot exist on its own for very long and will die after a short time outside the body.

Cells of the same type are generally found together in tissues. A **tissue** is a group of similar cells organised to do a particular job. For example, the muscle tissue in the wall of your stomach and gut is made from muscle cells. The nerve tissue in your brain and spinal cord is made from nerve cells.

In multicellular organisms, various tissues are arranged into a structure called an **organ**. An organ is a collection of specialised tissues that has a particular function. For example, a leaf whose main function is to make food contains food-making tissue, transport tissue, support tissue and lining tissue.

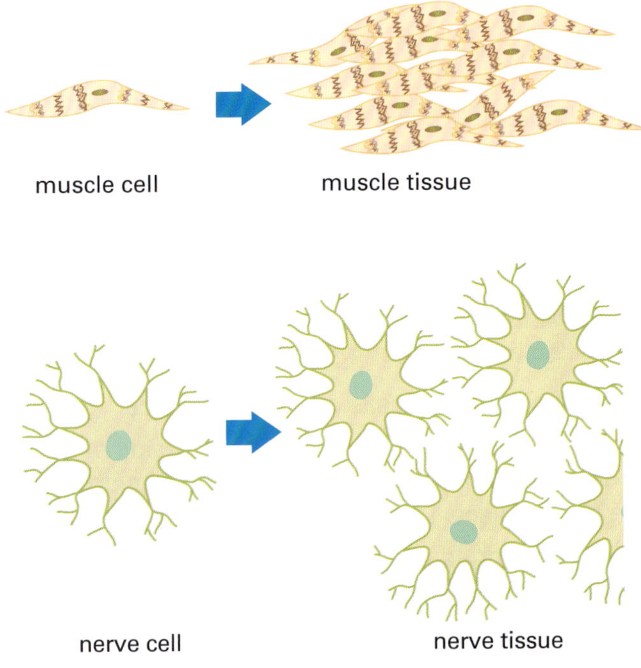

Figure 4.9 Many cells of the same kind combine to form tissues in the body.

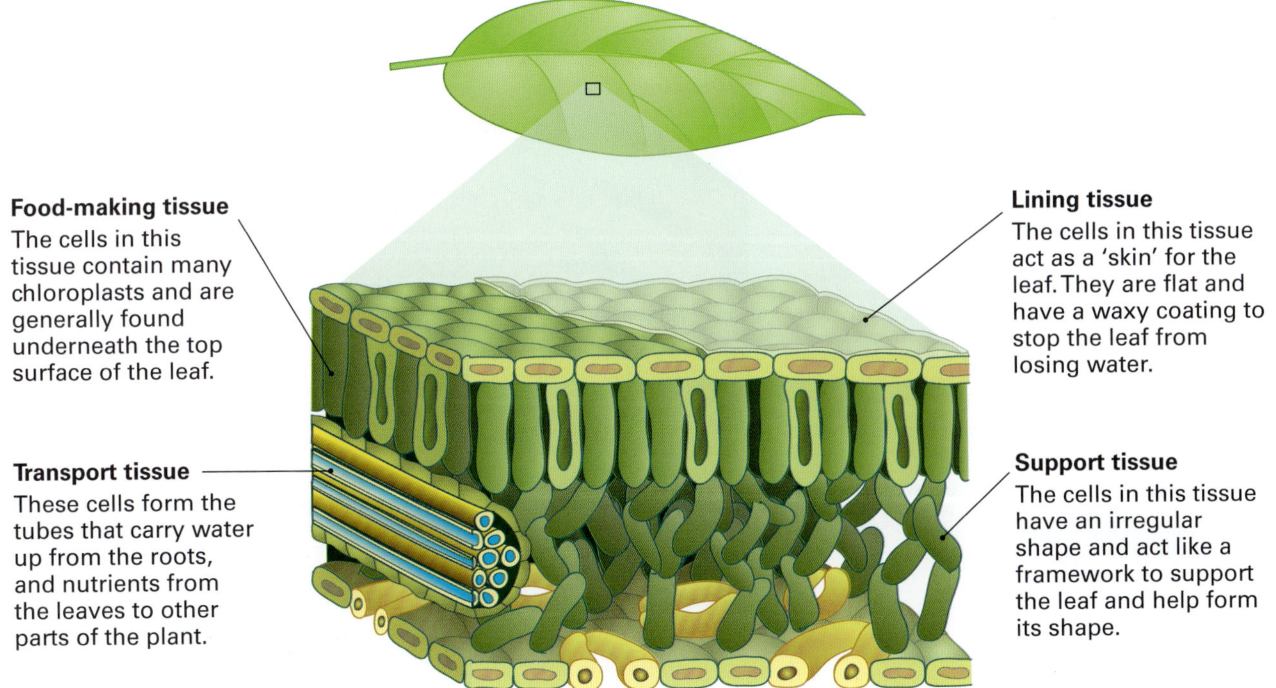

Figure 4.10 The leaf is an organ with many different tissues working together.

ISBN 978 1 4202 3735 1

You will need a microscope and slide, some prepared slides of various tissues, some clear nail polish and a leaf.

A Looking at tissues

Set up a microscope and ask your teacher for a prepared slide of a tissue.

Draw a sketch of the cells in a small section of the tissue (about six to ten cells).

Write down the name of the tissue (this will be written on the slide).

B Observing the cells on a leaf's surface

Brush some nail polish on the *underside* of a leaf, so that it covers an area about the size of a 20-cent piece. Let it dry for a few minutes.

Peel the dried nail polish from the leaf and look at it under a microscope. You will see a copy of the surface cells on the leaf.

Alternatively you can try and peel off the top layer of cells, stain them and view them under the microscope.

You will also see cells that form holes or pores in the surface of the leaf. Find out from the library what these pores are called. What is their function? You can see an image of the cells in the leaf surface on page 78.

The stomach is an organ whose function is to break down (digest) food. It contains glandular tissue that produces substances that chemically break down foods, muscle tissue that churns the food, and connective tissue that holds the other tissues together.

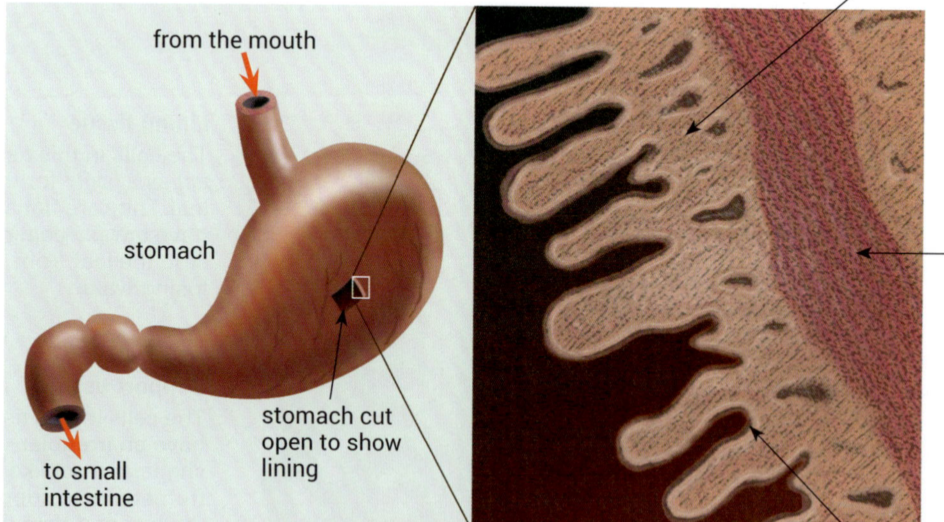

Connective tissue
This tissue is found between other tissues and helps to hold these tissues together.

Muscle tissue
The cells in this tissue contract and relax, thus helping to mix and move the digesting food in the stomach.

Gland tissue
The cells in this tissue make substances that help break down the food in the stomach.

Figure 4.11 The tissues in the stomach have a number of functions. Cells in the gland tissue make chemicals that help digest foods, and muscle tissue moves the stomach to help mix the food.

ISBN 978 1 4202 3735 1

CHECK

1 Draw up a table similar to the one below and list the features of plant and animal cells so you can compare them. One feature has been done for you.

Plant cells	Animal cells
Have a nucleus	Have a nucleus

2 A microscope lens has ×10 marked on it. What does this mean?

3 Copy the drawing of a cell below into your notebook. Use the information in the table above to determine whether it is a plant cell or an animal cell, then label the cell.

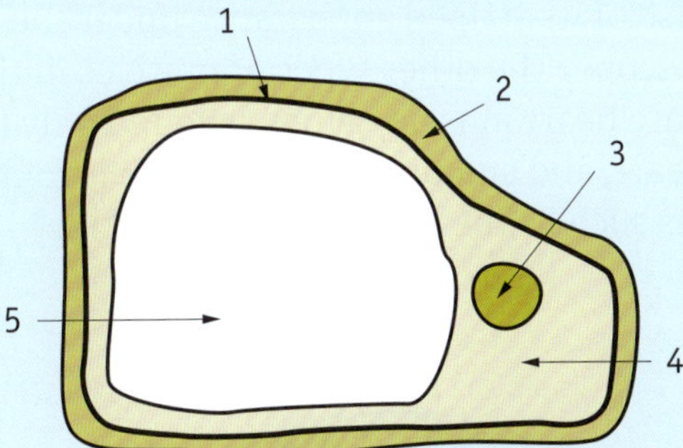

4 Describe the function of each of the five parts of the cell in Question 3.

5 Copy the following sentences into your notebook, then complete them using the words you have learnt in this section.

a Organisms are made of building blocks called _____.

b Cells in large organisms are called _____ cells, because they perform a particular function.

c The lens that you look through at the top of a microscope is called the _____.

d Organelles are found in the _____ of a cell.

e Chloroplasts are organelles that contain _____.

6 Explain the difference between a tissue and an organ. Give an example and use the words *cells* and *function* in your answer.

7 On page 80, the word *multicellular* was used. Explain what this word means.

8 Look at the diagram of the leaf on page 87. Make an inference for each of the following observations.

a The lining cells are very flat and fit together like tiles.

b There are many chloroplasts in the food-making cells.

c There are holes or pores in the underside of the leaf.

d The cells in the support tissue fit together like trusses in a house frame.

9 You are an illustrator for a Year 8 science textbook. Try to explain, using labelled drawings, how to make a wet-mount slide.

CHALLENGE

1 A microscope has two eyepiece lenses, ×4 and ×10, and three objective lenses, ×4, ×10 and ×40.

a What combination of lenses gives a ×160 magnifying power?

b What are the lowest and highest magnifying powers of the microscope?

c A specimen was photographed using the ×10 and ×10 lenses. On the photo the specimen measured to be 55 mm in diameter. What is the actual size of the specimen?

2 a What does the letter 'F' look like through a microscope?

b Through a microscope you observe a tiny insect moving diagonally across a slide, as shown in the diagram. Where should you place your finger to prevent it from escaping from the slide?

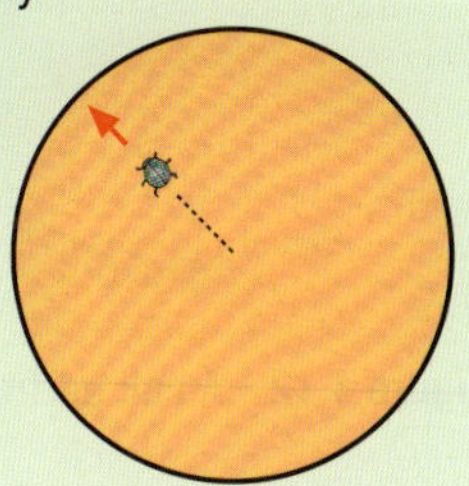

ISBN 978 1 4202 3735 1

4.2 Cell processes

A living cell is constantly active. Substances pass into and out of the cell through the cell membrane and, in plant cells, a cell wall. Chemical reactions occur in which large molecules are broken down to small ones and small molecules are built up to larger ones. For example, most of the matter in the cytoplasm of a cell is made of protein. In cell division and growth, the extra protein needed is built up from smaller amino acid molecules that pass into the cell from the blood or the liquid around the cell.

Cell respiration

The cells in your body may have to perform many of the following functions.

- All cells *make proteins* (e.g. enzymes) and other large molecules.
- Muscle cells *cause movement*.
- Nerve cells *send nerve impulses*.
- Many types of cells *divide*.

All of these functions require energy, and the cell's main source of energy is glucose. When glucose is broken down in **respiration** (RES-per-AY-shun), oxygen is used and carbon dioxide, water and energy are produced.

glucose + oxygen → carbon dioxide + water energy is released

This energy is used for muscle movement and nerve transmission, and in building large molecules such as proteins from smaller molecules.

Cell respiration occurs in organelles called **mitochondria** (might-oh-KON-dree-a). These tiny organelles vary in shape from round to sausage-shaped, depending on the type of cell, and are found in all cells that contain a nucleus. The number of mitochondria in a cell indicates its energy requirement. For example, a muscle cell contains up to 5000 mitochondria, while a skin cell may have fewer than 100.

Energy for muscle movement

Skeletal muscle is the type of muscle that usually moves bones. It is also called striated muscle because of its striped appearance under the microscope. The other types of muscles in the body are heart muscle, which is found only in the heart, and smooth muscle, which is found in organs such as the stomach and intestine.

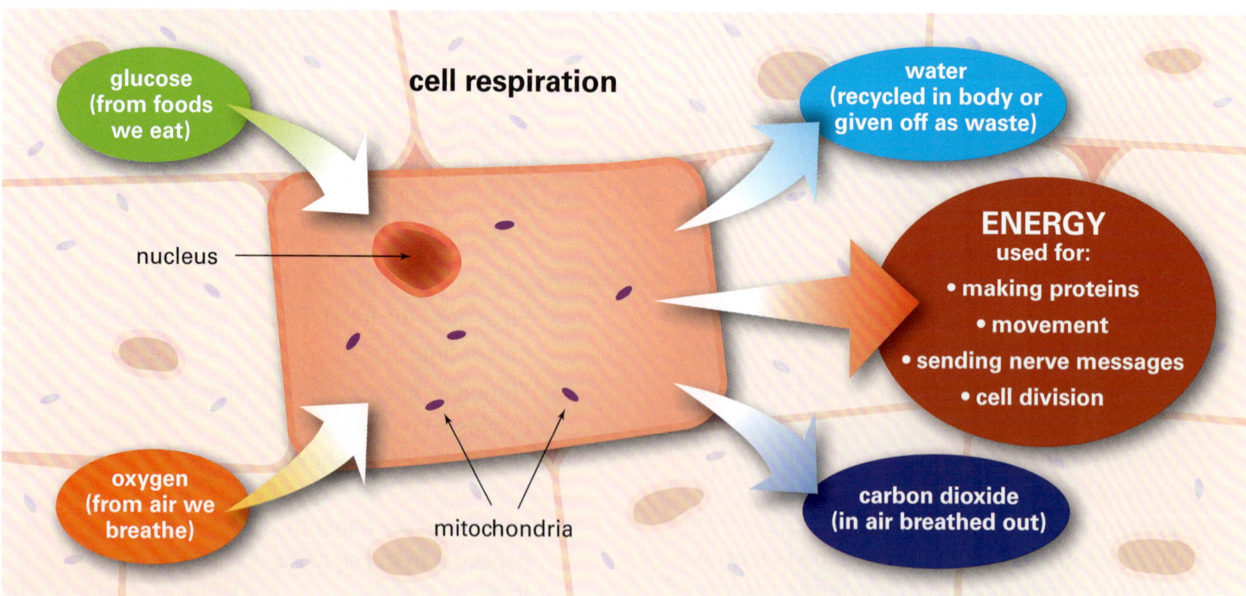

Figure 4.12 A cell performs a number of functions in the body. Most of these functions require energy.

ISBN 978 1 4202 3735 1

Skeletal muscle

When skeletal muscle contracts, it becomes shorter and thicker. You can feel this by placing your hand over your biceps muscle and bending your elbow.

Figure 4.13 shows the structure of skeletal muscle. Each muscle contains many muscle cells. These cells are different from other cells in that they are very long. Because of their length, muscle cells are often called *muscle fibres*.

Muscle cells are also different from other cells because they contain more than one nucleus. These are usually found on the surface of the cells.

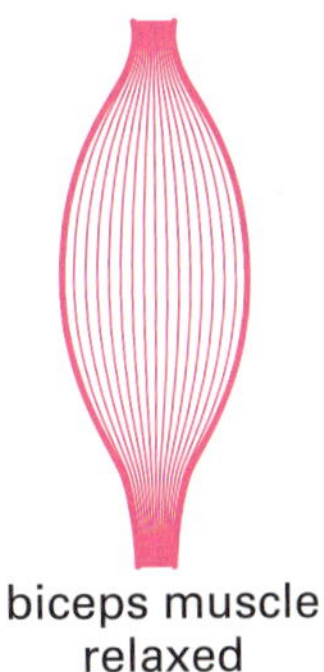

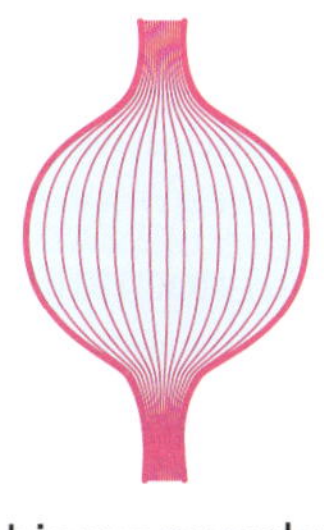

Muscle fibres are made up of tiny threads of protein called *fibrils*. When the muscle contracts, the strands in the fibrils slide over each other and become shorter. This process requires energy from cell respiration, and this is why muscle cells contain many mitochondria.

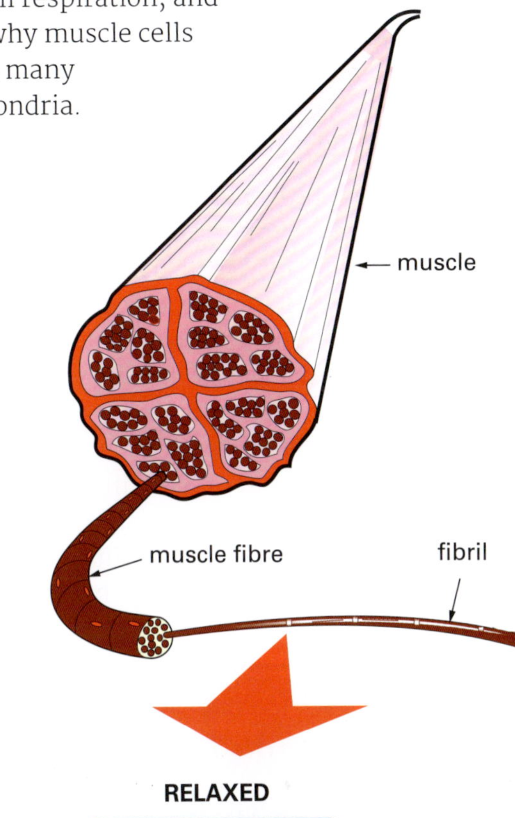

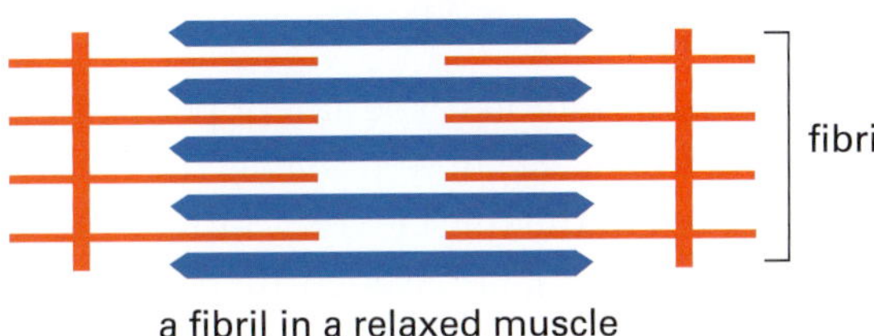

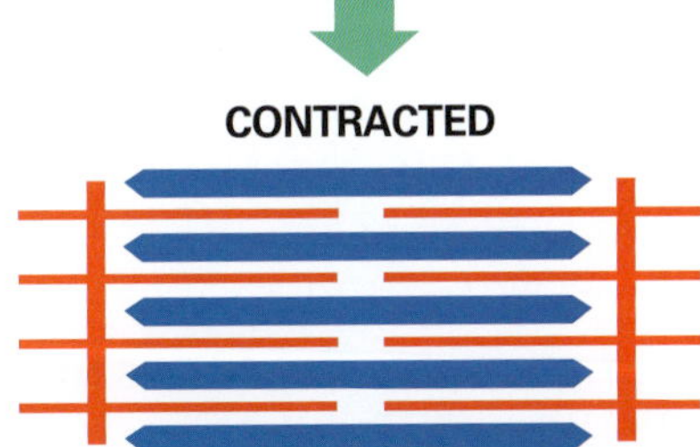

Figure 4.14 The structure of skeletal muscle (above). The model shows how a muscle fibril is thought to contract.

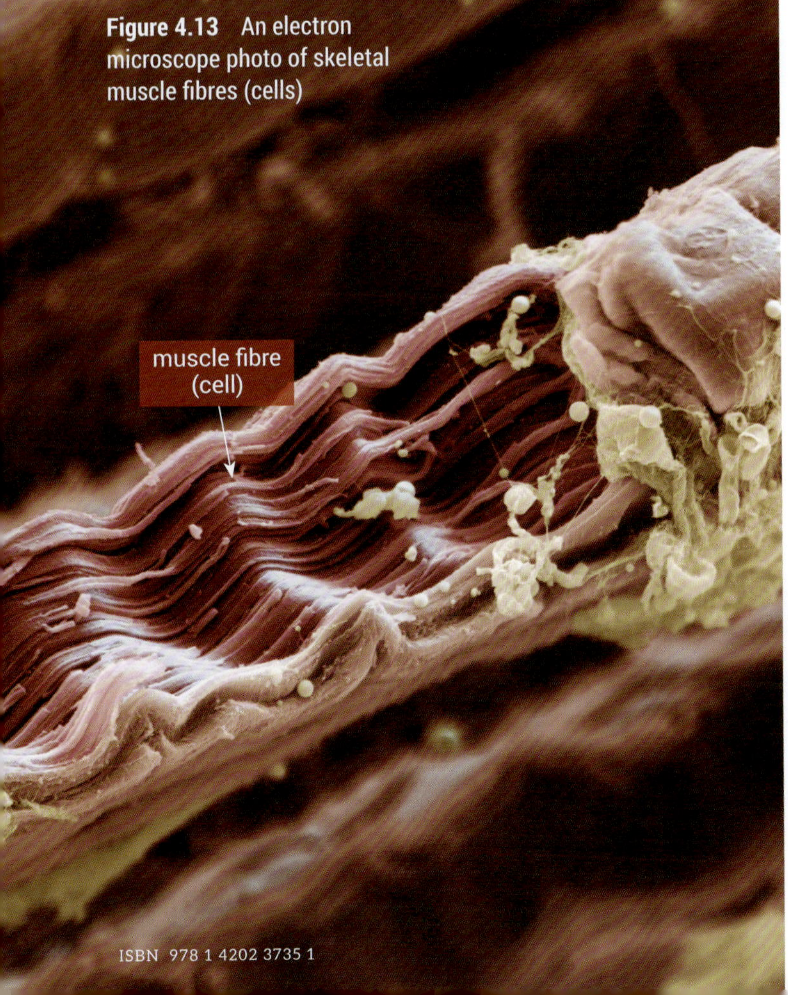

Figure 4.13 An electron microscope photo of skeletal muscle fibres (cells)

ISBN 978 1 4202 3735 1

Regular exercise builds up skeletal muscle. This is due to an increase in the number of blood capillaries that take blood to the muscle fibres, and an increase in thickness of the muscle fibres.

Figure 4.15 The muscles in a trained person contain many more capillaries and thicker muscle fibres than in an untrained person.

Muscles and energy

When glucose is broken down to produce energy in the mitochondria of cells, oxygen is normally used. This is called **aerobic respiration** (air-OH-bic). Aerobic means *with air*. However, muscle cells can also produce energy anaerobically, or *without air*. During vigorous muscular activity, the blood cannot supply the muscles with all the oxygen they need and glucose is broken down without oxygen. In this process, glucose is broken down to lactic acid, a molecule half the size of glucose. A much smaller amount of energy is produced anaerobically than aerobically.

After vigorous exercise, the blood brings more oxygen to the muscles and the lactic acid is gradually broken down to carbon dioxide and water. This is why your breathing and heart rates remain higher than normal for a short time after exercise.

The fitter you are, the more exercise you can do without feeling fatigued. This is because the muscles in a fit person have more blood capillaries which carry more oxygen to the tissues. Here the oxygen is used to break down the lactic acid.

Busting the myth—lactic acid and fatigue

In 1929 a now-famous experiment performed on a twitching frog's leg showed that as the lactic acid accumulated in the muscle cells, fatigue increased. As a result, lactic acid was identified as the cause of muscle fatigue.

However, Professors Graham Lamb and George Stephenson from La Trobe University believe that lactic acid is not the cause of muscle fatigue. It is a mistaken cause and effect! Lactic acid does build up in exercising muscles, but it is not the cause of fatigue.

Adenosine triphosphate, or ATP, is the energy carrier in cells. When ATP molecules split apart, energy is released for muscle activity. Professor Lamb has found that a low level of ATP inside muscle fibres may be one of the causes of fatigue. When the ATP drops to a critically low level, the muscle fibres stop contracting. When the energy usage drops, the concentration of ATP builds up again, and the muscles are able to do more exercise. Other substances such as potassium ions, phosphate ions and magnesium ions may also increase muscle fatigue.

Professors Lamb and Stephenson believe they and other researchers have busted the lactic acid–fatigue myth. 'This is a great lesson for young science students,' said Professor Lamb. 'Just because two variables correlate, it doesn't mean one is the cause and the other is the effect.'

Questions

1. Top athletes have their blood lactic acid levels checked after exercise. How does this indicate their level of fitness?
2. Use a dictionary to help you explain what the word *correlate* means. Then write a sentence using the word to help explain its meaning.
3. What does the sentence *It is a mistaken cause and effect!* mean?

ISBN 978 1 4202 3735 1

In this activity you are going to test the effect of temperature on muscle activity.

1 Pour some ice-cold water into a small plastic bucket or ice-cream container.
2 In your notebook, write a sentence (any sentence) of about 12 words.
3 Now put your writing hand and lower arm into the cold water for 2–3 minutes or for as long as you can.
4 Quickly wipe your hand dry with a towel and try to rewrite the sentence from step 2.

Describe the appearance of your hand after it has been in cold water. Suggest reasons for its appearance.

Suggest reasons for the difference between the two samples of handwriting.

EXPERIMENT 4.1 Muscles and exercise

PART A Temperature and muscles

The problem to be solved

In the activity above you saw the effect of temperature on muscle contraction. Use the results of this activity to design an experiment to *measure* the effect of temperature on muscle activity.

Designing the experiment

1 Work in a small group and discuss the tests you will do.
2 What variables will you control?
3 Write a draft of your experimental design, including safety issues and a materials list, and show it to your teacher.
4 After your teacher's approval, do the experiment and write a full report of your findings.

PART B Muscles and fatigue

The problem to be solved

In this part of the experiment you are going to use the equipment in the photo, or other equipment you think is necessary, to test one or more of the following questions.

a How much exercise can your middle finger do before it becomes fatigued?
b Can all fingers exercise at the same rate before they become fatigued?
c Does a short period of rest, say 10 seconds, between periods of exercise affect the time it takes for muscles to become fatigued?
d Do muscles become fatigued sooner in an unfit person than in a fit person?

Your report

Write a report of your findings, displaying your data so that other people can easily understand them. Your teacher may ask you to present your report to the class.

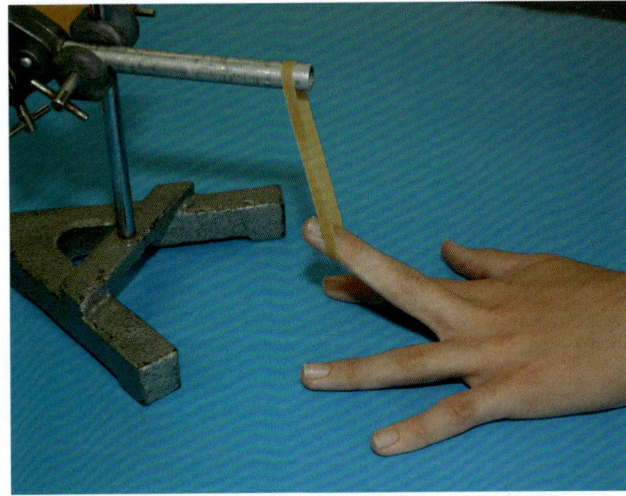

Figure 4.16 Adjust the height of the clamp so that your finger is raised as high as possible and the rubber band is stretched a little.

ISBN 978 1 4202 3735 1

Photosynthesis

Photosynthesis is a process that absorbs energy and occurs in organisms that contain chlorophyll. The energy source for this process is light, and the products are glucose and oxygen.

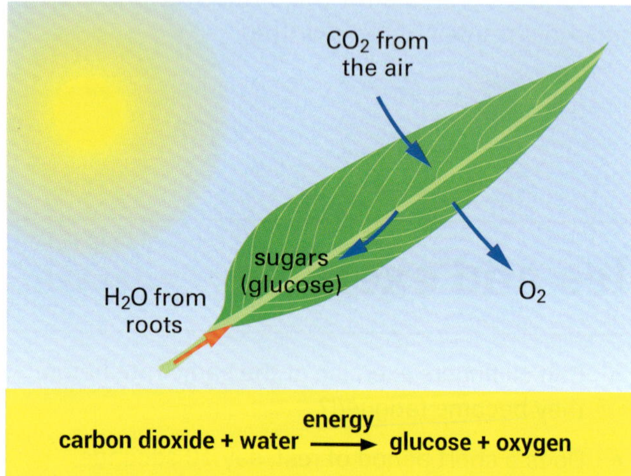

Figure 4.17 Photosynthesis uses light as its energy source. Glucose is made in the process.

The rate of photosynthesis depends on many factors, including the intensity and the colour of the light. White light from the sun is made up of the colours of the visible spectrum. Chlorophyll molecules absorb light strongly in the blue and red regions of the spectrum, and poorly in the green region (see Figure 4.19).

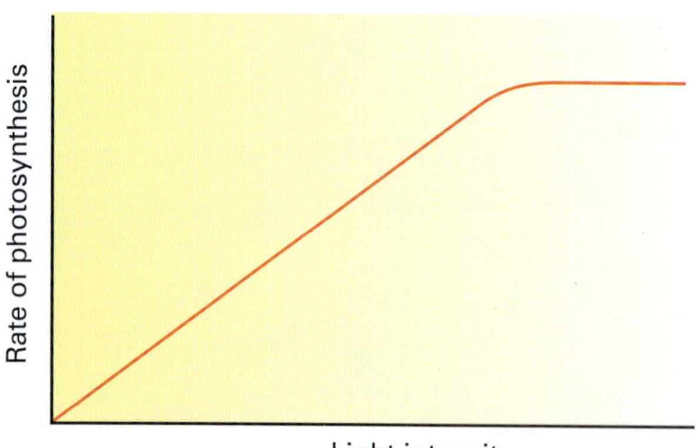

Figure 4.18 The effect of light intensity on the rate of photosynthesis

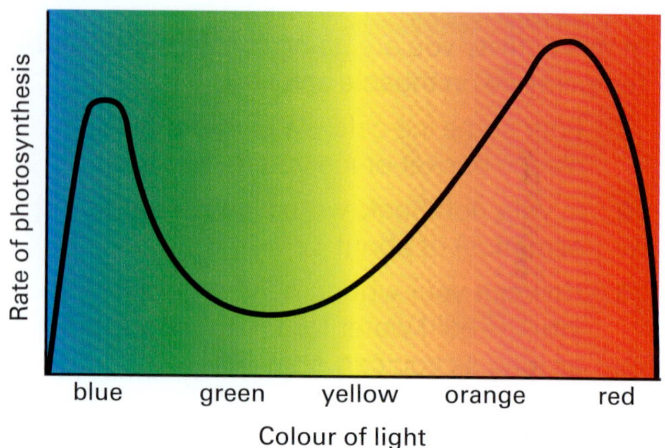

Figure 4.19 The effect of different coloured light on the rate of photosynthesis

If plants are placed in an area covered with a green plastic roof, they will grow poorly. This is because the green roofing allows only the green light in the sunlight to pass through to the plants. The rate of photosynthesis is much lower in green light than in light of other colours, causing poor growth in the plants.

The chlorophyll in plant cells is contained in the chloroplasts. Using an electron microscope, chloroplasts can be seen to contain *granules*. These can be stained with iodine, which indicates that they contain starch. When plants are kept in the dark for some time, the chloroplasts no longer contain granules. This is further evidence that the granules contain starch that is made by photosynthesis.

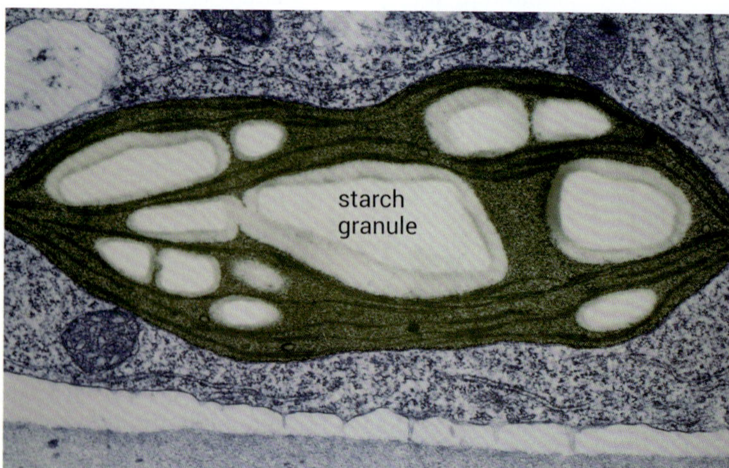

Figure 4.20 A chloroplast viewed through an electron microscope. The white patches inside it are starch granules.

ISBN 978 1 4202 3735 1

INVESTIGATION 4.2 Observing chloroplasts

Aim

To observe chloroplasts in aquarium plants.

Materials

- piece of pondweed, e.g. elodea
- spirogyra
- microscope
- 2 microscope slides and cover-slips
- iodine stain (10 g potassium iodide in 100 mL water, then add 5 g iodine)
- aquarium water

Teacher note: A videoflex microscope camera can be used to show students the chloroplasts and other cell structures.

Method

1 In Investigation 4.1 on page 85 you saw the structures in a plant leaf. In this investigation you will make more detailed observations.

2 Choose a young leaf from a pondweed stem and mount it on a slide with a drop of aquarium water. Add a cover-slip.

3 Observe the leaf under low power and then higher power.
Identify the cells in the leaf. Then identify the round green chloroplasts. You might see the chloroplasts moving. This is caused by the movement of the cytoplasm in the cell.

4 Place a strand or two of spirogyra on a slide and add a drop of iodine stain. Then add a cover-slip and let it stand for a minute or two. Use low power to try to identify the spiral chloroplast.

5 Use high power and look for darkly stained granules in the chloroplast. These are where starch is stored.

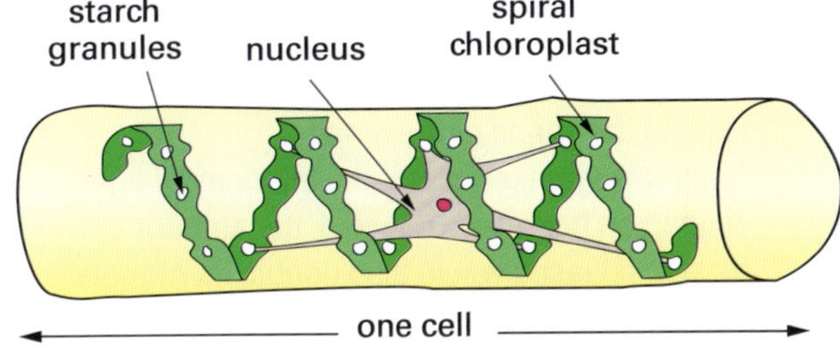

CHECK

1 Describe the four functions of body cells that require energy.

2 Cell respiration is said to be the reverse of photosynthesis. Explain what this statement means.

3 Mitochondria and chloroplasts are cell organelles.
  a What are organelles and where in the cell are these two organelles found?
  b Do all cells contain these organelles? Do all cells have the same numbers of these organelles? Explain both answers.
  c Mitochondria vary in size between 2 μm and 5 μm. What is their size in mm?

4 How is a skeletal muscle cell different from other cells in your body?

5 Describe how skeletal muscle contracts. Why are there many more mitochondria in muscle cells than in skin cells?

6 What happens in muscle cells when they have to contract vigorously and there is not enough oxygen available?

ISBN 978 1 4202 3735 1

7 What type of respiration occurs in each of the following situations?
  a a 20-metre sprint for the bus
  b doing housework
  c doing 20 chin-ups
  d walking the dog

8 Describe how the muscles of body builders are different from your muscles.

9 What are the advantages to the body of muscles using energy supplied anaerobically?

CHALLENGE

1 The drawing below shows part of a plant cell that has been enlarged many times. Use the magnification to calculate the actual size of the nucleus and the chloroplast in micrometres.

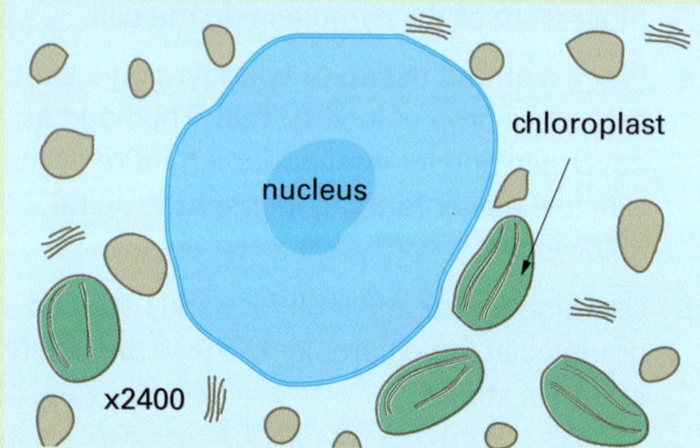

2 Look at the graph, Figure 4.18 on page 94.
  a Explain the shape of the graph.
  b Antonia suggested that if the intensity of light shining on her plants doubled, the amount of oxygen released would also double. Is Antonia correct? Explain.

3 A light was shone onto a freshwater plant, and the rate of photosynthesis was measured at different temperatures. The experiment was repeated with a second light of a different intensity. The graph at the top right shows the results.
  a What is the aim of the experiment?
  b Which light was of higher intensity? Why?
  c Which variables were changed? Which would have been controlled?
  d What was measured in the experiment?
  e Draw the apparatus that might have been used.
  f Write a conclusion for the experiment.
  g Suggest why the slope of the light B graph decreases above 10 °C.

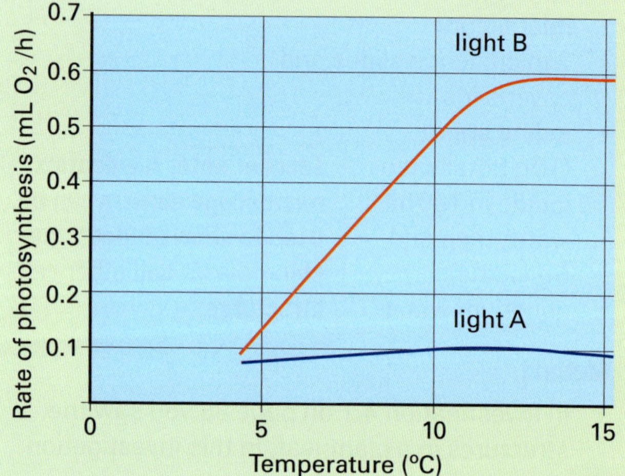

4 Three people agreed to take part in an experiment to test the amount of lactic acid in their blood after vigorous exercise. The three people exercised in the gym every day for 30 days. The table below shows their results.

	Blood lactic acid (mg/100 mL of blood)		
Day	**Person A**	**Person B**	**Person C**
0	72	70	90
5	40	72	45
10	32	28	32
15	26	28	–
20	25	24	46
25	21	22	28
30	21	21	22

  a How does the training affect the amount of lactic acid in the blood?
  b Suggest two inferences to account for the lower lactic acid in the blood after regular vigorous training for 30 days.
  c Account for the results of person C.

ISBN 978 1 4202 3735 1

4.3 Investigating cells

Alex decided to buy his mother some flowers for helping him with his science project.

He placed them in a container on the kitchen table and wrote a note on a card.

When his mother came home from work many hours later the flowers had wilted.

Alex's flowers wilted because the cells in the plant lost water to the air. Instead of being tight and rigid, the cells became loose and floppy, like a fully inflated balloon that has lost some of its air. Water particles passed out of each cell, through the cell membrane and the cell wall. This movement of particles is called **diffusion**.

In Investigation 4.2 on page 95 you added iodine stain to your slide of spirogyra. After a few minutes the iodine had passed into the cell and the chloroplasts had become stained. The iodine had diffused into the cell from the outside.

Why did the iodine diffuse into the cell? Particles are constantly moving and bumping into other particles. In doing this they tend to move away from each other and spread out. The iodine particles moved from outside the cell where they were concentrated to inside the cell where they were less concentrated. Eventually the iodine particles spread evenly throughout the water inside and outside the cell.

water diffuses out of cells

Cells are full and tight → Cells are soft and floppy

Figure 4.21 Water diffused out of the cells in the stem of Alex's flowers and the flowers wilted.

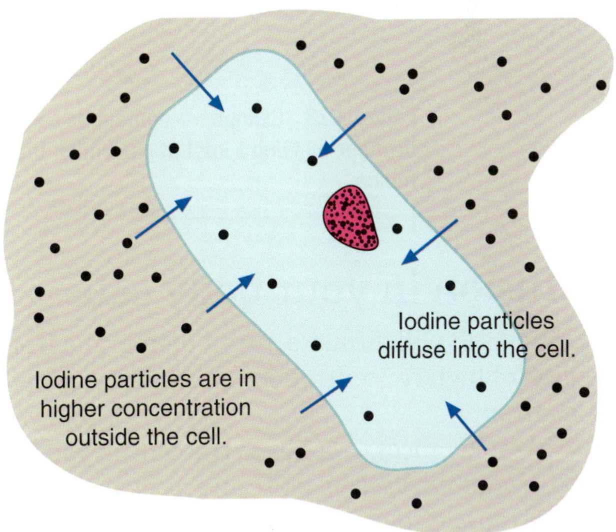

Figure 4.22 The iodine particles diffuse through the cell membrane and into the cell.

ISBN 978 1 4202 3735 1

EXPERIMENT 4.2 A model for diffusion

Your task

To use a model to explain how various substances diffuse through membranes.

Your task is to use lengths of cellophane tubing to design an experiment to show that substances such as salt and sugar diffuse through membranes.

Risk assessment and planning

- Work in a group to discuss and design your tests. Draw up data tables where necessary.
- You will use silver nitrate solution to test for salt. Silver nitrate is toxic and it stains skin and clothes. Wash any spills immediately with plenty of water. Use disposable gloves and wash your hands thoroughly at the end of the lesson.
- Check with your teacher about how to dispose of the waste liquids.

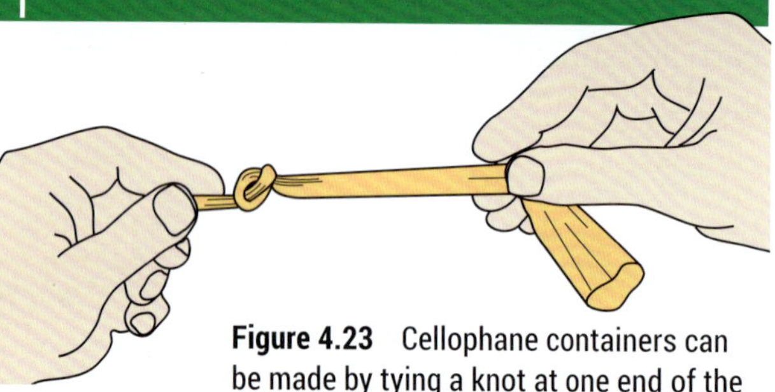

Figure 4.23 Cellophane containers can be made by tying a knot at one end of the tubing.

Helpful hints

1. The cellophane tubing can be made into a container to hold a solution as follows: Wet a piece of cellophane tubing with tap water. Tie a knot in one end. Rub the other end with your fingers until the tubing opens.
2. Try using a salt or sugar solution in the tubing and distilled water in the beaker. Then see if the salt or sugar diffuses into the beaker. (You could also try the reverse of this and see if salt or sugar diffuses into the tubing.)
3. You will need to leave your set-up overnight.
4. The presence of salt can be tested with a drop of dilute silver nitrate solution. The presence of sugar can be tested with a Clinistix or Benedict's solution.
5. Make a list of the equipment you will need. Your teacher will need to supply you with:
   - 10% glucose solution or 20% sodium chloride solution
   - lengths of cellophane tubing (20 cm long)
   - 5% **silver nitrate** solution in a dropper bottle.

Writing your report

Write a report of your findings. How does this model explain how substances might move into and out of cells?

Diffusion in your body cells

Sugar and salt will diffuse across a membrane from a region of high concentration to a region of lower concentration.

This same process applies to the cells in your body. In your muscles, glucose and oxygen diffuse into the muscle cells from the capillaries that surround them. This is because the glucose and oxygen are constantly used up as the muscle cells contract. Hence the amounts of these two substances decrease in active muscle cells.

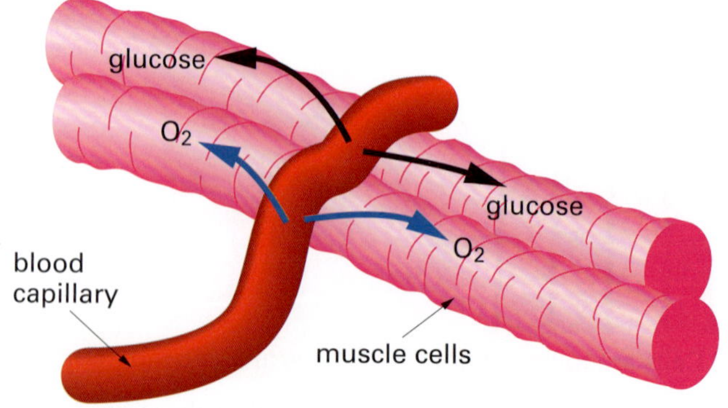

Figure 4.24 Glucose and oxygen diffuse into the cells from the blood because they are used up during respiration.

ISBN 978 1 4202 3735 1

Glucose and oxygen diffuse into the muscle cells from the capillaries that surround them because the concentrations of glucose and oxygen are higher in the blood than in the cells.

Carbon dioxide is produced in respiring muscle cells as a waste product. In active muscles the concentration of CO_2 increases in the cells. As the concentration of CO_2 becomes greater than in the blood, it diffuses out of the muscle cells and is carried away to be expelled by the lungs. Other wastes from cellular activities also build up in the cells and diffuse out into the blood.

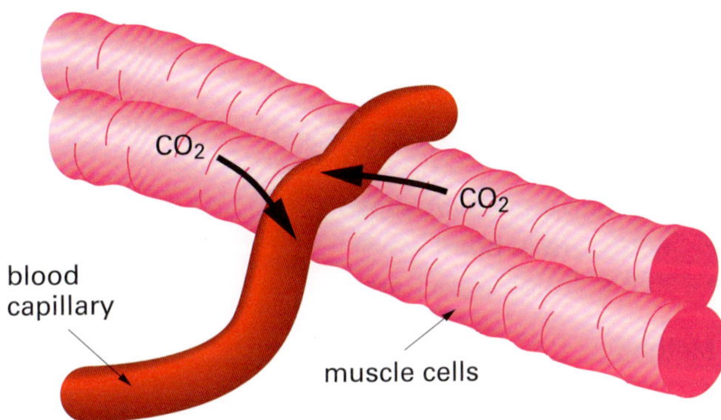

Figure 4.25 During respiration carbon dioxide builds up in the muscle cells and diffuses out into the blood.

Osmosis

If a potato cube is left in distilled water for about an hour, it becomes hard. If another potato cube is left in salt water, it becomes soft. These results are due to a special case of diffusion of water molecules through the cell membranes of the potato cells. This process is called **osmosis** (oss-MOW-sis).

Osmosis occurs in all cells because the cell membranes are *semipermeable*. This means that the cell membrane allows water particles to move freely across it but slows down the movement of dissolved substances, such as salts and sugars.

The cells in the potato cube contain about 90% water. When the potato is placed in distilled water (100% water), water particles diffuse into the cell from where they are at a higher concentration to where they are at a lower concentration. This makes the potato firm and larger. This is why vegetables, such as carrots and celery, that are a little soft become firm and crisp when you put them in water.

When the potato is placed in the salt solution (only 80% water), the water diffuses out because the water is now at a higher concentration inside the cell. The cells become soft and smaller because there is a movement of water particles from inside the cell to the outside.

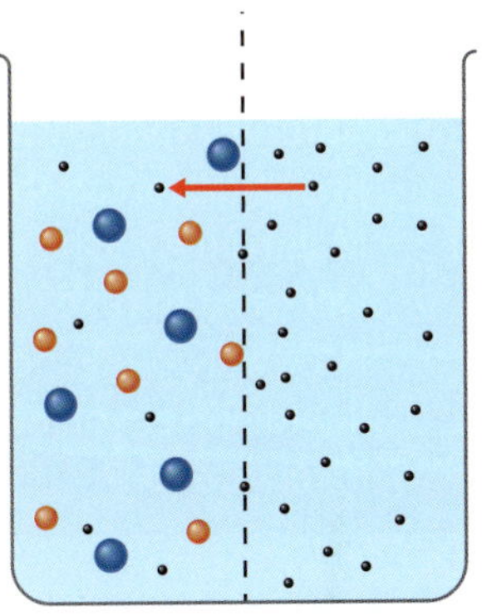

Figure 4.26 A semipermeable membrane allows smaller particles such as water to pass through, while larger particles such as glucose cannot.

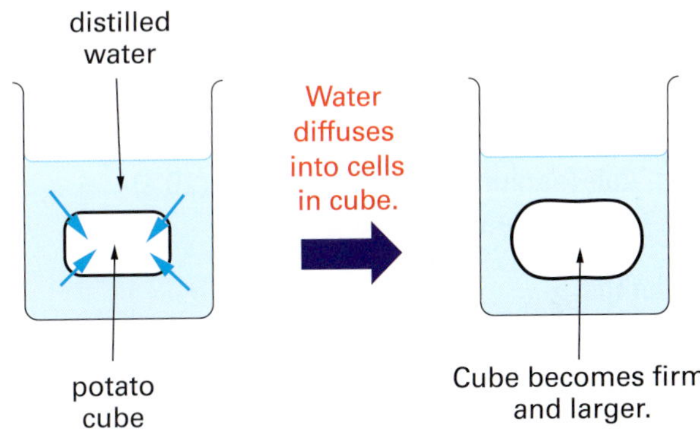

Figure 4.27 A potato cube becomes firm when placed in distilled water because water diffuses into the cells.

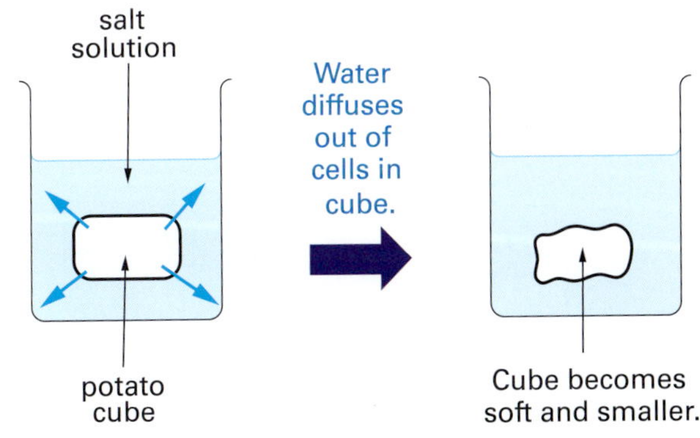

Figure 4.28 A potato cube becomes soft when placed in salt water because water diffuses out of the cells.

ISBN 978 1 4202 3735 1

Osmosis—the movement of water

Aim

To observe the effects of the movement of water into and out of cells.

Risk assessment and planning

- Carefully read through the two parts of this investigation so that you know what to do.
- Make a list of the safety precautions you will need to take.
- How will you dispose of the solids and liquids after you have finished the investigation?

PART A

Materials

- fresh potato
- distilled water
- salt (sodium chloride) solution (20%)
- 2 small beakers or jars

Method

1 Peel the potato and cut two 2 cm cubes from it.

2 Place one cube in a beaker of distilled water and the other in a beaker of the salt solution.

3 Leave the potato cubes in the liquids for at least an hour, or overnight. While you are waiting, go on with Part B.

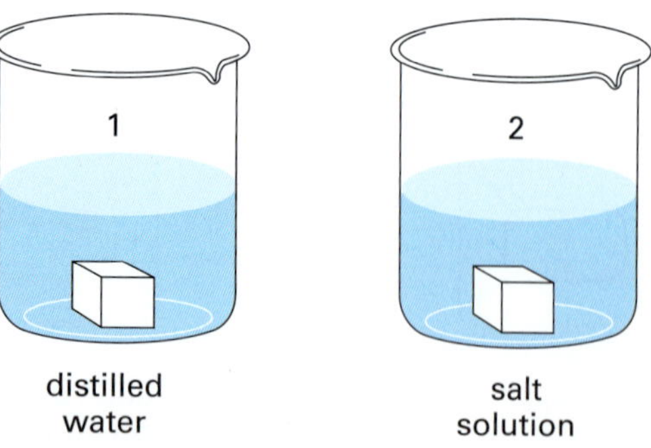

4 Take the potato cubes out of the liquids. How do they feel? Record your observations.

Discussion

1 Account for the texture and feel of the potato cubes in each beaker. Could you return the salt potato cube to its original condition? How?

2 Repeat the experiment but this time weigh the potato cubes before and after.

PART B

Materials

- microscope
- 2 slides and cover-slips
- spirogyra in pond water
- salt solution from Part A

Method

1 Use the diagrams below to prepare two slides of spirogyra to observe under the microscope.

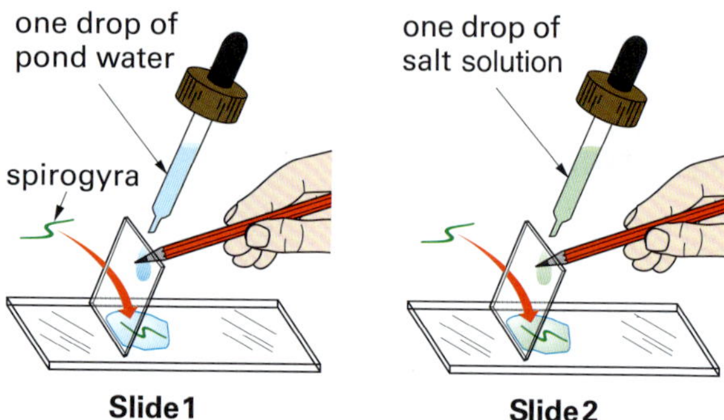

2 Look at each slide under a microscope. Draw the cells of spirogyra on each slide.

Discussion

1 The movement of water through cell membranes is responsible for the results in Part A and Part B. Suggest why this is so.

2 Use your results to explain why wilted carrots or celery will become crisp again when you soak them in water.

3 How do the results explain why freshwater fish do not have to drink water?

ISBN 978 1 4202 3735 1

1 A small crystal of copper sulfate was placed in a beaker of distilled water. The beaker was observed 4 hours later. The diagrams show the results. In terms of particles, explain what happened in the beaker.

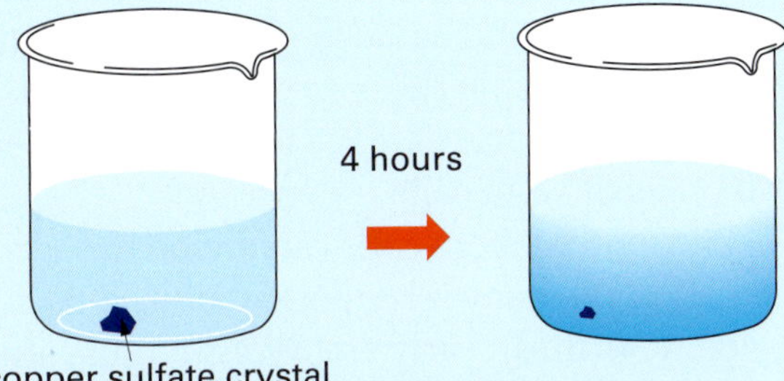

2 A piece of limp celery is placed in a glass of water. After 30 minutes the celery is firm and crisp. When this piece of celery is placed in sea water it becomes limp after 30 minutes. Explain what has happened in each case.

3 What is a semipermeable membrane? Design a test to show that glucose will diffuse through cellophane tubing but starch will not. (Hint: Starch turns black when iodine solution is added to it, and Benedict's solution turns red when glucose is present.)

4 Explain why a cube of apple after being placed in distilled water for an hour is firmer than the original cube.

5 Some red blood cells were placed on a microscope slide and a drop of liquid was added. When the slide was observed under a microscope, the cells had shrivelled up. Was the liquid that was added to the blood cells distilled water or salt solution? Explain your answer.

1 Starch solution was poured into two pieces of cellophane tubing that had each been tied at one end. One piece of tubing was placed in a beaker of lukewarm distilled water. Saliva was added to the second tubing, which was then placed in a second beaker of lukewarm distilled water.

The set-ups were left for a few hours. Then the liquids in each beaker and piece of tubing were tested with iodine and Benedict's solution. (Note: Benedict's solution turns red when glucose is present in the solution.)

a What results would you expect in each of the set-ups? Explain your answer.
b What was the aim of the experiment?
c Why were the pieces of tubing placed in lukewarm water?

2 An amoeba is a protist that lives in still water such as lakes, ponds and dams. It has a vacuole that is able to contract (contractile vacuole). Water that diffuses into the cell is expelled to the outside by this contractile vacuole.

contractile vacuole

nucleus

The following data show the number of contractions of the vacuole when the amoeba was placed in different solutions.

Solution	No. of contractions per 10 minutes
distilled water	60
pond water	40
dilute sugar	10
solution A	none

a Explain why the number of contractions is less in the dilute sugar solution than in distilled water.
b Suggest why there are no contractions in solution A.
c Predict what would happen to the amoeba if it did not have a contractile vacuole.

ISBN 978 1 4202 3735 1

SCIENCE AS A HUMAN ENDEAVOUR

Stem cell research

Figure 4.29 'Superman' Christopher Reeve

Christopher Reeve, the star of the film *Superman* (1978), was paralysed when he fell from a horse in 1995. He was confined to a wheelchair, and until his death in 2004 he lobbied politicians to approve stem cell research to find a cure for people with spinal cord injuries. But what are **stem cells**?

Stem cells are unspecialised cells that can develop into any one of over 200 cell types in the body. Stem cells are responsible for renewing tissues, growing into new cells to replace old ones and repairing damage. Scientists see the possibility of using these stem cells to treat diseases and repair damage to the body. There are two types of stem cells.

- Pluripotent stem cells can divide to form any cell in the body. Pluripotent stem cells can come from an early embryo, a foetus, amniotic fluid, the placenta or umbilical cord blood.
- Tissue-specific stem cells are also known as adult stem cells. They are found in tissues throughout the body, including in muscle, fat, skin, hair follicles, bone marrow, brain and spinal cord, the lining of the gut and lungs, and joint fluid. These stem cells only grow into the type of tissue they are found in.

Using stem cells

Recently scientists discovered that a mature adult cell could be induced to (made to) copy the behaviour of a pluripotent stem cell. This means that an adult cell could be used to grow any tissue in the body. These cells are called induced pluripotent stem cells (iPS cells). Imagine how diseases could be cured or damaged bodies repaired using iPS cells. Maybe scientists could grow a new liver rather than using transplants!

Is it ethical?

Much research, especially in the early days of stem cell discovery, has been done into stem cells using donated embryos that remain after IVF procedures. However we need to ask: is this *ethical*?

With new scientific developments there are always people who are for it, and those who are against. In this case, some right-to-life and religious groups are against the use of embryos, as they believe the embryos have a right to live even though they are unborn. On the other side, the potential medical benefits are enormous.

ISBN 978 1 4202 3735 1

As Christopher Reeve once asked, 'Is it more ethical for a woman to donate unused embryos that will never become human beings, or to let them be tossed away as garbage when they could help save thousands of lives?'

Corner discussion

1 Put the following signs in the four corners of the room: 'agree', 'disagree', 'unsure but I think I agree' and 'unsure but I think I disagree'.

2 *Do you think leftover human embryos should be used for stem cell research?* Move to the corner that applies to you.

3 People in each corner now try to convince the people in the two unsure corners to join them. Everyone should give a reason.

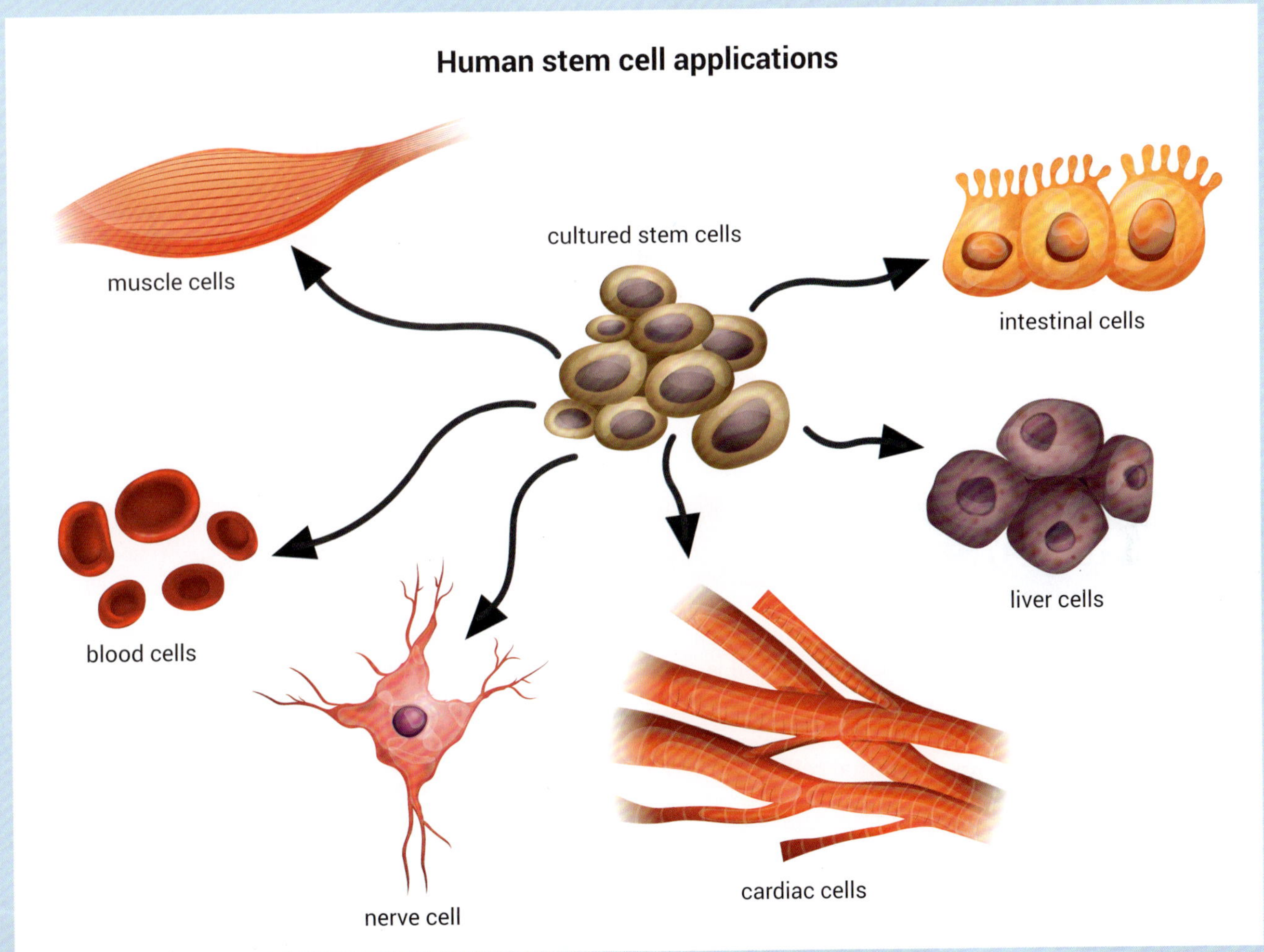

Figure 4.30 Can you think of a disease that could be cured with the help of each of these cells?

ISBN 978 1 4202 3735 1

MAIN IDEAS

Copy and complete these statements to make a summary of this chapter. The missing words are on the right.

1 All organisms are made of _____. There are many different types of cells, with different shapes and different functions.

2 A _____ can be used to identify the various parts of a cell: the nucleus, cell membrane, cytoplasm and organelles. A _____, chloroplasts and vacuoles can be observed in plant cells.

3 Cells of the same type are generally found together in _____. Each type of tissue has a particular _____ in an organism.

4 Tissues are arranged in structures called _____ in multicellular organisms. Each tissue has a specific function in an organ.

5 _____ uses glucose and oxygen and produces energy for cell functions such as cell division and for making _____.

6 _____ respiration occurs in cells that have a good supply of oxygen. Certain cells such as muscle cells can produce energy _____ (without oxygen).

7 In plants, _____ occurs in organelles called chloroplasts. The rate of photosynthesis depends on the _____ and its colour.

8 Substances move into and out of cells by the process of _____.

proteins
cells
aerobic
cell wall
photosynthesis
microscope
function
organs
tissues
cell respiration
light intensity
diffusion
anaerobically

CH•6 REVIEW

1 A cell is observed under a microscope to have a nucleus, cytoplasm and organelles. The cell is:
- **A** definitely an animal cell.
- **B** definitely a plant cell.
- **C** either a plant cell or an animal cell.

2 Which one of the following statements about cells is *false*?
- **A** Plant cells have large vacuoles.
- **B** A nerve cell is an example of a specialised cell.
- **C** All cells are rectangular or brick-shaped.
- **D** Plant cells have cell walls.

3 Match the cell part in the list with the correct description below.

cell membrane, cytoplasm, vacuoles, chloroplast, nucleus, cell wall

- **a** an organelle that is involved in the process of photosynthesis
- **b** the jelly-like material that fills a cell
- **c** the part of the cell that controls its activities and keeps it alive
- **d** a covering that controls the movement of materials into and out of a cell
- **e** a thick, tough layer that helps support and protect the cell
- **f** liquid-filled spaces found in some cells.

ISBN 978 1 4202 3735 1

4 A microscope has a ×10 objective lens and a ×4 eyepiece lens. How big would an object 0.05 mm in diameter appear through the microscope?

5 Photosynthesis is a process that requires energy, while respiration is a process that produces energy.

a Explain what this sentence means.

b Why is respiration important for an organism like you?

6 Explain how a unicellular organism is different from a multicellular organism.

7 Why is your stomach called an organ? Use the words ***cells*** and ***tissues*** in your answer.

8 The two cells in the drawing below are found in different tissues in your body. The cell from tissue A is box-like and makes a watery substance called mucus. The cell from tissue B is very flat. Infer the function of each tissue and where it might be found in your body.

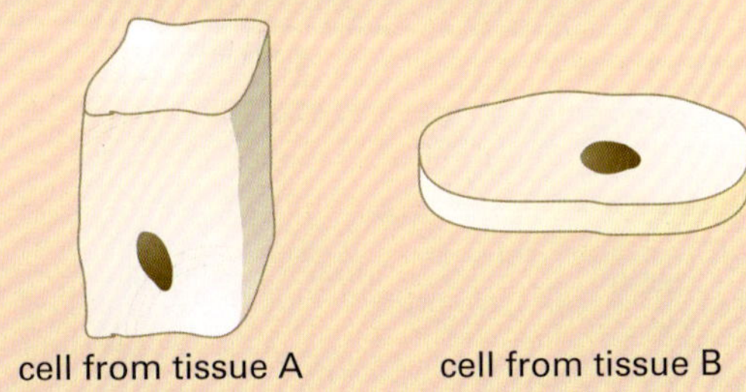

9 Sugar solution was poured into a thistle funnel and the top of the funnel was covered with a semipermeable membrane. The funnel was then turned upside down, placed into a beaker of distilled water and left for 2 hours. The diagram above right shows the set-up.

a Why did the liquid rise in the funnel?

b Predict what would have happened if the liquid in the funnel had been a more concentrated sugar solution.

c Predict the results if the experiment was repeated with distilled water in the funnel and sugar solution in the beaker.

d Use your knowledge of the process that occurred in this experiment to explain why a stick of celery becomes limp if it is placed in salty water.

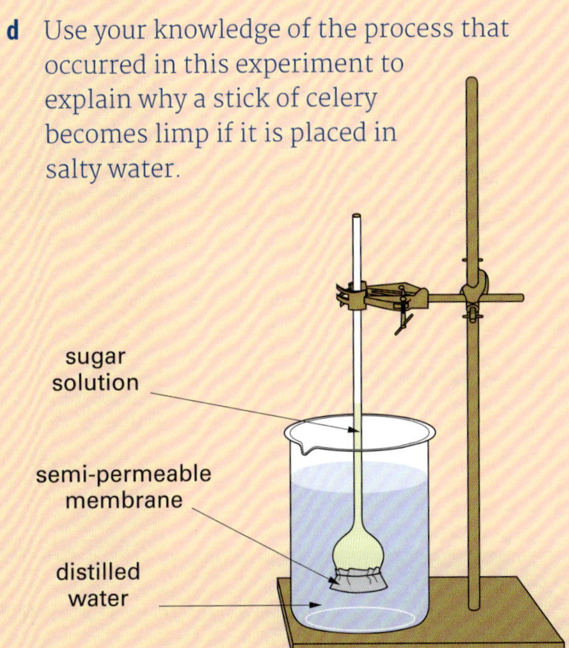

Microscope licence test

LAB REVIEW

You will be working in pairs and assessing each other's work in this practical test of microscope skills.

You will be given—a microscope, microscope slide and cover-slip and a small piece of newspaper that contains a few letters. Your teacher will also give you an assessment grid to help you assess your partner's task.

Your task—to make a wet-mount slide of some letters on a small piece of newspaper without any air bubbles or excess water, and then draw it under the microscope.

The test—your teacher and the class will discuss what you have to do to pass the licence test. Your partner will then assess the quality of your wet-mount slide and drawing and record your results on the assessment grid. Remember, you can only pass or fail this test. If you fail you must repeat the test until you pass.

Your teacher may issue you with a microscope operator's licence when you pass the test.

Check your answers on page 296.

ISBN 978 1 4202 3735 1

IN THIS CHAPTER

Science Understanding

- identify the differences between chemical and physical changes
- identify evidence that a chemical change has taken place
- investigate simple reactions and identify some different reaction types
- consider the issue of whether we should act now to reduce global warming due to carbon dioxide emissions

Science Inquiry Skills

- use science equipment and chemicals safely, without hurting yourself or others

CH•5 Chemical reactions

ISBN 978 1 4202 3735 1

GET STARTED: *EXPLORE*

Your teacher will give you a sparkler, some matches and safety glasses. Alternatively your teacher may burn a sparkler as a demonstration.

> Record as many observations of the sparkler as you can (including measuring its mass). Call this *Observations before lighting*.
>
> Put on the safety glasses and light the sparkler.

> Record your observations of the burning sparkler. Call this *Observations of a burning sparkler*.

> When the sparkler has finished burning, make another list of observations. Call this *Observations of a burnt sparkler*.
>
> **Caution:** Let the sparkler cool for 5 minutes before you find its mass.

> In small groups, discuss the changes that occurred in the burning sparkler. Make a list of these changes.

Keep your notes as you will use these later in the chapter.

ISBN 978 1 4202 3735 1

5.1 Physical and chemical properties

Suppose you are designing a surfboard. It has to float and has to support a surfer standing on it and manoeuvring it in the waves.

Secondly, you have to decide on what type of materials you are going to use. To do this you have to consider the **properties** of the materials that are available to use. A material's properties include its strength, flexibility, colour, hardness and so on.

You could use wood for the surfboard. It floats, is strong and easily shaped, but it is heavy and easily cracked. Foam also floats and is easily shaped, but it is not strong and would break in the surf. However, foam that is covered with fibreglass is strong, floats, is easily shaped and is light. The properties of fibreglass and foam make them suited to the uses of a surfboard.

Good-quality bicycle crash helmets are made from high-density foam (a plastic) because the foam absorbs the impact of a crash and protects the head from injury. Builders' nails are made from galvanised steel because the steel is hard and relatively cheap to buy and the galvanising stops them rusting.

Physical properties such as colour, strength and hardness are easy to observe. However, *chemical properties* can only be observed during chemical reactions. For example, you might not know that hydrogen gas is flammable until you put a match in it. In the next two sections you will investigate chemical reactions and the chemical properties of substances.

INVESTIGATION 5.1 Properties of materials

Aim

To list the properties of some common materials and to identify the materials from which supermarket packaging containers are made.

Materials

- at least 10 different manufactured items; for example, pencil, pen, piece of lab glassware, piece of clothing, shoe, lunchwrap, watch
- four packaged items that were purchased at the supermarket

ISBN 978 1 4202 3735 1

Structure	Materials	Property of materials	Alternative materials

Risk assessment and planning

- Before you start, draw up a data table like the one above for Part A. You will need to use the whole width of a page in your notebook for it.
- For Part B, work in a group of four or five people. Carefully read through the steps and the questions and conclusions, then design a data table for your results.

PART A

Method

1 For each of the items supplied to you, list the material(s) it is made from.
Beside each material, write down some of the properties of the material that make it suitable for use in the item.

2 List alternative materials from which each item could be made.

PART B

Method

1 Inspect each supermarket item and list the material(s) from which the packaging is made.

2 List the properties of each material that make it suitable for use in that packaging.
You may also want to consider the cost of the material, visual appeal, ability to be recycled and ease of manufacture.

3 When your group has finished with the packaged items, swap them with another group. Aim to observe at least 10 items.

Discussion

1 For each of the 10 items you observed, describe a property of the material that is important for the people who *sell* the product.
Then describe a property of the material that is important for the people who *buy* the product.

2 From your 10 items, list four. Beside each, write down two alternative materials in which this product could be packaged.
What advantages and disadvantages do these alternative materials have?

ISBN 978 1 4202 3735 1

CHECK

1 a Name *four* properties of materials.
b The table below shows five common materials. Copy the table and list *two* important properties for each material. The first one is done for you.

Material	Properties
aluminium	lightweight, will not rust
paper	
glass	
copper	
nylon thread	

2 Each of the drinking containers below is made from a different material. What sort of contents are they designed for? Give reasons for your answers.

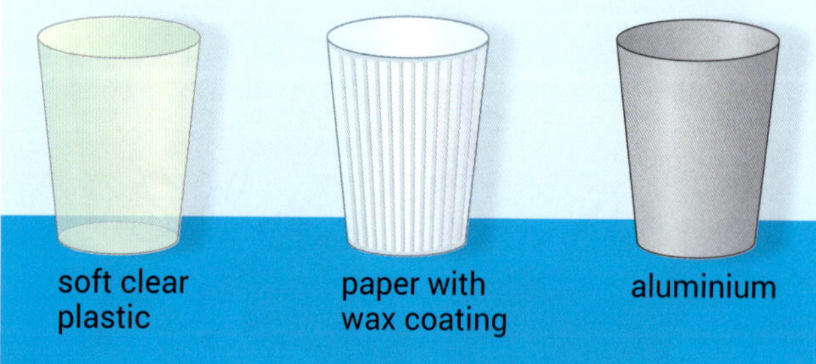

3 Which properties of glass make it useful for drink containers? Why are glass containers banned from swimming pools, concerts and football games?

4 Styrofoam is used as a packaging material for computers, televisions and stereos. Why is this material used? What are the disadvantages of using it? What other materials could be used?

5 Which of the following are *physical* properties and which are *chemical* properties?
a Carbon dioxide is heavier than air.
b Stainless steel doesn't rust.
c Glass breaks easily.
d Hydrogen gas is flammable.
e Oxygen is a colourless, odourless gas.
f Most plastics are not biodegradable.
g Styrofoam floats in water.

CHALLENGE

1 Suppose you are going to make each of the following items.
a clothes pegs
b container to store food in the freezer
c container to carry water on a bushwalk
d screws for the decking at a beach house
e container to heat food in a microwave oven

Which of the following materials would you use to make each of the structures above? Give reasons for your choices.

steel
stainless steel
brass
clear soft plastic
hard coloured plastic
hard clear plastic
pottery

2 What properties allow you to tell the difference between the substances in each of the following pairs?
a rubber and aluminium
b chalk and snow
c soft drink and water
d salt and sugar
e wood and plastic

3 The plastic polyvinyl alcohol is soluble in water. It is used to make packages for detergents, insecticides etc. Suggest some other uses for this plastic.

ISBN 978 1 4202 3735 1

5.2 What is a chemical reaction?

You have already investigated **physical changes**, such as water freezing to form ice. In a physical change the substance changes some of its properties, but the substance itself is still the same. Water is still water whether it is a liquid or a solid. Physical changes can usually be reversed, for instance ice can melt to become liquid water again.

You have observed that a burning sparkler gives off light, heat, sound and gases, and that the burnt sparkler looks quite different from the original one. New substances are also formed, with black material left on the wire and smoke produced. The changes that occurred in the burning sparkler are called chemical changes and are caused by a **chemical reaction**. Chemical reactions cannot easily be reversed. You cannot get the sparkler back once you have burnt it.

There are many types of chemical reactions and some of these will be explored in this chapter.

Signs of a chemical reaction

The signs for recognising when a chemical reaction has occurred are listed below. Sometimes only one of these things happens; at other times several things happen.

Signs of a chemical reaction

1 Gas is produced.
2 Permanent colour change occurs.
3 Solid is formed.
4 Heat, light, sound or electricity is produced.

Figure 5.1 An Alka-Seltzer tablet in water releases carbon dioxide gas. The tablet contains sodium bicarbonate (baking soda) and citric acid, which react together. This is an acid reaction.

Figure 5.2 A chemical reaction occurs when a match burns, producing heat and light. First the phosphorus in the match head reacts with oxygen in the air, then the wood in the match burns leaving a new substance—black carbon. This is a combustion reaction.

ISBN 978 1 4202 3735 1

Figure 5.3 A solid is sometimes produced in a chemical reaction (precipitation reaction) when two clear solutions are mixed. In this example, the new yellow substance formed is called lead iodide.

Figure 5.4 Rusting is a very slow chemical reaction (combination reaction). Iron reacts with water and oxygen in the air to produce red-brown iron oxide, called rust.

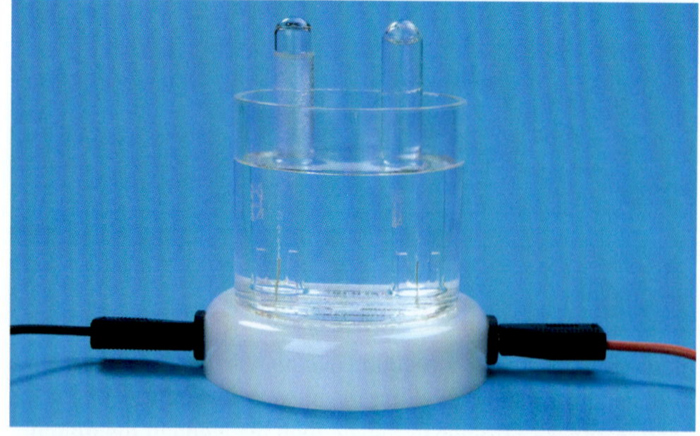

Figure 5.5 Water can be decomposed into hydrogen and oxygen gas using electricity.

Types of chemical reactions

Chemical reactions can be classified into various types. Here are some of the more common reactions you should be able to identify.

Acid reactions

When a substance reacts with an acid, it is called an *acid reaction*.

Example: A metal such as magnesium reacts with acid to produce hydrogen gas.

As you work through the rest of this chapter, try to identify some of these common reaction types.

Combustion reactions

A **combustion** reaction occurs when something burns. In combustion a substance reacts with oxygen in the air to produce heat and light (a flame), and new substances are produced.

Examples: Gas burning in the stove; candle wax burning.

Precipitation reactions

When a solid forms during the mixing of two solutions this is called a precipitation reaction. The solid is called the **precipitate** (pre-SIP-it-ate).

Example: Formation of lead iodide solid when colourless/clear solutions of lead nitrate and sodium iodide are mixed. See Figure 5.3.

Combination reactions

A *combination* reaction is the opposite of decomposition. Substances join together to create new substances.

Examples: During **rusting** iron reacts with water and oxygen in the air to form iron oxide. Hydrogen reacts explosively with oxygen to form water. See the photo on the chapter opening page.

Decomposition reactions

In a **decomposition** reaction substances break up to form two or more new substances.

Example: Passing electricity through water breaks down (decomposes) the water to form oxygen gas and hydrogen gas.

ISBN 978 1 4202 3735 1

Observing reactions

Aim

To observe a variety of chemical reactions.

Risk assessment and planning

Carefully read the instructions for both parts. Which safety precautions will be necessary?

In your notebook, design and draw up a large data table to record your observations. Note: There are seven different reactions in Part A.

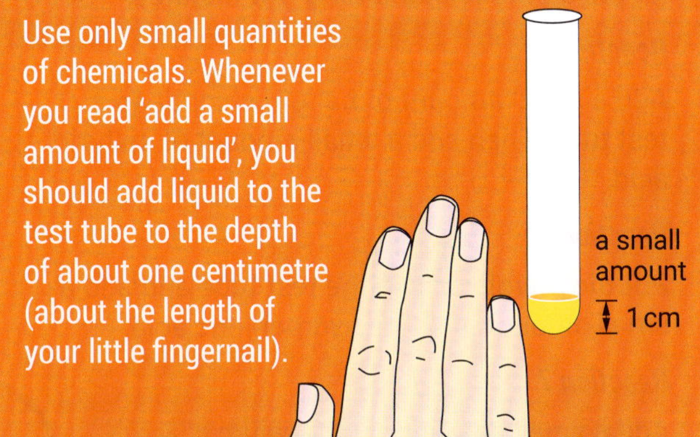

PART A

Materials

- dilute **hydrochloric acid** (1 M)
- dilute **sodium hydroxide** solution (1 M)
- dilute **sulfuric acid** (1 M)
- sodium thiosulfate (hypo) crystals
- 2 small pieces of magnesium ribbon
- **copper sulfate** solution (0.1 M)
- **barium chloride** solution (0.1 M)
- marble chip
- potassium permanganate solution (0.001 M)
- hydrogen peroxide (3% solution)
- 7 test tubes and test tube racks
- test tube holder

Wear safety glasses

Point the test tube away from yourself and anyone else

piece of magnesium

hydrochloric acid

Method

1 Number the seven test tubes.

2 Carry out the seven reactions below.
For each reaction, describe the changes you see, noting any of the four signs that a reaction has occurred. (Feel the test tube to detect any change in temperature.)
Draw up a data table for your results. Put what you did on the left-hand side, and what you observed on the right.

SAFETY GLASSES MUST BE WORN

Reaction 1

Put a piece of magnesium in the test tube and add a small amount of dilute hydrochloric acid. Don't forget to point the test tube away from yourself and other people, and wear safety glasses.

Reaction 2

One-third fill the test tube with copper sulfate solution. Add a piece of magnesium.

Reaction 3

One-third fill the test tube with water. Add about a spatula of hypo crystals and shake to dissolve.

ISBN 978 1 4202 3735 1

Reaction 4

Add a small amount of copper sulfate solution to a test tube. Then add a small amount of sodium hydroxide solution (caustic soda) and let the test tube stand for a few minutes.

Reaction 5

Add a small amount of barium chloride to a test tube. Then add a small amount of sulfuric acid.

Reaction 6

Put the marble chip in the test tube and add a small amount of dilute hydrochloric acid.

Reaction 7

Add a small amount of potassium permanganate solution to a test tube. Then add 2 or 3 drops of hydrogen peroxide.

Discussion

1 For each reaction, list which of the signs of a chemical reaction (see page 111) occurred.
2 In which test tube reactions did you notice a temperature change?
3 Sort the reactions into groups so that each group contains similar types of reactions.

PART B

Materials

- lemon
- strip of copper (10 mm × 80 mm)
- strip of magnesium or aluminium or a large steel nail
- millivoltmeter or multimeter
- 2 connecting wires with alligator clips

Method

1 Clean the two pieces of metal with steel wool until they are shiny.
2 Push the metal strips into the lemon about 10 mm apart (not touching!).
3 Connect the wires to the metal strips and to the multimeter, as shown in Figure 5.6. Record your observations.

Discussion

1 Was there a chemical reaction? How do you know?
2 After some time the multimeter will read 0. Suggest why.
3 Take the metal strips out of the lemon. Are there any signs that a chemical reaction has occurred?

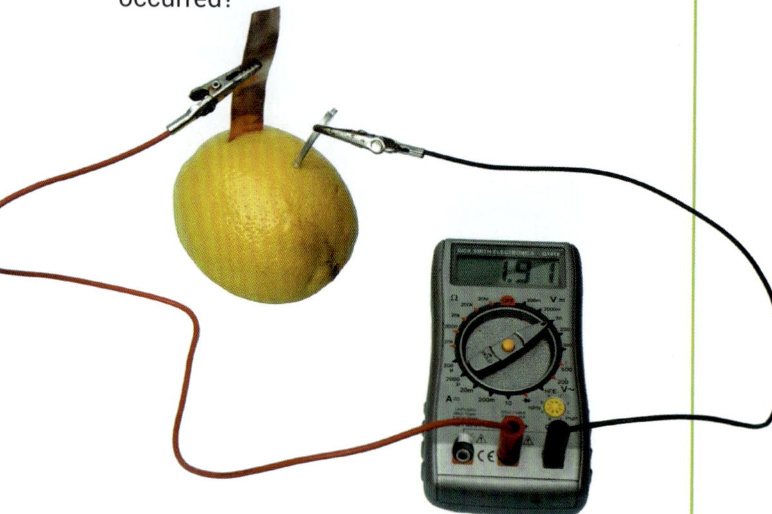

Figure 5.6 Connecting the lemon to the multimeter

PART C

SAFETY GLASSES MUST BE WORN

Do this activity outdoors.

You need a clear plastic bottle and a stopper or cork that fits well. Put about three heaped teaspoons of baking soda in the bottle. Moisten the stopper with water. Get about half a cup of vinegar. Now be prepared to work quickly.

While your partner holds the bottle and stopper, pour the vinegar into the bottle. Put the stopper in immediately, but not too tight! Make sure you point the bottle away from people or things that could break.

ISBN 978 1 4202 3735 1

Reactants and products

You can divide the substances in any chemical reaction into two groups—**reactants** and **products**. The reactants are the substances you start with and which react with each other. The products are the new substances produced in the reaction. For example, in Investigation 5.2 on pages 113–114, vinegar reacted with baking soda to produce bubbles of gas that caused fizzing. So in this case the vinegar and baking soda are reactants, and the gas (called carbon dioxide) is the product.

There is a shorthand way of describing what happens in a chemical reaction. It is called a **chemical equation**. Instead of saying baking soda reacts with vinegar to produce carbon dioxide, you write:

baking soda + vinegar → carbon dioxide

(reactants) → (products)

The reactants are on the left-hand side of the equation, and the product is on the right-hand side. Sometimes there is only one reactant and sometimes there are two or more. The same goes for the products. The reactants and products can be solids, liquids or gases. Sometimes it is hard to know exactly what the reactants and products are. For example, when wood burns, it reacts with oxygen (an invisible gas present in the air), and produces another invisible gas—carbon dioxide.

When do reactions occur?

Chemical reactions occur only under certain conditions.

First, you need to have the right substances present. Iron will not rust without air and water. This is why rusting does not occur on the moon—there is no air or water.

Second, you quite often need heat to cause a chemical reaction. To fry an egg, you need to heat it in a pan. To get dough to rise you need to put it in a warm place; and to get a sparkler to burn, you have to heat it with a match.

Third, you sometimes need electricity to cause a chemical reaction. This is what happens when you charge a battery. The electricity produces chemical reactions that store energy.

Forming solids

Two of the reactions in Investigation 5.2 (on pages 113–114) formed solids. When you added sodium hydroxide to copper sulfate (in Reaction 4), a pale blue solid called a precipitate formed.

In Reaction 5 a precipitate formed when you added sulfuric acid to barium chloride solution. This precipitate is barium sulfate. The equation for this reaction is:

barium chloride + sulfuric acid → barium sulfate + hydrochloric acid

Figure 5.7 When charging a battery, the electricity causes chemical reactions that store energy.

ISBN 978 1 4202 3735 1

Reaction rate

Some reactions are slow, like rusting. Others occur quickly, like the ones you observed in Investigation 5.2. Explosions are chemical reactions that occur very quickly and produce large amounts of heat, light and sound.

The speed of a reaction is called its **reaction rate**. A slow rate means the reaction takes a long time. A fast rate means it takes only a short time. In Experiment 5.1 below you can investigate the way in which the temperature affects the reaction of acid on marble chips.

Figure 5.8 An electrical spark was thought to have caused this fire. This occurred because the oil reacted explosively with the oxygen in the air. This is an example of a very fast reaction.

EXPERIMENT 5.1 Reaction rate

This experiment is a little different from the others you have done. In this one you have to work in a small group to write the method.

Planning

The aim of this investigation is to see how temperature affects reaction rate.

- For the reaction use a marble chip in some hydrochloric acid.
- Try three temperatures—room temperature, ice water and hot tap water (although you might have other ideas!).
- Pour the hydrochloric acid into the test tubes and sit them in hot or cold water for at least 5 minutes before you add the marble chip.
- Make it a fair test! Use the same amount of acid and the same size marble chip in all test tubes.

Getting started

1 Write the title, aim, planning and safety check, materials and method.
2 Show your write-up to your teacher before you start. Your teacher has to sign your investigation.
3 Make sure you have designed data tables to record your results.

Writing your report

1 Remember to include results, discussion and a conclusion in your report.
2 Work in a group to compile the report but write your own to show to your teacher.

ISBN 978 1 4202 3735 1

Wanted reactions

All life depends on chemical reactions. Plants use the chemical reaction called photosynthesis to make food. Chemical reactions occur in your bodies to digest the food you eat. And further reactions are needed to produce energy for your body cells. Fuels such as coal and oil are also used to provide energy; for example, petrol is burnt in the engine of a car.

Most of the materials we use every day are made by chemical reactions. For example, synthetic rubber, plastics, detergents, fertilisers, paints and some medicines are made from chemicals found in oil. And the chemical reactions that occur in batteries allow us to produce electricity anywhere.

Figure 5.9 A bushfire is a large, uncontrolled chemical reaction.

Unwanted reactions

Some chemical reactions are unwanted. For example, house fires and bushfires are uncontrolled reactions that cause loss of life and property. Rusting and other forms of corrosion cause millions of dollars worth of damage. Tooth decay is caused by chemical reactions between your teeth and acids in food.

Reactions that occur in homes, cars and factories all create products that can harm our environment if not carefully controlled. Every year we learn more about the properties of materials we use. For example, we now know that CFCs (chlorofluorocarbons), the gases once used in aerosols and refrigerators, can destroy the ozone layer by reacting with it.

SCIENCE AS A HUMAN ENDEAVOUR

Painting the bridge

The Sydney Harbour Bridge is the largest steel bridge in Australia. Construction of the bridge started in 1924 and it took 1400 men eight years to complete.

The 52 800 tonnes of steel in the bridge would cover an area of about 60 football fields. But there is one major problem with a steel bridge—it rusts. This is an unwanted chemical reaction.

To stop the steel reacting with the air, the steel is painted. The first painting in 1930 took 272 000 litres of paint to give the bridge three coats of paint. This amount of paint would fill seven petrol tankers. The bridge is continually being painted to stop it rusting.

ISBN 978 1 4202 3735 1

1 Copy and complete these sentences.
 a Two types of change that substances undergo are _____ and _____ changes.
 b When a chemical reaction takes place _____ _____ are formed.
 c Chemical reactions cannot easily be _____.
 d Fizzing means that a _____ has been produced.
 e _____ and light may be produced in a chemical reaction.
 f In the reaction hydrogen + oxygen → water, the reactants are _____ and _____. The product is _____.

2 How do the properties of an egg change when it is fried?

3 Draw up two columns using the headings *Physical changes* and *Chemical changes*. Put the following in the correct columns:
 a ice-cream melting
 b an apple rotting
 c milk going sour
 d an egg being poached
 e dynamite exploding
 f clothes drying on a clothes line
 g a toaster element becoming red-hot
 h gas burning in a stove
 i sugar dissolving in water
 j a lamp filament glowing

Think of one more example for each column.

4 The teacher heated some orange-coloured crystals in a test tube. The crystals crackled and turned to a green powder. Give two reasons why you think a chemical reaction occurred.

5 Maya said that an Alka-Seltzer tablet fizzing in water is a chemical reaction, but sugar dissolving in water is not. Do you agree with her? Give reasons.

6 Give three examples of wanted reactions and three examples of unwanted reactions.

7 a Explain what a precipitate is.
 b Give an example of a precipitation reaction.

8 a Explain the difference between a combination reaction and a decomposition reaction.
 b Give an example of each type of reaction.

1 Go back to your results for Investigation 5.2 (on pages 113–114). Write equations to describe what happened in Part A Reactions 2 and 4.

2 Use your knowledge of chemical changes to explain why batteries go flat.

3 When nitric acid is added to a piece of copper, a brown gas (nitrogen dioxide) is formed. A blue substance called copper nitrate is also formed, plus some water. Write a chemical equation for the reaction.

4 Ian investigated the reaction that occurs when yeast is added to dough. He measured the height that the mixture rose up a graduated icy-pole stick. Here are his results:

Temperature (°C)	15	20	25	30	35
Distance up stick (cm)	0	0	5	10	16

Draw a line graph of 'Distance up stick' against 'Temperature'.
 a Explain the shape of your graph.
 b The dough did not rise until the temperature reached 25 °C. Write an inference to explain this.

ISBN 978 1 4202 3735 1

5.3 Some common gases

Surrounding the Earth is a thin layer of air called the **atmosphere**. This atmosphere protects us in several different ways. It causes meteors to burn up before hitting the Earth. It also contains a gas called ozone that protects us from harmful ultraviolet radiation. Beyond the atmosphere is space.

The air in our atmosphere is a mixture of colourless gases. These gases can be separated from each other and used in various ways depending on their properties. For example, nitrogen is used to make fertilisers, and liquid nitrogen is used to cool things. The unreactive gas argon is used in light bulbs to stop the filament from burning.

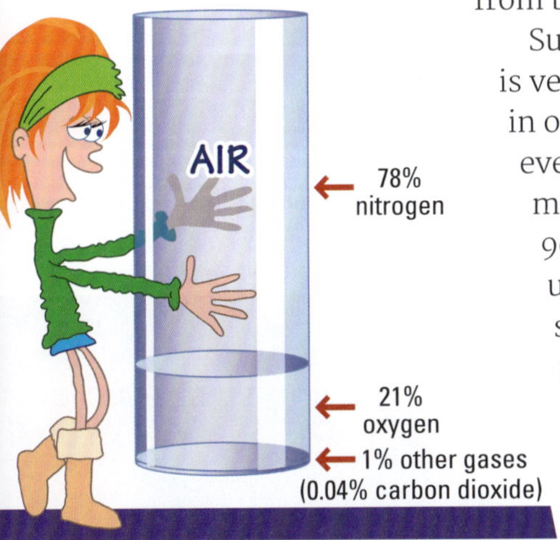

Surprisingly, there is very little hydrogen in our atmosphere, even though it makes up about 90% of the total universe. The sun and stars are mainly hydrogen, and so are the outer planets.

Oxygen

We cannot live without oxygen—our bodies need a constant supply of it. Oxygen is needed to get energy from food in the process of respiration. We use about 20 litres every hour when we are resting, and much more when we are active.

Oxygen is used to help people breathe in difficult situations such as after car accidents and in hospitals during operations. It is also used in places where there is not enough oxygen to breathe normally. For example, jet plane pilots, mountain climbers and firefighters carry a supply of oxygen.

Figure 5.10 Mountain climbers carry oxygen at high altitudes where the air is thin to avoid getting altitude sickness.

Oxygen (O_2) is a colourless gas that has no smell. It is very reactive, meaning it reacts with many substances. Combustion (or burning) is the process in which oxygen combines rapidly with other substances, producing light and heat. Corrosion of metals, for example rusting, is mostly due to metals reacting slowly with oxygen in the air.

In space there is no oxygen, so rockets have to carry their own supply in the form of liquid oxygen. In an oxy-acetylene torch, acetylene gas is burnt with oxygen to produce temperatures above 3000 °C.

Figure 5.11 In an oxy-acetylene torch, heat from the burning gas melts metal.

ISBN 978 1 4202 3735 1

Hydrogen

In the early part of the 20th century, airships were a popular form of transport. They were filled with hydrogen to give them lift. In 1937 a large German airship called the *Hindenburg* caught fire as it landed on the east coast of the United States. The hydrogen in the airship reacted violently with the oxygen in the air, and 35 of the 97 passengers and crew were killed. After this accident, airships quickly lost their popularity.

Hydrogen (H_2) is a colourless gas that is at least 15 times lighter than any other substance. This is why it was used in the early airships and balloons. It is still used today to fill weather balloons, but modern airships use helium gas because it is not flammable.

In the presence of a flame, hydrogen reacts so rapidly with the oxygen in the air that it explodes.

hydrogen + oxygen → water

This same reaction is used in the rocket engines of the space shuttle. Some people believe hydrogen will be used as a fuel in the future because when it burns in air, the only product is water, which does not cause air pollution.

Hydrogen is used in industry to make many different everyday materials. For example, the flow chart below shows how it is used to make margarine from vegetable oil.

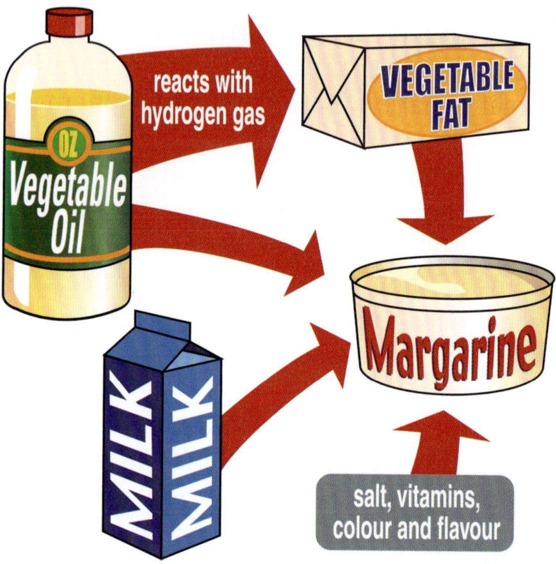

Figure 5.12 How hydrogen is used in making margarine. Some margarines do not contain milk.

Figure 5.13
The airship Hindenburg crashes in flames in 1937.

ISBN 978 1 4202 3735 1

INVESTIGATION 5.3 Making hydrogen

Aim

To make hydrogen gas and test it.

Materials

- test tube rack
- 2 test tubes
- 1 cm of magnesium ribbon
- dilute **hydrochloric acid** (1 M)
- taper and matches
- test tube holder

Risk assessment and planning

Read the investigation carefully.

- Why is it essential that you wear safety glasses when doing this experiment?
- Why do you turn the collecting test tube upside down over the first test tube?

Method

1. Put a test tube in the test tube rack and one-third fill it with dilute hydrochloric acid.
2. Put the magnesium ribbon into the acid. Then use a test tube holder to hold an empty test tube upside down over the mouth of the first test tube, as shown.
3. Ask another person to light a taper. Then remove the top test tube, tilt it upwards slightly and *immediately* put the burning taper near its mouth. A 'pop' or 'squeak' indicates that the gas in the tube is hydrogen.

   What do you observe on the inside of the test tube?
4. If necessary, repeat the experiment.

Discussion

1. What evidence is there that a chemical reaction occurred in the test tube?
2. After a while no more hydrogen is produced. Why is this?
3. How can you explain the presence of water droplets inside the mouth of the test tube after you 'pop' the hydrogen?
4. Why did you tilt the test tube upwards before bringing the burning taper near?
5. Suggest why you didn't hold the collecting test tube in your hand when you tested for hydrogen.

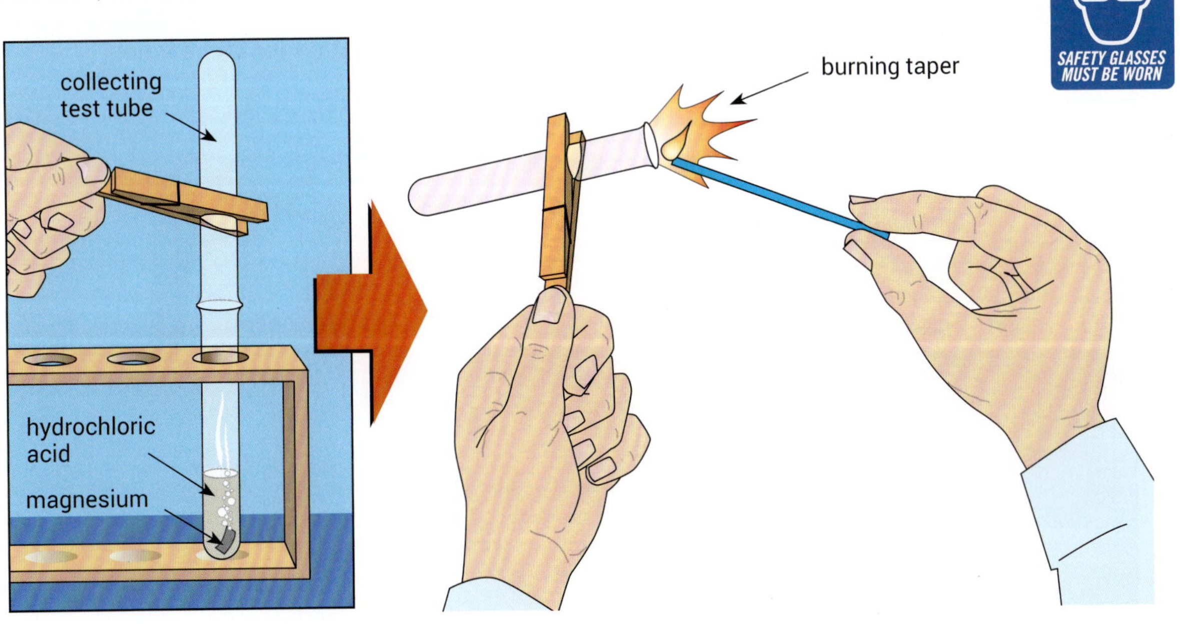

ISBN 978 1 4202 3735 1

INVESTIGATION 5.4 Making oxygen

Aim

To make oxygen gas and test it.

Materials

- one-holed stopper fitted with U-shaped piece of glass tubing as shown
- manganese dioxide powder
- hydrogen peroxide
- 2 test tubes
- test tube rack
- test tube holder
- wooden splint and matches

Risk assessment and planning

This investigation is similar to the previous one except that you will use a piece of U-shaped tubing to collect the gas. This time you have to write the method.

Read through the investigation carefully and use the diagram as a guide to write a step-by-step method to make oxygen gas. (It is best to work in a small group for this.)

Show the method to your teacher before proceeding.

Hints for writing the method

1 Use a pea-sized amount of manganese dioxide powder in the test tube.
2 Use a small amount (see page 113) of hydrogen peroxide.
3 To test for oxygen gas, light a wooden splint, let it burn for a short while then blow it out, so that it still has a glowing tip. Immediately put the glowing splint into the test tube. The splint should burst into flames.

Discussion

1 Why is it that the splint only glowed in air, but burst into flames in oxygen gas?

2 The black manganese dioxide powder was not a reactant in the reaction. It speeds up the reaction but is not used up.
Predict what would happen if you placed the manganese dioxide remaining in the test tube into a new test tube containing hydrogen peroxide. Give a reason for your answer.

3 In the reaction that produced oxygen, water was also produced. The hydrogen peroxide was used up. Which were the products in the reaction? Which were the reactants? Try to write an equation for the reaction.

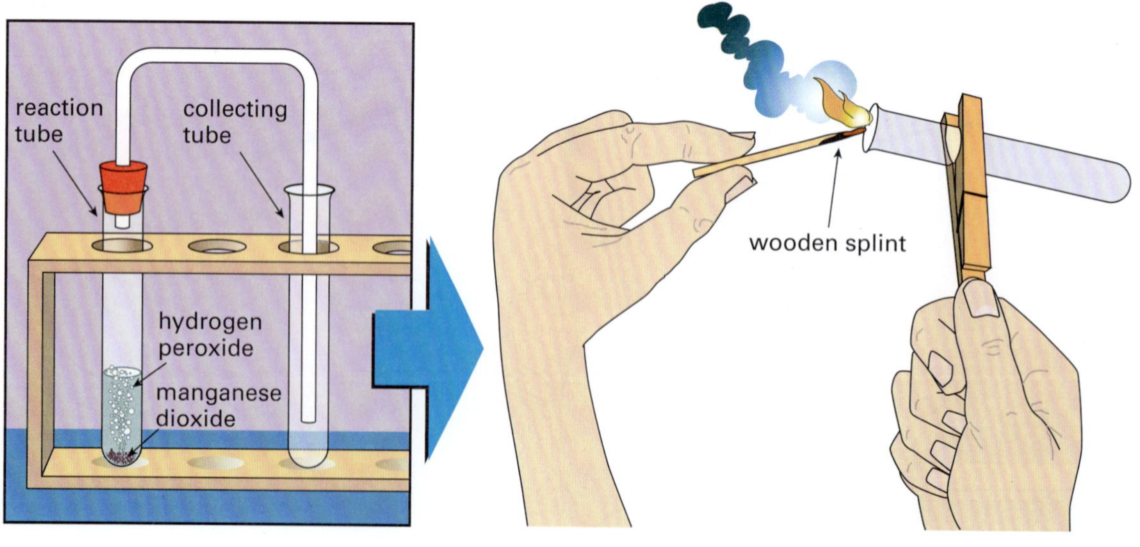

ISBN 978 1 4202 3735 1

Carbon dioxide

In Part C of Investigation 5.2 on page 114, when baking soda and vinegar were mixed, the gas produced built up pressure in the bottle and the cork was blown out. This gas was carbon dioxide.

Carbon dioxide (CO_2) is another invisible gas. It is found in air, but there is not very much of it—only 0.04% of the total of all gases. Despite this, carbon dioxide is a very important gas and has many uses.

Drinks and bread

When yeast and sugar are mixed together in warm water, a chemical reaction occurs. Yeast is a type of fungus that changes sugar into alcohol and carbon dioxide. This is why beer and wine bubble when they are fermenting. The bubbles contain carbon dioxide gas. Yeast is also used in making bread. The small 'holes' in a slice of bread are produced by bubbles of carbon dioxide, which expand in the hot oven.

Carbon dioxide is put into soft drinks under pressure. When you open the can or bottle, the carbon dioxide escapes, causing bubbles.

Fire extinguishers

Carbon dioxide has two important properties—things will not burn in it, and it is much more dense than air so it sinks to the ground. This means it can be used in fire extinguishers. The carbon dioxide is under pressure and when the trigger is pressed, the gas rushes out. It forms a 'gaseous blanket' over the fire and stops air getting to it. Without air, the fire soon goes out.

Dry ice

When carbon dioxide is cooled to about −80 °C it freezes to a white solid called *dry ice*. It is called dry ice because it changes directly to carbon dioxide gas without forming a 'wet' liquid. Dry ice is much colder than water ice, so it is used to keep things cold and dry. It is also used in cloud seeding. In this process tiny particles are dropped from aircraft into clouds to produce rain.

Figure 5.14 Carbon dioxide is used in fire extinguishers because it does not allow things to burn and is heavy. It sinks to the ground to smother the fire.

Plants, burning and breathing

Green plants use up carbon dioxide from the air in the process of photosynthesis.

carbon dioxide + water + sunlight → plant food + oxygen

There are many ways in which the gas is put back into the air. For example, a lot of carbon dioxide goes into the air from power stations, which burn coal, and from cars, which burn petrol. We also breathe out carbon dioxide produced in our bodies during the process of respiration.

food/fuel + oxygen → carbon dioxide + water + energy

The total amount of carbon dioxide throughout the world stays about the same. But in recent years it has increased slightly, and people are worried that this could lead to global warming due to the **greenhouse effect**.

ISBN 978 1 4202 3735 1

INVESTIGATION 5.5 Making carbon dioxide

Aim

To make carbon dioxide gas and investigate its properties.

Risk assessment and planning

Read through the four parts and discuss with your teacher which ones you will do and how much time you will spend on this investigation.

PART A Making and testing carbon dioxide

Materials

- 3 test tubes and test tube rack
- stopper
- one-holed stopper fitted with U-shaped piece of glass tubing as shown
- taper and matches
- drinking straw
- limewater (calcium hydroxide solution)
- 2 or 3 marble chips (calcium carbonate)
- dilute **hydrochloric acid** (1 M)

Method

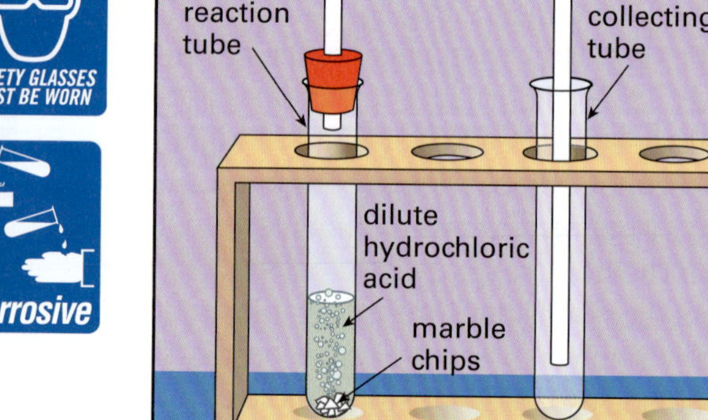

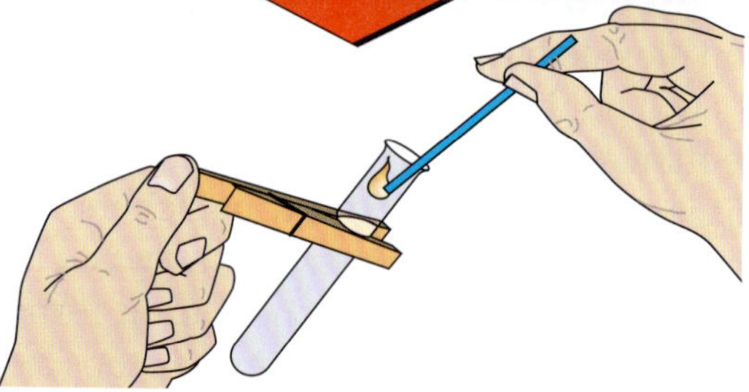

1. Set up the apparatus as shown. Make sure the collecting tube is dry. Put two or three marble chips into the reaction tube.
2. Add some hydrochloric acid to the reaction tube, then quickly fit the stopper and tubing. Bubbles of carbon dioxide gas will form. This gas will go to the collecting tube.
3. After about 3 minutes remove the collecting tube and put a stopper in it. Replace it with another tube one-third full of limewater. Allow the gas to bubble through the limewater while you do step 4.
4. Light the taper, remove the stopper from the collecting tube, and carefully put the taper into the tube as shown.

   Record what happens.
5. Now go back and observe the limewater from step 3.

   Record your observations.

   Has there been a chemical reaction? How do you know?
6. Tip out the limewater and wash out the test tube. Then pour in some fresh limewater. Blow gently through a straw into the limewater.

   What do you observe?

   What can you infer from this observation?

ISBN 978 1 4202 3735 1

PART B Fire extinguisher

Materials

- 3 birthday candles of different lengths
- large ice-cream container (taller than the longest candle)
- baking soda
- vinegar (acetic acid)
- beaker or jar
- teaspoon
- matches
- Blu-Tack

Method

1 Use Blu-Tack to attach the three candles to the bottom of the ice-cream container, as shown. Light the candles.

2 Put 2 teaspoons of baking soda in a beaker. Then cover the baking soda with vinegar. Tilt the beaker over the container as shown, but don't pour out any froth or liquid.

What do you observe?

Write an inference to explain what happened.

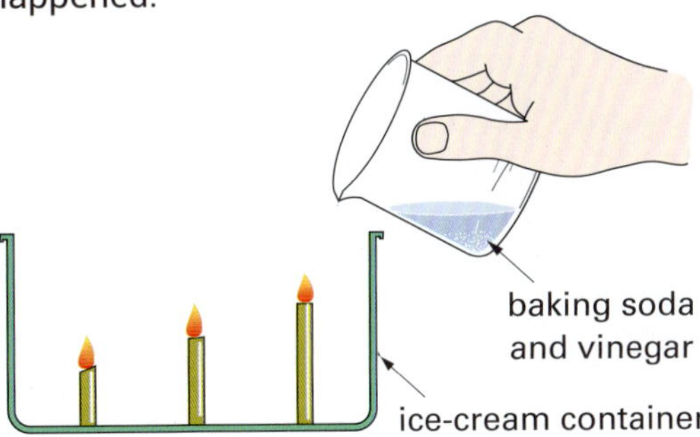

PART C Yeast action

Materials

- 3 large test tubes
- sachet of powdered yeast
- 2 teaspoons of sugar
- teaspoon
- four 250 mL beakers or jars

Method

1 Dissolve 2 teaspoons of sugar in about 100 mL of lukewarm water in a glass jar or beaker.

2 Add about half a sachet of powdered yeast and stir. Put the beaker in a warm place.

3 Prepare three water baths as follows:

Beaker 1—fill with cold water (add an ice cube if necessary).

Beaker 2—fill with slightly warm water.

Beaker 3—fill with hot tap water.

4 Pour equal amounts of the yeast mixture into the three test tubes. Then place a test tube in each water bath.

5 Observe the test tubes for 10–15 minutes and note any evidence of a chemical reaction.

Record your observations. In which test tube did the most vigorous reaction occur?

Under which conditions does yeast work best?

PART D Sultana bouncers

Materials

- 4 sultanas
- soda water or lemonade
- glass or jar

Method

1 Half fill the glass with soda water.

2 Place the sultanas in the glass.

Observe what happens to the sultanas. Record your observations.

In a group try to come up with an explanation of why the sultanas move up and down. (Hint: Do the sultanas sink or float in ordinary water?)

Why do the sultanas stop bouncing after some time?

ISBN 978 1 4202 3735 1

SCIENCE AS A HUMAN ENDEAVOUR

Carbon dioxide—friend or foe?

There is a very small amount of carbon dioxide (CO_2)in the atmosphere—about 0.04%. This is the amount of the gas measured in 2016. In 1915, about 100 years ago, the amount was about 0.03%. The difference between the 1915 and 2016 amount doesn't seem all that much. However, scientists now believe that there is already too much CO_2 in the atmosphere. In fact, there is now more CO_2 in the atmosphere than there has ever been. It was in the 1950s that the amount of carbon dioxide became higher than at any other time in the last 400 000 years.

CO_2 and the greenhouse effect

The Earth absorbs sunlight during the day and warms up. At night, the heat is released into the air and it becomes cooler. However, if you were on the moon, the nights would be extremely cold. This is because the moon does not have an atmosphere to trap some of the heat.

The gases in the Earth's atmosphere act like the glass in a greenhouse. The sunlight passes through the glass and heats up the plants and soil in the greenhouse. At night, the heat inside the greenhouse is trapped by the glass. This is the greenhouse effect.

Carbon dioxide traps more heat than the other common gases in the atmosphere. This is why it is called a *greenhouse gas*.

CO_2 and global warming

Many people are concerned about the amount of CO_2 in the atmosphere. Scientists have suggested that the CO_2 produced by power stations, industries, and cars and trucks is trapping too much heat in the atmosphere. This has resulted in an increase in the average temperature around the world, and is called **global warming**. Global warming is now considered to be a very real danger to life on this planet.

It is predicted that global warming will cause flooding as the sea level increases due to the melting of ice in the polar regions. Other problems may also occur—ocean warming and coral death, extinction of animal and plant species, changing climates, droughts, cyclonic weather and the spread of tropical diseases.

Figure 5.15 The Earth acts as a greenhouse. Heat is trapped in the atmosphere by carbon dioxide and other gases. As greenhouse gases increase, so does the Earth's temperature.

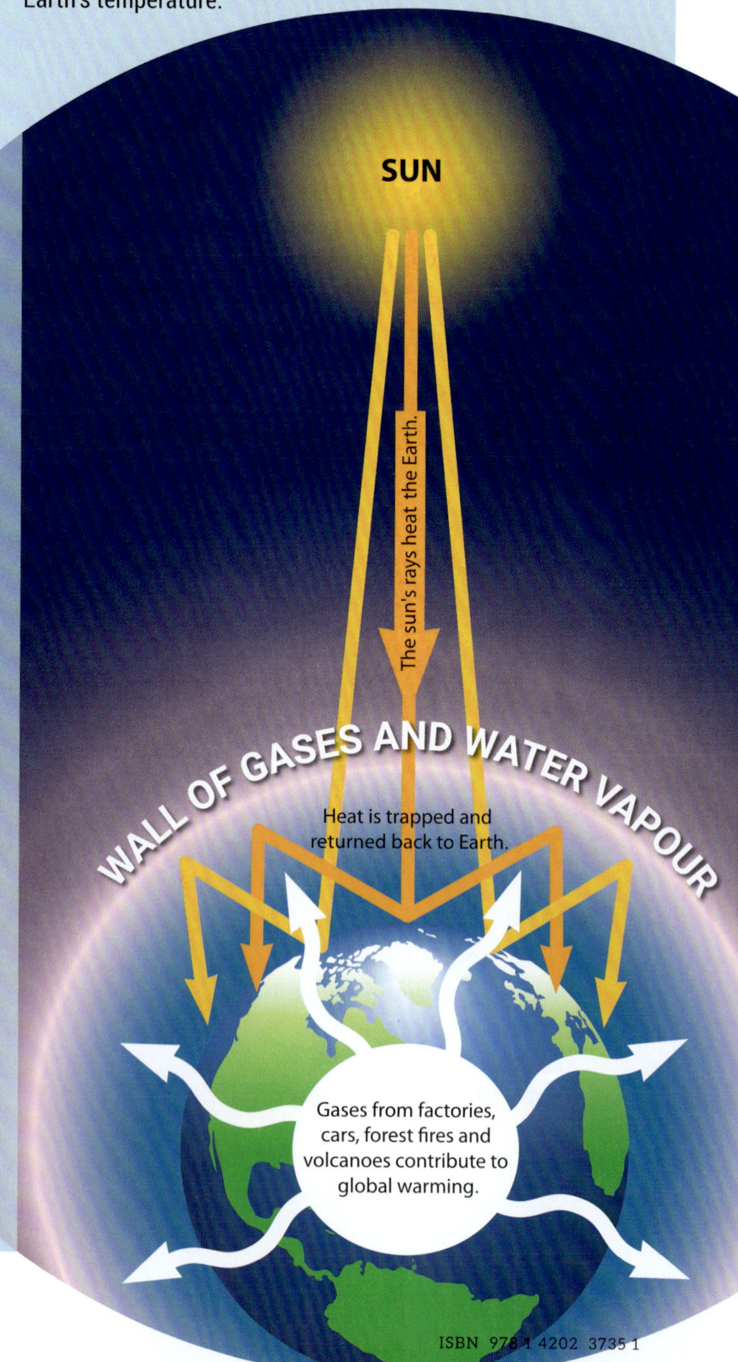

ISBN 978 1 4202 3735 1

ACTIVITY

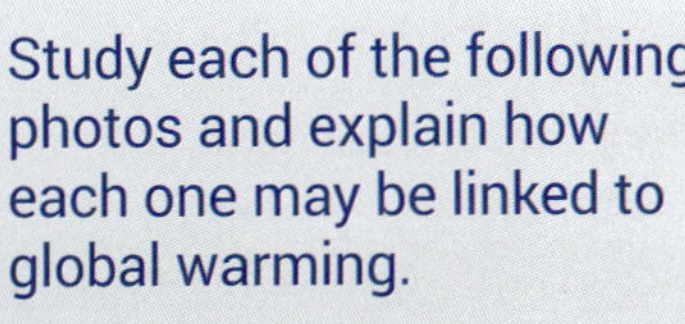

Study each of the following photos and explain how each one may be linked to global warming.

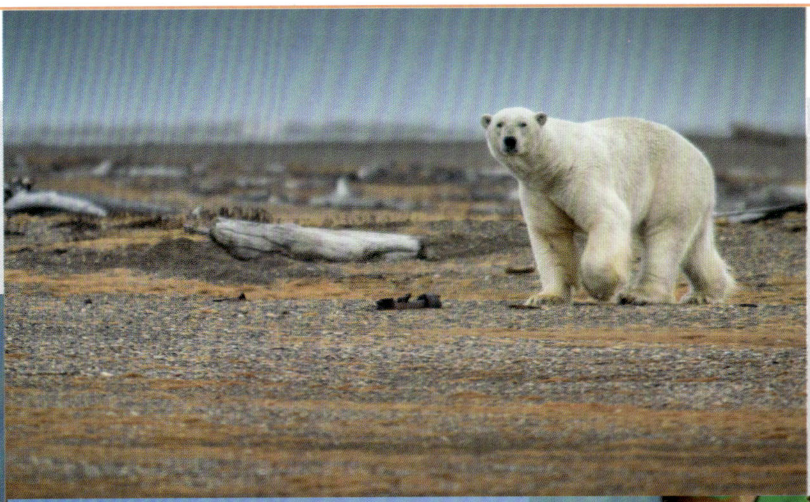

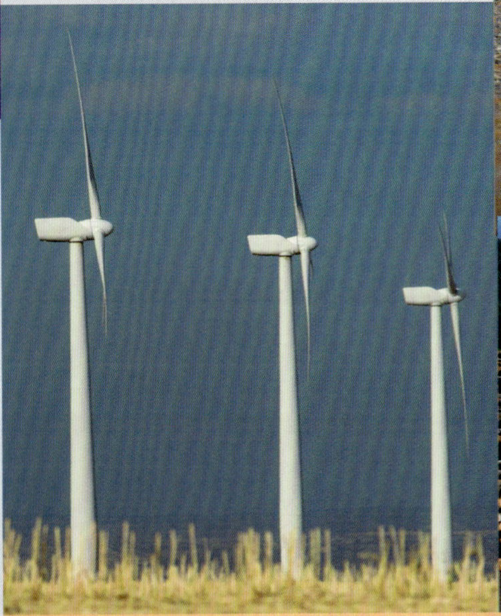

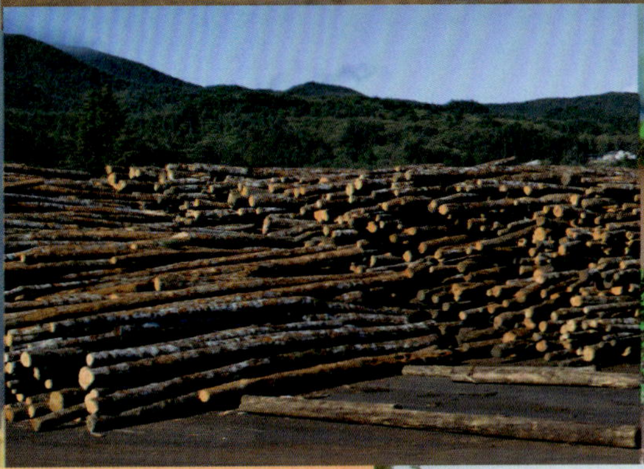

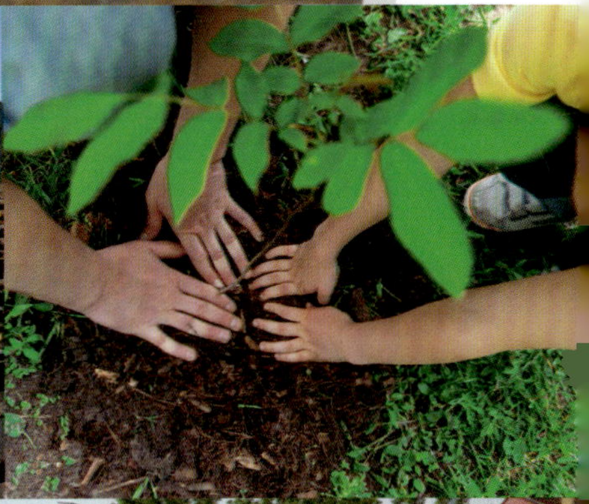

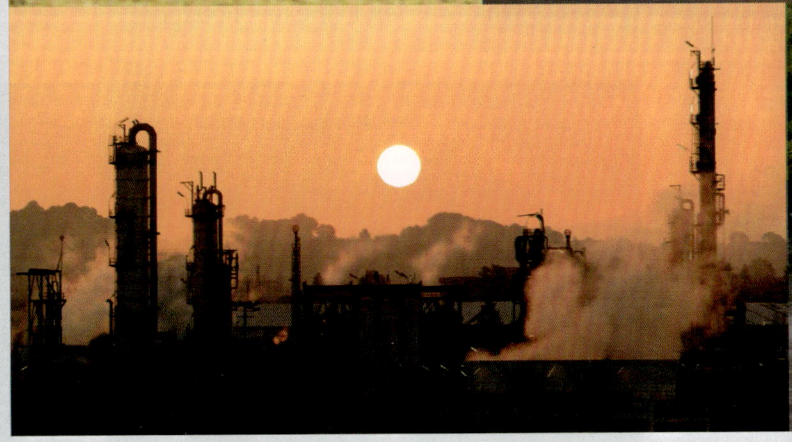

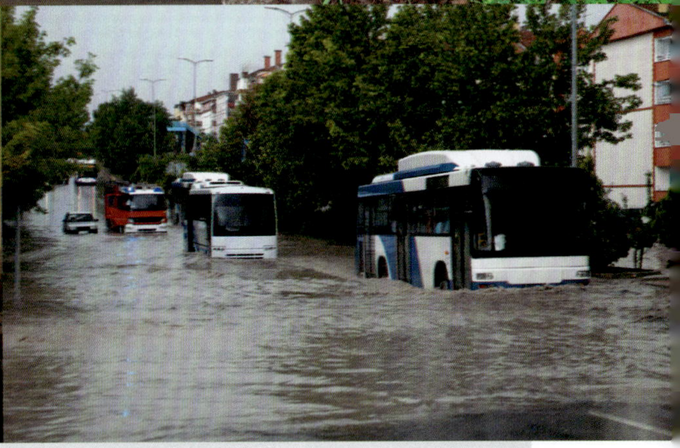

EXPLORE ONLINE

Search the internet to research answers to the following questions:

1 Explain how the greenhouse effect is good for Earth, but can also have negative effects.

2 Make a list of some of the predicted effects of global warming on Australia.

3 Identify some things that you can personally do to reduce greenhouse gas emissions in your daily life.

4 What is Australia as a nation doing to reduce its greenhouse gas emissions?

5 Did you find any examples of how technology can help reduce greenhouse gas emissions? If so explain your answer using examples.

6 Imagine the government asks you to invent a new technology that can help slow or stop global warming. Describe what your technology is and how it would work. Use a diagram to show your invention.

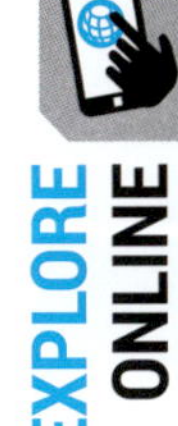

ISBN 978 1 4202 3735 1

CHECK

1 Why are modern airships filled with helium rather than hydrogen gas?

2 Why is oxygen so important to life on Earth?

3 What chemical reaction occurs during combustion?

4 Some oxygen dissolves in water. Why is this important for some living things?

5 Which is heavier—hydrogen gas or oxygen gas? How do you know?

6 You have a test tube containing a colourless, odourless gas. How could you test whether it is hydrogen or oxygen?

7 Use Figure 5.12 on page 120 to describe in a few sentences how margarine is made.

8 Write a short paragraph comparing and contrasting the properties of hydrogen and oxygen.

9 Which of the following statements about carbon dioxide are true and which are false? Rewrite the incorrect ones.

- a About 40% of the air around us is carbon dioxide.
- b Carbon dioxide is always a gas.
- c Carbon dioxide is more dense than air.
- d Carbon dioxide helps burning.
- e Humans breathe out carbon dioxide.
- f Vinegar reacts with baking soda to produce carbon dioxide.
- g Carbon dioxide is produced during photosynthesis.

10 List three ways of making carbon dioxide.

11 Which two properties of carbon dioxide are illustrated in this diagram?

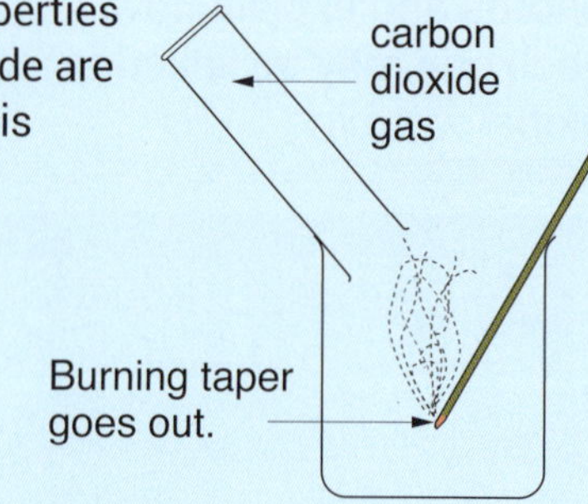

12 Which two substances are needed for fermentation to occur?

13 A bottle of soft drink goes 'flat' if it is left open. Why?

14 This equation describes what happened in Investigation 5.5, Part A.

calcium carbonate + hydrochloric acid
→ calcium chloride + water + carbon dioxide

- a What are the reactants?
- b What are the products?
- c What was left in the reaction tube?

Figure 5.16 This image of the Eagle Nebula, taken by the Hubble Space Telescope, shows massive pillars of hydrogen gas in which stars are born. Hydrogen is one of the most common elements in the universe.

ISBN 978 1 4202 3735 1

CHALLENGE

1 If you pass electricity through water, the gases hydrogen and oxygen are produced. Write an equation to describe this reaction.

2 How could you distinguish between a bottle of water and a bottle of hydrogen peroxide?

3 Tania burnt some magnesium ribbon in oxygen. A white powder called magnesium oxide was left.

- a Write an equation to describe the reaction.
- b Tania found that the magnesium oxide had a greater mass than the magnesium she started with. How could she explain this?
- c What type of reaction is this?

4 Merali and Wayne put a candle in a saucer and added some water. They then lit the candle and put a glass upside down over the candle. The candle burnt for a while then went out, but they were surprised when the water rose inside the glass. How can you explain why this happened?

5 Why are some people worried about the amount of carbon dioxide in the Earth's atmosphere?

6 Work out a way of showing that the gas bubbles in a bottle of soft drink are carbon dioxide.

7 When carbon dioxide is bubbled through limewater, the solution goes cloudy. Suggest why this happens.

8 Fizzy drink powders contain baking soda and tartaric acid. When this powder is added to water it fizzes.

- a Using what you have learnt in this chapter, suggest what chemical reaction has occurred here.
- b Why does it fizz only when water is added?

9 How can you explain the 'fizzy' taste of sherbet? It is a mixture of baking soda, citric acid and sugar.

10 When making bread it is important to let the dough rise slowly in a warm place instead of putting it straight into a hot oven. Suggest a reason for this.

11 The 'mist' in the photo below is produced by adding dry ice directly to water.

- a What is dry ice?
- b Explain what happens when dry ice is added to water. Is the change a physical or chemical one?
- c Why does the 'mist' stay low and sink along the bench.

Use the internet to find information on the common gases listed below. For example, how do we extract them from the air, and how do we use them?

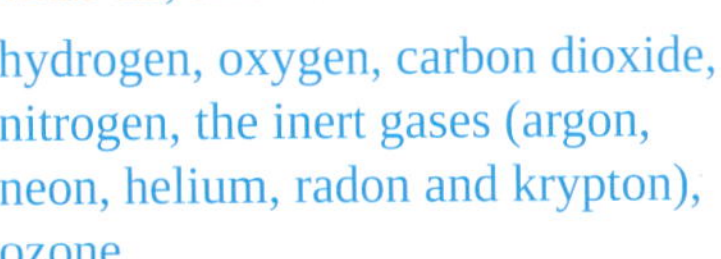

hydrogen, oxygen, carbon dioxide, nitrogen, the inert gases (argon, neon, helium, radon and krypton), ozone

You could also search for more general topics such as *gases in atmosphere* or *composition of atmosphere*.

Summarise your information in a large table.

ISBN 978 1 4202 3735 1

MAIN IDEAS

Copy and complete these statements to make a summary of this chapter. The missing words are on the right.

1 The ________ of a material depend on its properties.

2 Chemical changes (reactions) produce new substances with ________ different from those of the original substances.

3 A chemical reaction may produce a gas, a colour change, a solid called a ________, heat, light, sound or ________.

4 A ________ change is one in which no new substances are formed. It can usually be reversed.

5 A chemical ________ shows what you start with (the reactants) and the substances produced (the ________) in a chemical reaction.

6 A chemical reaction occurs only under certain conditions. ________ and electricity may be needed.

7 Some reactions occur slowly, while others occur quickly. The speed of a reaction is called its ________.

8 Our atmosphere is a mixture of ________, about $^{4}/_{5}$ nitrogen and $^{1}/_{5}$ ________. It also includes small amounts of other gases, e.g. carbon dioxide, ozone and hydrogen.

9 A ________ reaction occurs when a substance burns by reacting with oxygen.

10 In a ________ reaction substances break up to form two or more new substances. This is the opposite of a ________ reaction.

precipitate
uses
combination
properties
equation
decomposition
rate
gases
heat
oxygen
combustion
products
electricity
physical

CH•5 REVIEW

1 Which *one* of the following is a chemical reaction?
- **A** melting a block of ice
- **B** burning magnesium ribbon
- **C** magnetising a piece of iron
- **D** expanding air by heating it

2 Which one of the following is *false*?
- **A** Heating sometimes causes a reaction.
- **B** Some reactions produce electricity.
- **C** Reactions take only a few seconds.
- **D** Reactions sometimes produce colour changes.

3 Read the descriptions below of four reactions carried out by Emilia. For each one, decide which of these four signs she used to tell there was a chemical reaction:
- colour change
- precipitate formed
- gas produced
- heat produced

- **a** I mixed two clear solutions, and a white solid settled to the bottom of the tube.
- **b** After a while the test tube felt warm.
- **c** When I added water the mixture fizzed.
- **d** When I added the acid it turned red.

ISBN 978 1 4202 3735 1

4 The two gases that make up most of the atmosphere are:
- **A** nitrogen and hydrogen.
- **B** nitrogen and oxygen.
- **C** oxygen and carbon dioxide.
- **D** carbon dioxide and hydrogen.

5 Here are some reactions that take place in the home. List them from the fastest reaction to the slowest.
- **a** paint drying
- **b** fruit rotting
- **c** gas burning in a stove
- **d** a cake baking
- **e** a metal can rusting

6 When Kim heated some sugar in a test tube, it turned to black carbon. She also noticed some drops of water around the mouth of the tube. Write a word equation to describe the reaction that probably occurred.

7 Which *one* of the following reactions produces the gas hydrogen?
- **A** baking soda + vinegar
- **B** hydrogen peroxide + manganese dioxide
- **C** marble chips + dilute hydrochloric acid
- **D** magnesium + dilute hydrochloric acid

8 Some food was left in a closed plastic container on the kitchen bench. After 2 weeks, the lid started to bulge. What do you infer from this?

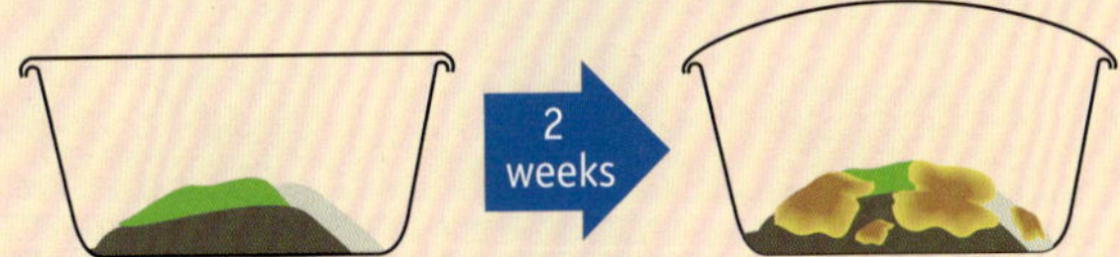

9 Below are four statements, each with a reason. In which case is the wrong reason given?
- **A** Hydrogen is used as a rocket fuel because hydrogen is the lightest gas known.
- **B** A flame should not be used when making hydrogen, because hydrogen and air form an explosive mixture.
- **C** Pure oxygen is used in an oxy-acetylene torch, because substances burn better in pure oxygen.
- **D** Carbon dioxide is used in fire extinguishers because things will not burn in it.

Questions 10 and 11 refer to the following information.

Ian was doing a science experiment. He took a piece of bread and broke it into two equal halves. He then added 10 drops of water to each half and placed each in a similar, sealed glass jar. He put one jar in a dark cupboard and the other on a well-lit window sill. After a week Ian found that mould had grown on the bread placed in the dark cupboard, but none had grown on the bread in the light.

10 What problem was Ian investigating?
- **A** What effect does moisture have on the growth of bread mould?
- **B** What effect does light have on the growth of bread mould?
- **C** What is bread mould?
- **D** Where does bread mould come from?

11 Below are four factors that may or may not influence the growth of bread mould.

Factor 1: the amount of water added to the bread
Factor 2: the temperature in each jar
Factor 3: the type of glass each jar is made of
Factor 4: the amount of light falling on the jars

For each factor choose either A, B, C or D.
- **A** This factor was deliberately varied by Ian.
- **B** This factor could influence the growth of mould, but Ian made special allowance for it so that it would not affect the results.
- **C** This factor could influence the growth of mould, but Ian did not allow for it in his experiment.
- **D** This factor is not important and is unlikely to influence the growth of mould.

Check your answers on page 297.

ISBN 978 1 4202 3735 1

IN THIS CHAPTER

Science Understanding

- give everyday examples of the way heat energy can be transferred by conduction, convection and radiation
- investigate energy transformation in devices
- apply particle theory to explain heat
- discuss an example of how a scientific theory was rejected

Science Inquiry Skills

- measure temperature accurately using a thermometer
- perform a controlled experiment to see which absorbs more radiation—a shiny silver can or a dull black one
- discuss the reliability of experimental results with others
- design experiments to solve everyday problems, e.g. determining which type of material keeps you warmest in winter

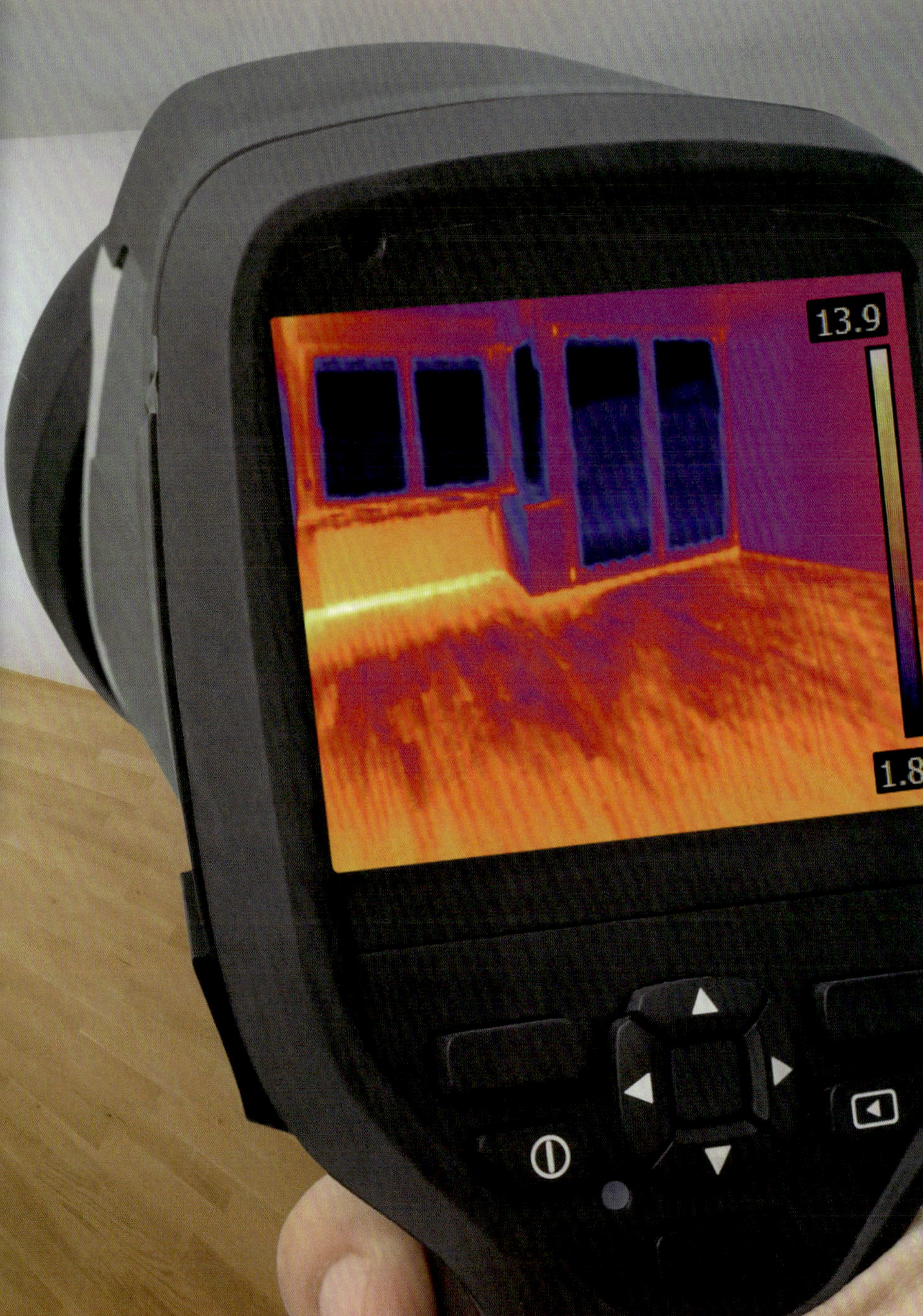

CH•6 Heat energy

ISBN 978 1 4202 3735 1

GET STARTED: *QUESTION*

The photo shows a glassworker. The molten glass inside the furnace is at a temperature of more than 1000 °C. The furnace and glass give off a huge amount of heat.

- What do you notice about the end of the metal rod in the furnace?
- List the items of protective clothing the worker is wearing.
- Suggest why the glassworker's head wear is silver-coloured.
- Describe two ways in which the heat moves from the furnace to the glassworker.

ISBN 978 1 4202 3735 1

6.1 Heat and temperature

Heat is very important in our lives. Our bodies function best at a temperature of about 37 °C. If we get too hot or too cold, we feel uncomfortable. If our body temperature rises too far above normal or too far below normal, we can die.

We use fans, heaters and air conditioners to keep us comfortable. The walls and ceilings of our homes are insulated to keep heat in during winter and out during summer. We use heat for cooking food and heating water. Heat is used by industries to make new materials such as glass, steel and plastics. Our cars produce heat when they burn petrol. Heat from burning coal is used to generate electricity. But what is heat? And how is it different from temperature?

Several hundred years ago, people thought of heat as a special fluid called *caloric* that flowed in and out of objects as they were heated or cooled. An American named Benjamin Thompson, who later moved to Germany and became Count Rumford, showed that this caloric idea was incorrect. He observed that when holes were drilled in brass to make cannons, so much heat was produced that water had to be poured over the cannons to cool them. From this Rumford inferred that it was the movement of the drills that made the cannons hot. The kinetic energy of the drill had been converted into heat energy. People soon realised that some heat is always produced when energy changes from one form to another. In other words, **heat** is a form of energy, and it is therefore measured in joules (J).

Heat and temperature are not the same, but there is a connection. **Temperature** is a measure of how hot or cold something is. It is measured in degrees Celsius (°C) using a thermometer.

You have probably played with sparklers. Each spark is actually a tiny piece of white-hot metal, and its temperature may be as high as 800 °C. (The temperature of boiling water is only 100 °C.) However, if a spark falls on your hand, you don't even feel it. This is because each spark contains only a small amount of heat energy. Some of this heat energy is transferred to your skin, but the resulting temperature rise is so small that you usually cannot detect it.

Figure 6.1 Drilling brass cannons produced considerable heat. From this, Count Rumford inferred that heat is a form of energy.

Figure 6.2 Each tiny spark has a high temperature but contains very little heat energy.

ISBN 978 1 4202 3735 1

Heat and the particle theory

We can use the particle theory (Chapter 2) to explain heat. When you heat an object, the particles in it move more rapidly and therefore have more energy. This is why the temperature is higher. When the particles lose energy and move more slowly, the temperature is lower.

Look at the diagram below. When a hot object comes into contact with a cold object, heat flows from hot to cold until both objects are at the same temperature. The rapidly moving particles in the hot object transfer some of their energy to the particles in the colder object. The larger the temperature difference, the faster the transfer. Cool objects in warm places take in energy from their surroundings. For example, an ice block melts quickly on a hot day. Warm objects such as a cup of hot coffee lose heat energy to their cooler surroundings.

Did you know?

If the temperature of a substance was lowered to −273 °C, its particles would have no energy at all and would therefore be completely still. This temperature is called *absolute zero*, and scientists have come close to this in some experiments.

Figure 6.3 Particle theory

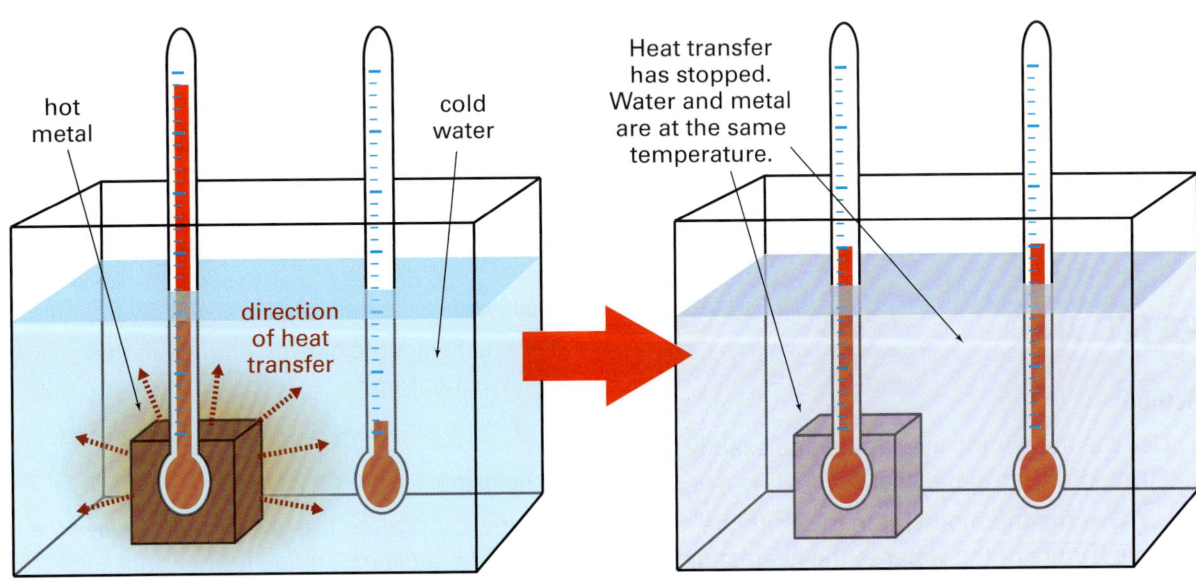

Figure 6.4 Heat moves from the hotter object to the cooler one until both reach the same temperature.

ISBN 978 1 4202 3735 1

Heat and temperature

Aim

To find answers to these questions:

A Does the mass of a substance influence how much its temperature rises?

B Does the type of substance influence how much its temperature rises?

Materials

- hotplate or burner, tripod and gauze
- 250 mL beaker
- thermometer
- 100 mL measuring cylinder
- stopwatch
- **olive oil**
- paper towel

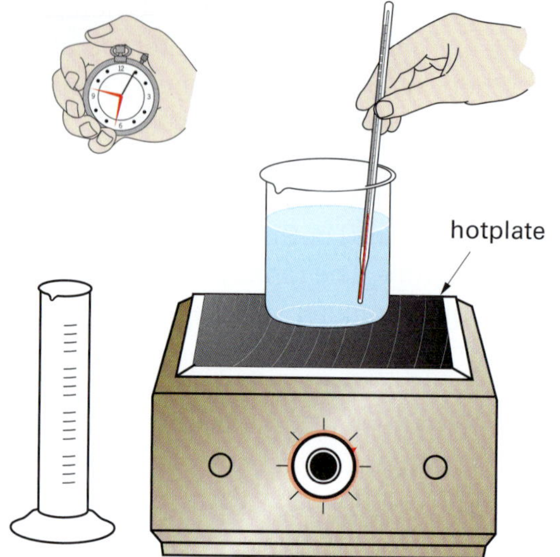

Risk assessment and planning

Read through both parts of the investigation.

- Discuss with your teacher the safest way to handle the hot beaker.
- How is Part B different from Part A?
- Suggest why a hotplate is used in this experiment instead of a Bunsen burner.

Draw up a data table like the one below.

	Temperature of the liquid (°C)		Rise in temperature (°C)
	before heating	after heating	
50 mL water			
100 mL water			
60 mL olive oil			
120 mL olive oil			

PART A

Method

1 Use the measuring cylinder to add exactly 50 mL of water to the beaker.

Use the thermometer to measure the temperature of the water to the nearest degree. Record this in the data table.

2 Adjust the hotplate or the burner to medium heat. *Leave it at the same setting throughout the experiment.* This is to make sure that the heater supplies heat at a constant rate.

3 Place the beaker of water on the hotplate for exactly 2 minutes. Then remove the beaker from the hotplate, stir the water *gently* with the thermometer and read the temperature.

Record this temperature in the data table.

Calculate and record the rise in temperature.

4 Empty the beaker, cool it under running water, and dry it.

5 Add 100 mL of water to the same beaker and measure the temperature before and after heating for 2 minutes.

Record your results in the data table.

Discussion

1 Which variable did you change in this investigation?

2 Which variables did you keep the same?

Conclusion

Write an answer to the question *How does the mass of a substance influence how much its temperature rises?*

ISBN 978 1 4202 3735 1

PART B

Method

Repeat Part A, but this time use olive oil—60 mL and 120 mL. (60 mL of olive oil has the same mass as 50 mL of water.)

Record all results in the data table.

Discussion

Which variable did you change going from Part A to Part B? Which variables did you keep the same?

Conclusion

Write an answer to the question *How does the type of substance influence how much its temperature rises?*

From Investigation 6.1 you can conclude that:

A the same amount of heat will raise the temperature of 50 mL of water twice as much as it raises the temperature of 100 mL of water.

B the same amount of heat will raise the temperature of olive oil more than it raises the temperature of an equal mass of water. In other words, olive oil heats up more quickly than water does. The bar graph below shows the amounts of heat needed to warm 1 gram of various materials by 1°C.

You could also predict that if you supply twice as much heat to water or olive oil, you raise the temperature twice as much.

To summarise, the amount of heat gained or lost by an object depends on three variables:

- its mass
- the temperature change
- what it is made of.

You can calculate the amount of heat that is transferred if you know these three variables.

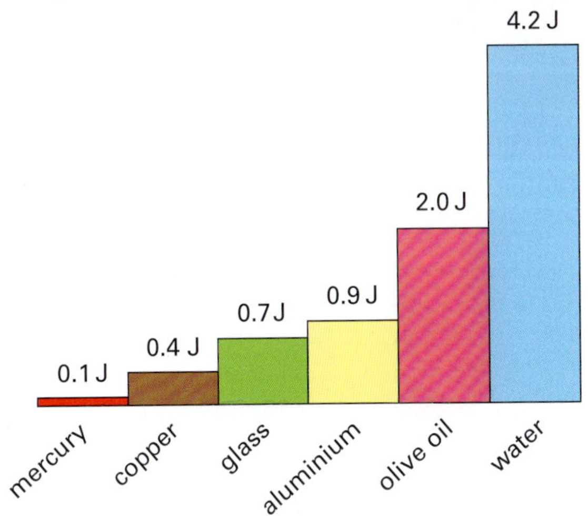

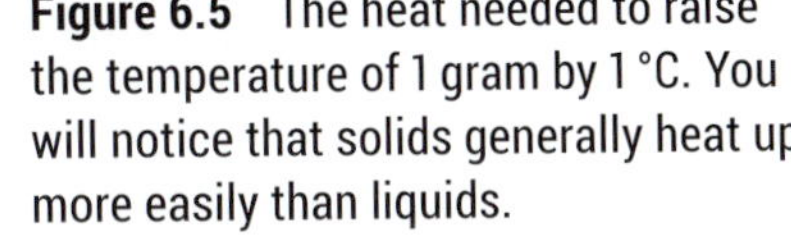

Figure 6.5 The heat needed to raise the temperature of 1 gram by 1 °C. You will notice that solids generally heat up more easily than liquids.

SKILLBUILDER

Calculating

The amount of heat energy needed to raise the temperature of 1 gram of a substance by 1°C is called its *specific heat capacity*. For example, the specific heat capacity of water is 4.2 joules per gram per °C. To calculate the heat needed to change the temperature of something, you can use this mathematical formula:

heat (J) = mass (g) × specific heat capacity × change in temperature (°C)

So, the heat needed to raise the temperature of 50 mL (50 g) of water by 10 °C can be calculated as follows:

heat = 50 g × 4.2 × 10 °C
= 2100 J

Use the specific heat capacities from the bar graph above to answer these questions.

1 How much heat is required to raise the temperature of:
   a 100 mL of water by 10 °C?
   b 60 mL of olive oil by 10 °C?
2 A 5-gram block of aluminium was heated from 30 °C to 100 °C. How much heat energy was needed?
3 How much heat is given out when 60 grams of copper cools from 100 °C to 20 °C?
4 Using your results from Investigation 6.1, calculate the amount of heat transferred to 50 mL and 100 mL of water, and 60 mL and 120 mL of olive oil. Is the amount of heat transferred the same for all?

ISBN 978 1 4202 3735 1

SCIENCE AS A HUMAN ENDEAVOUR

How a theory was rejected

ACTIVITY

1. Measure 50 mL of cold water into a beaker and record its temperature. Then measure out 50 mL of hot water and record its temperature.
2. Add the hot water to the cold water and stir carefully with a thermometer. Record the temperature.
3. Use what you have learnt about heat to explain your results.

To explain the results of the activity, did you write a ***hypothesis*** that heat is transferred from a hotter substance to a colder one? Scientists test a hypothesis like this by carrying out more experiments. Other scientists then repeat these experiments. If all the experiments support the hypothesis, and none have shown it to be false, the hypothesis is accepted as a *theory*. This theory is accepted until new evidence is found that doesn't support it.

In 1783, Antoine Lavoisier proposed the theory that heat is a special fluid called *caloric* that flows from warmer to colder objects. Unfortunately Lavoisier was beheaded in the French Revolution in 1794. Four years later Count Rumford did his experiment with drilling brass cannons (see page 134), and suggested that the caloric theory was incorrect. However, scientists were reluctant to reject the caloric theory as it successfully explained many experiments.

In 1842 James Joule carried out a famous experiment that meant the caloric theory had to be rejected. He made the apparatus shown below in which a falling weight turned a paddle in a tank of water. The friction caused by the paddle caused the temperature of the water to rise slightly. From this he was able to show that heat is just a form of energy and can be explained in terms of the motion of particles.

Questions

1. How does a hypothesis become a theory?
2. Draw a time line showing the events outlined on this page, as well as Dalton's particle theory in 1803. How long did the caloric theory last?
3. Why was the caloric theory rejected?

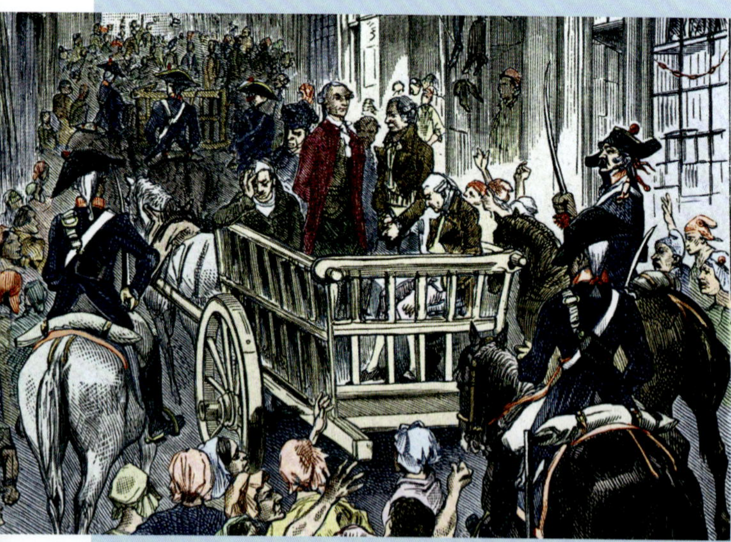

Figure 6.6 Lavoisier, in the red coat, is carried to the guillotine during the French Revolution.

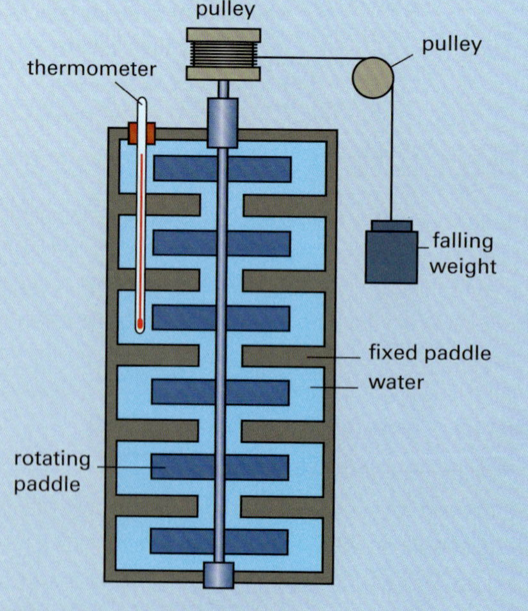

Figure 6.7 Joule's experiment

ISBN 978 1 4202 3735 1

CHECK

1 Decide which of the following statements are true and which are false. Rewrite the false ones to make them true.
  a Heat is a form of energy.
  b When you strike a match, you convert kinetic energy into heat energy.
  c When an energy change occurs, some heat energy is always produced.
  d A block of ice contains no heat energy.
  e Heat is measured in degrees Celsius.
  f As an object becomes hotter, its particles move more rapidly.
  g Heat travels from cold objects to hot objects.

2 When Faith used an electric hair dryer to dry her hair, the hair dryer became quite hot. What energy change has occurred?

3 a Which is hotter—a cup of water at 50 °C or a bathtub full of water at 50 °C?
  b Which contains more heat energy?

4 A cold saucepan is put into a sink containing hot dishwashing water.
  a What will happen to the temperature of the saucepan?
  b What will happen to the temperature of the water?
  c Does heat flow from the water into the saucepan, or from the saucepan into the water?

5 Josh heats two identical iron nails together until they are red hot. He drops one into 50 mL (50 grams) of water and the other into 60 mL (50 grams) of olive oil. If both liquids are at the same temperature to start with, predict which will be hotter 1 minute after the nails have been dropped in. Explain your answer.

6 Explain why heat energy can be considered a form of kinetic energy.

7 Eva had four identical beakers containing different amounts of water, as shown. She heated them for different lengths of time and none of them boiled.

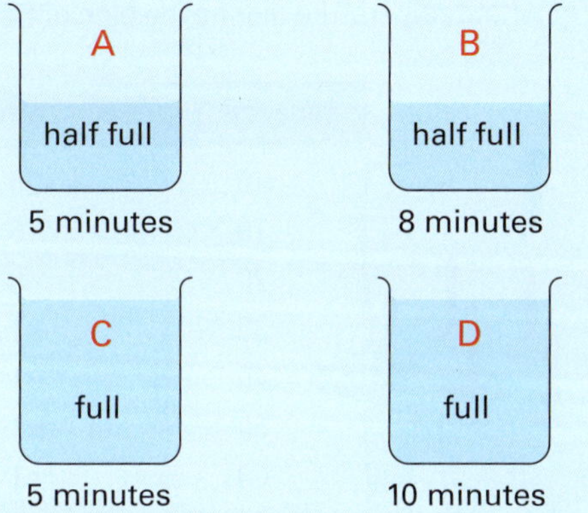

  a Which beaker of water received the most heat?
  b Which beaker would you predict had the highest temperature after heating?
  c Which beaker would have the lowest temperature after heating?

8 Samples of 50 grams of aluminium, copper, glass and water all initially at 20 °C are heated for 5 minutes on a hotplate with a constant setting. Predict the order (from highest to lowest) of the final temperatures of each sample. See the bar graph on page 137.

9 Suggest why mercury is used in thermometers. (Hint: See the bar graph on the previous page.)

10 On a hot summer's day the dry sand at the beach can be almost unbearable to stand on, while the water is cool. Try to explain this temperature difference.

ISBN 978 1 4202 3735 1

CHALLENGE

1 Ramone was making some ice blocks from fruit juice, and decided to investigate their temperature as they cooled in the freezer. He put a thermometer in one of them and measured the temperature every 10 minutes. His results are listed in the table below.

a Was heat being added to the ice blocks during Ramone's experiment, or being taken away from them?

b Plot Ramone's results on a line graph.

c Suggest a reason for the flat part of the graph between 40 min and 60 min.

d Suggest a reason for the flat part between 90 min and 100 min.

e At what temperature did the ice blocks freeze?

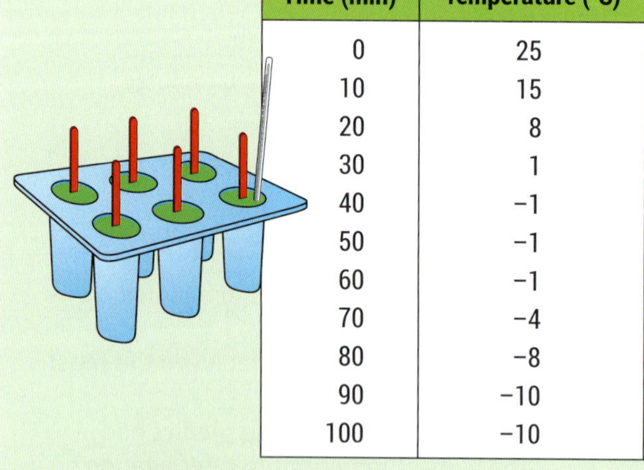

Time (min)	Temperature (°C)
0	25
10	15
20	8
30	1
40	−1
50	−1
60	−1
70	−4
80	−8
90	−10
100	−10

2 Harry did an experiment and drew these diagrams to show his method.

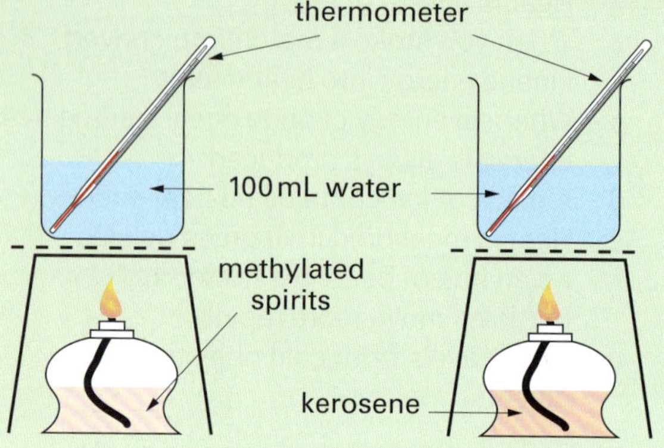

Heat for 5 minutes.

a Which variables did Harry control in this experiment?

b Which variable did he purposely change?

c Which variable did he measure?

d What do you think was the aim of Harry's experiment?

3 Nicky measured the temperature of a saucepan of hot water as it cooled. She plotted her results as shown in the graph below.

a What was the temperature of the water after 10 minutes? After 40 minutes?

b Why is Nicky's graph steep to start with but flatter near the end?

c What do you think the temperature of the room was when Nicky did her experiment? Explain your answer.

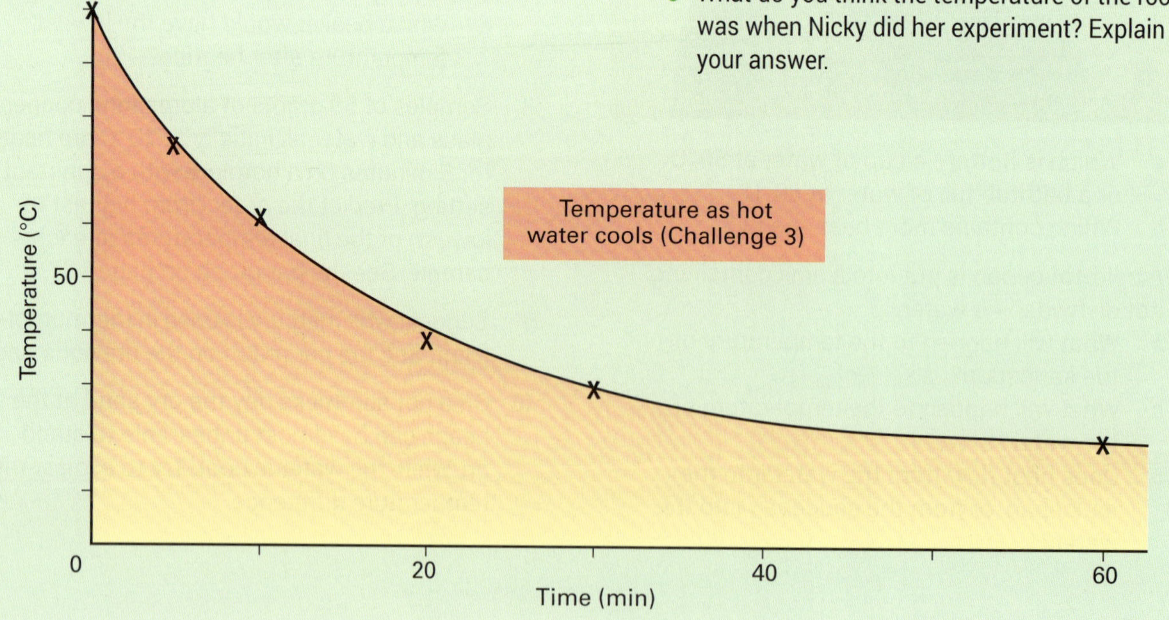

ISBN 978 1 4202 3735 1

6.2 Heat transfer

Heat energy can be transferred in three different ways. The direction in which it flows is always from a higher temperature to a lower temperature.

Conduction

ACTIVITY

1. You will need a glass rod about 20 cm long, and a metal rod the same length and thickness as the glass one.
2. Use wax or grease to stick a paperclip about 5 cm from the end of each rod. Then lay the rods across a tripod so that the paperclips hang down as shown.
3. Heat the end of both rods equally.

Which paperclip falls off first? What does this mean?

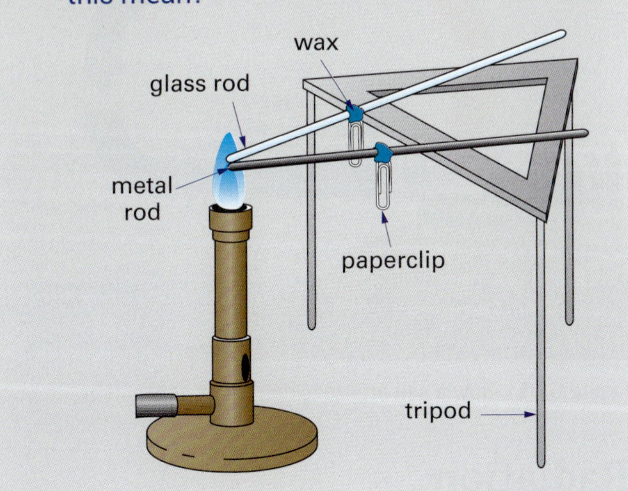

A metal rod in contact with a hot flame quickly becomes hot. The heat is transferred along the rod by the process of **conduction**. The particles in the end of the rod gain energy from the flame. This causes them to vibrate faster and collide more energetically with each other. This process continues like a chain reaction from particle to particle along the rod. As a result, heat energy is transferred from the hot end of the rod to the cooler end.

As you saw in the activity, some solids conduct heat better than others. Substances that conduct heat well are good **conductors**, and most metals are good conductors. Substances, such as glass, that are poor conductors of heat are called **insulators**. Most plastics are poor conductors of heat, so those that do not melt easily are used to make handles for saucepans and frying pans.

Insulating handles allow you to pick up hot objects without the heat being conducted to your hand. Plastic foam is a good insulator, and is used in the walls of refrigerators to keep heat out.

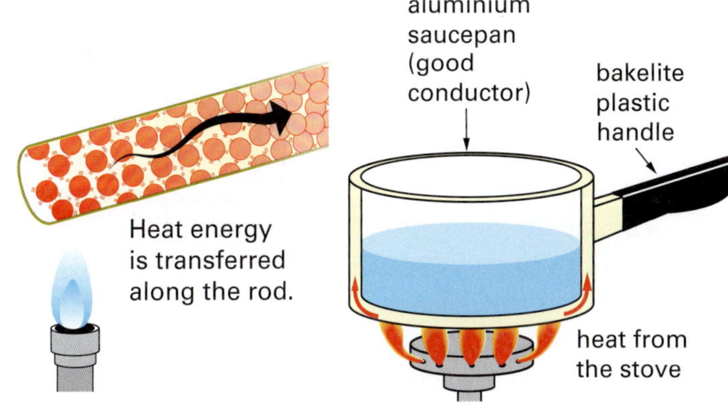

Figure 6.8 Conduction

Figure 6.9 Metal conducts heat and plastic does not.

ISBN 978 1 4202 3735 1

Convection

1 Fill a large beaker with water and allow it to stand until the water is completely still.
2 Carefully drop half a teaspoon of used tea leaves down one side of the beaker, making sure not to disturb the water.
3 Heat underneath the tea leaves as shown.
   - Suggest why the tea leaves rise.
   - Draw a diagram showing the movement of the tea leaves.

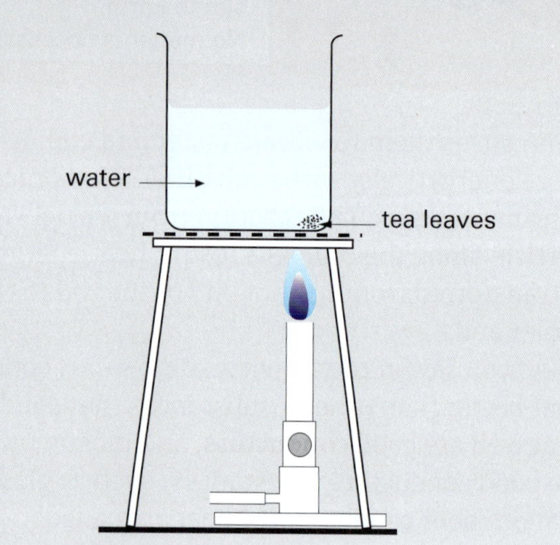

The movement of the tea leaves in the activity above demonstrates the movement of heat energy by the process of **convection**. This process can be explained using the particle theory. When water particles at the bottom of the beaker are heated, they gain more energy and move more rapidly. Because of this, they move further apart than the particles above them. Hence, the warm water near the bottom is less dense than the water above it. This warmer water therefore rises, and colder water moves in to take its place. This movement of particles is called a *convection current*.

Hot water systems work by convection. The heater at the bottom warms the water, which moves upwards as the cool water takes its place, setting up convection currents. The hot water is drawn off from the top.

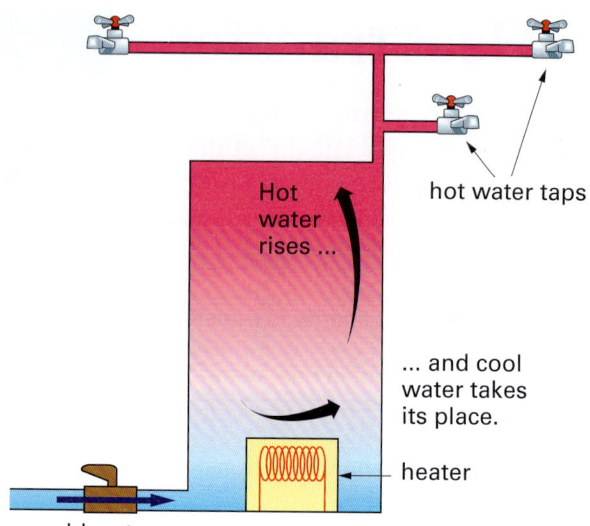

Figure 6.10 A hot water heater works by convection.

Convection currents also occur in air. When you turn on a heater in winter, the warm air rises above the heater. A convection current is then set up as the cooler air sinks. The same thing happens on a larger scale to form a sea breeze. During the day, the land is warmer than the sea. Warm air rises above the land, and cooler air blows in from the sea to take its place.

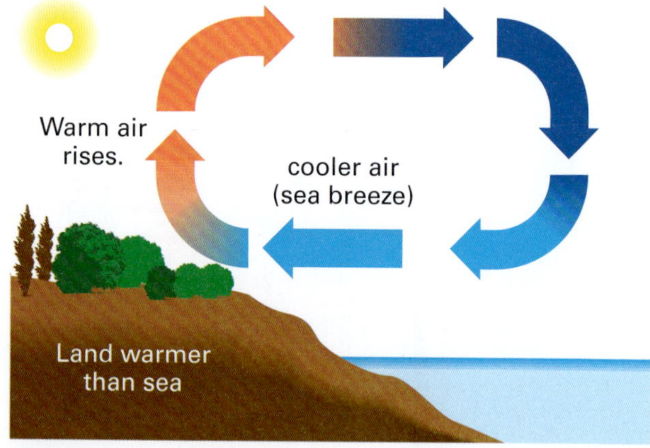

Figure 6.11 How a sea breeze is caused by convection

Radiation

The sun's rays heat the Earth. However, the space between the sun and the Earth does not contain matter, so heat energy cannot be transferred by the processes of conduction or convection.

ISBN 978 1 4202 3735 1

Instead, the sun transfers heat energy by the process of **radiation**.

All objects transfer some heat by radiation. The hotter the object, the more heat it radiates. The radiation itself is not hot, but when it is absorbed by an object, it causes the particles in the object to move more rapidly, thus heating it.

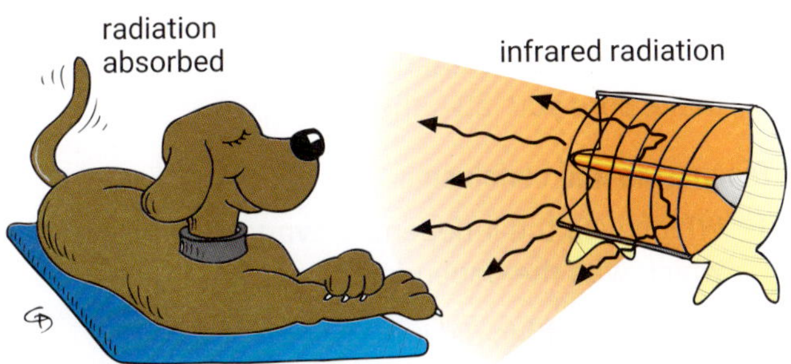

Figure 6.12 The curved silver mirror at the back of an electric radiator reflects the radiation.

Warm objects radiate heat mainly in the form of *infrared radiation*, which we cannot see but which can be detected by special infrared scanners. People are usually warmer than their surroundings and give off more infrared radiation. This is why infrared scanners are used at night by air–sea rescue helicopters to help find people who are lost.

If a metal object becomes hot enough, it will glow, giving off visible light as well as infrared radiation.

Light, infrared and other forms of radiation all travel as extremely high-speed waves, which can pass through a vacuum, such as the vacuum of space. In a vacuum, the speed of radiation is 300 million metres per second or 3×10^8 m/s. Different types of radiation have different wavelengths. The waves travel in straight lines, and they can be *reflected, absorbed* or *transmitted* by matter.

All of these properties of radiation are applied in microwave cooking. Microwaves are *reflected* off metals, so they reflect from the inside of the oven onto the food being cooked. The microwaves penetrate the food to a depth of between 2 and

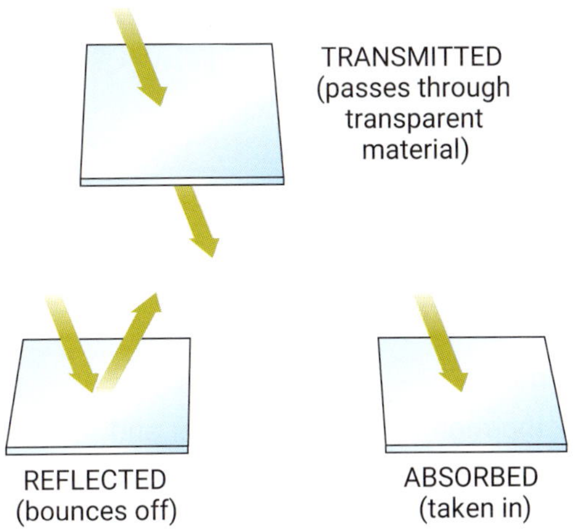

Figure 6.13 Heat and light can be transmitted, reflected or absorbed.

4 cm, where they are *absorbed*. This causes the molecules in the food (mainly water molecules) to move more rapidly, and hence the food heats up. The heat is then transferred to other parts of the food by conduction. This process may continue for a short time after the food has been removed from the oven. The microwaves are *transmitted* through the glass dish, which remains relatively cool because it absorbs very little radiation. (Microwaves will also pass through paper and most plastics.)

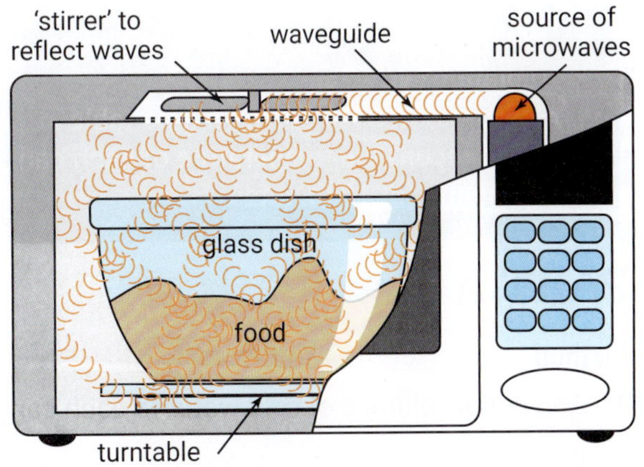

Figure 6.14 How a microwave oven works. You can see through the glass door, but the microwaves cannot pass through the metal lattice behind the glass.

ISBN 978 1 4202 3735 1

INVESTIGATION 6.2 Which absorbs more radiation?

Aim

To compare the amount of radiation absorbed by a shiny silver can and a dull black can.

Materials

- 2 thermometers or datalogger and 2 temperature probes
- portable spotlight or electric radiator
- 2 metal cans—one shiny silver and one dull black

Notes

1 **Instead of using a spotlight, you could use a microscope lamp, or you could put the cans in direct sunlight.**

2 **If you use empty food cans, you could blacken one by holding it in the smoke from a burning candle. Painted soft-drink cans work well.**

3 **To cut down on heat loss by convection, you need lids.**

Risk assessment and planning

Read the Method carefully and discuss with your teacher what equipment you will use.

- What safety precautions will be necessary?
- Which can do you predict will absorb more radiation? Why?

In your notebook design a data table in which to record the temperature of each can every minute for 15 minutes.

PART A

Method

1 Add equal volumes of cold water to each can.

2 Position the spotlight or radiator at an equal distance from each can.

3 Record the initial temperature of the water in each can. (These should be the same.)

4 Turn on the spotlight and at the same time start timing.

Record the temperature in each can every minute for 15 minutes.

5 Plot the temperature for each can on a single graph. (A datalogger will do this for you.) You could use a different colour for each can, but make sure you label the two curves.

Discussion and conclusion

1 Which was the independent variable and which was the dependent variable?

2 Which variables did you control?

3 Which can absorbed more radiation? How do you know? Was your prediction correct?

4 Look at your graph. What does the slope of each line tell you about the warming rate of the can?

5 Based on the results of this experiment, write a generalisation saying how the amount of heat absorbed by an object depends on the type of surface.

6 Could the experiment be improved? If so, how?

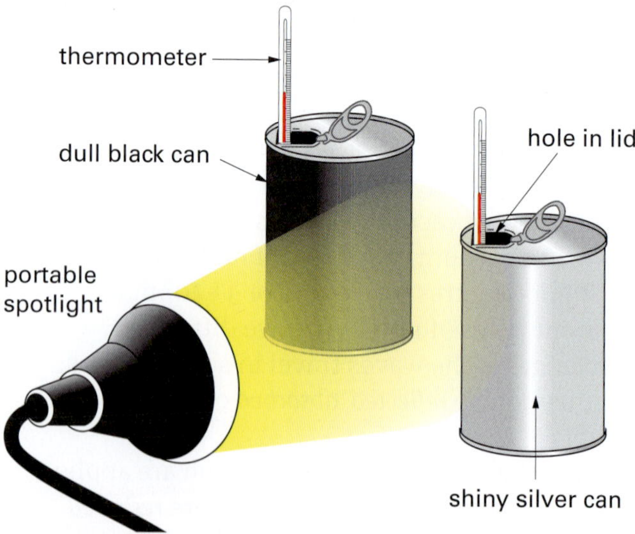

PART B Inquiry

Design a similar experiment to find out which can *cools* more quickly.

ISBN 978 1 4202 3735 1

Absorbing and emitting radiation

Dark-coloured surfaces are better absorbers of radiation than light-coloured ones. This is because light-coloured surfaces reflect more of the radiation. Bright shiny surfaces are the best reflectors and the poorest absorbers. This is why aluminium foil is used in ceilings and walls of houses to reflect heat. This is also why the absorbing panels of solar water heaters are painted black so that the copper pipes inside them absorb as much of the sun's radiation as possible. Dark-coloured cars become much hotter than light-coloured cars when left in the sun. And dark-coloured clothes are hotter in summer than light-coloured clothes.

Figure 6.15 Dark colours absorb heat better.

All objects emit (give out) infrared radiation if they are at a higher temperature than their surroundings, but some radiate heat more readily than others. Dark-coloured objects radiate heat more effectively than light-coloured objects. Rough surfaces also radiate heat more effectively, due to their greater surface area.

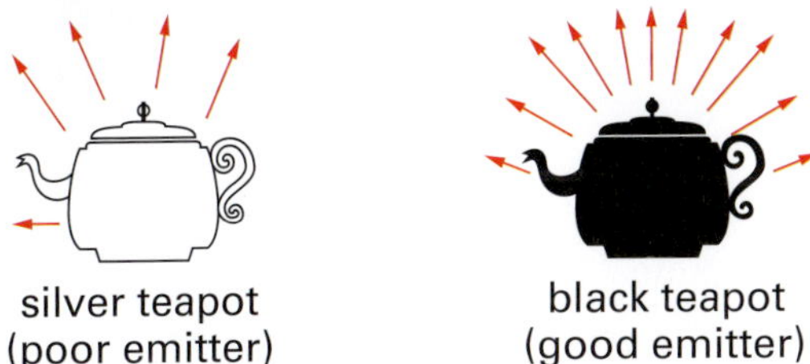

Figure 6.16 Dark colours emit heat better.

Figure 6.17 The cooling fins on an air-cooled motorcycle engine have a large surface area to increase radiation of heat.

Controlling heat transfer

An object that is warmer than its surroundings will lose heat until it is the same temperature as its surroundings. Similarly, an object that is cooler than its surroundings will gain heat from its surroundings. We use insulators to control this transfer of heat.

You may have seen birds fluffing up their feathers on cold days. This is to trap air between their feathers. Because air is an insulator, it slows down the loss of heat from the bird's skin to the surrounding cooler air. Woollen jumpers, sleeping bags and the batts used to insulate houses also work by trapping air.

The wetsuits worn by surfers and divers are made of neoprene (see Figure 6.18). A thin layer of water warmed by body heat is trapped between the suit and the diver's skin. This water helps to prevent the diver's body heat from escaping.

Eskys and thermos flasks (see page 151) are insulated containers that keep food and drink at the temperature we want it—either hot or cold. Pizza delivery people put their pizzas in special insulated boxes to keep them warm.

ISBN 978 1 4202 3735 1

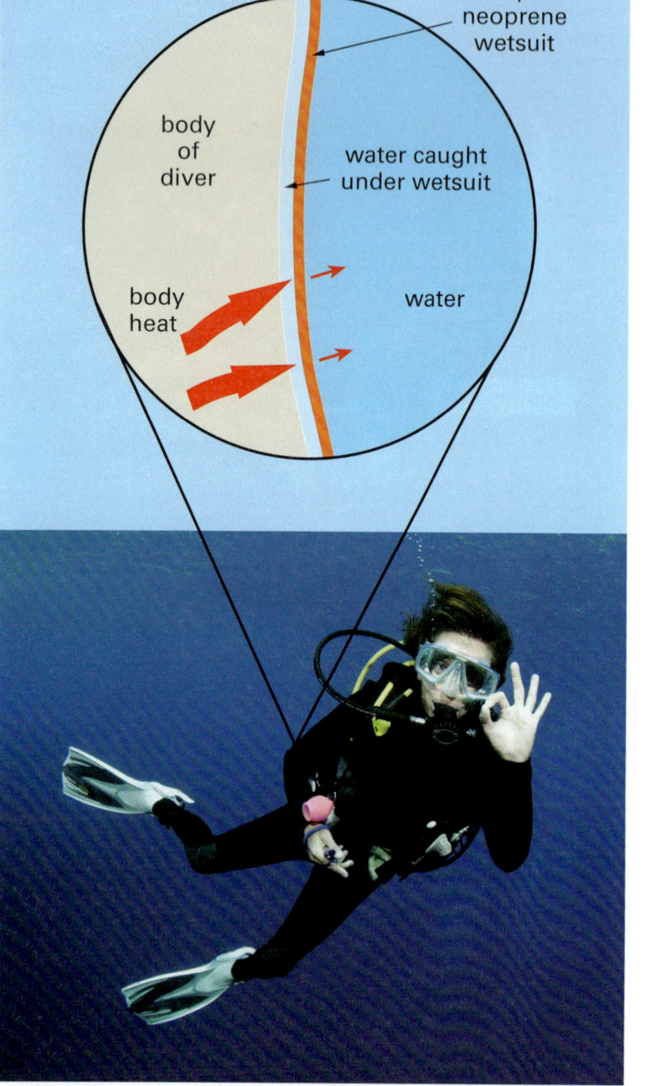

Figure 6.18 Divers' wet suits insulate their bodies using a layer of water.

Figure 6.19 Food is carried in insulated containers to keep it hot.

We insulate the walls and ceilings of our homes. This keeps them cool in summer by preventing heat from coming in from outside. It also keeps them warm in winter by preventing heat from escaping (see page 155).

Figure 6.20 This silver insulation helps to stop heat loss from pipes by reflecting heat back into the pipes.

Figure 6.21 Home insulation keeps heat out in summer and in during winter.

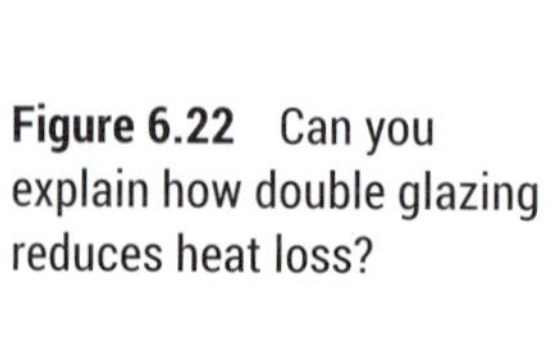

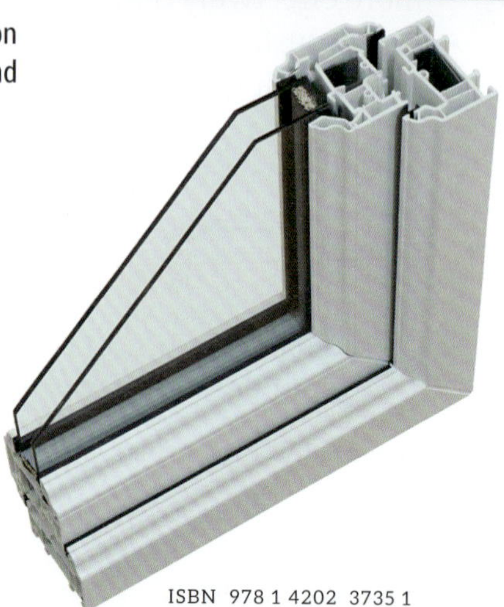

Figure 6.22 Can you explain how double glazing reduces heat loss?

ISBN 978 1 4202 3735 1

Which is the best insulator?

The problem to be solved

Your task is to design an experiment to solve the problem *Which type of material keeps you warmest in winter?*

Designing your experiment

The design of the experiment is up to you, but here are some questions to guide you.

- What will you use as a model for a human body? One idea is to use a can filled with hot water.
- How will you clothe your model bodies? What materials will you use? Some possibilities are wool, cotton, nylon, polyester, flannelette. How many model bodies will you need?
- How will you measure the temperature? How often will you do this, and for how long will you continue the experiment?
- How will you make your test fair (as in Chapter 1)? You are varying the type of clothing, but what other variables are there? How will you keep these other variables constant?
- It would be a good idea to use an *experimental control*—a model body with no clothes. You can then compare it with the clothed bodies to see how effective the different types of clothes are.
- If possible, repeat your experiment to improve the accuracy of your measurements. If you get the same results, then you can be more confident your conclusion is correct. If someone else repeats your experiment and gets the same results, you can be even more confident. Results like this are said to be **reliable**.

Results

How will you record and display your results? If a datalogger is available, you could use it. Would a graph be useful?

Writing your report

Look carefully at your results and write a report of your findings, giving your answer to the problem. You could take a digital photo of your set-up to include in your report. Could you improve your design? How?

Inquiry

Instead of keeping something warm you often want to keep something *cold*. Design an experiment to find out which is the best insulator for this.

ISBN 978 1 4202 3735 1

CHECK

1 What is the advantage of a copper bottom on a saucepan?

2 The four beakers shown are identical and contain the same volume of water at 80 °C. After 10 minutes the temperature of each is measured again.

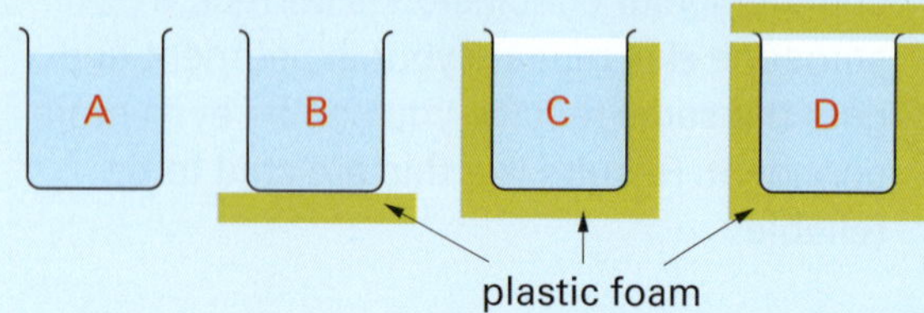

a Which beaker do you think will be the hottest after 10 minutes? Why?

b Explain why the water in B will probably be a little warmer than that in A.

c What happens to the heat energy lost from the beakers?

3 A gas or electric oven is more correctly called a convection oven. Why? Draw a diagram showing how it works by convection.

4 You put a can and a glass bottle of ginger beer into the refrigerator at the same time. They both contain 375 mL and are both at room temperature. Predict which one will cool more quickly. Why?

5 Refrigerators and freezers are painted white. Yet the coils at the back are painted black. Why is this?

6 Predict the effect that a chocolate coating would have on the rate at which an ice-cream melts. How could you test your prediction?

7 Polar bears have white fur and black skin. In winter their fur is fluffed up, and in summer it sits down flat. Suggest how these adaptations allow the bears to control their temperature.

8 In a supermarket the doors of vertical refrigerators must be kept shut, yet the freezer unit in the foreground of the photo has no lid. How can you explain this?

9 There are similarities and differences in the way light, heat and sound are transmitted, reflected and absorbed.

a Can heat travel through space where there is no air? What about light and sound?

b Can heat, light and sound be reflected? Give examples.

c Can heat, light and sound be absorbed? Give examples.

ISBN 978 1 4202 3735 1

CHALLENGE

1 What colour would you paint large petrol storage tanks? Why?

2 Look at the cartoon showing Jason thinking about how heat travels. Is he right? How would you answer him?

3 Ansa and Tammy filled two paper cups, one with water and the other with soil. They placed them in the refrigerator overnight. The next morning they took both cups, put them in the sun, and measured their temperature every 15 minutes. Here are their results:

Time	Water	Soil
9.00 am	10 °C	10 °C
9.15 am	10 °C	11 °C
9.30 am	12 °C	13 °C
9.45 am	13 °C	16 °C
10.00 am	14 °C	20 °C
10.15 am	15 °C	25 °C
10.30 am	15 °C	30 °C

a Plot these results on a graph.

b Which absorbs heat more readily—water or soil?

c During the day, which becomes hotter—the land or the sea?

d Where would a glider pilot look for thermals (rising air)—above land or above a lake?

4 Hot-air balloons work by using a burner that heats the air below the balloon. How does this make the balloon rise?

5 Use the particle theory to explain the following.

a Conduction occurs much more rapidly in solids than in gases.

b Convection currents can occur in liquids and gases, but not in solids.

6 Design an experiment to compare the insulating properties of four different house bricks. Try it if you have time.

7 One end of a long glass rod is heated to 100 °C and the other end is cooled to 0 °C.

a What will happen to the temperature at each end if the rod is left at room temperature?

b Sketch graphs to illustrate the temperature changes at the two ends of the rod.

8 Using what you have learnt in this chapter suggest four ways of preventing:

a heat loss from your house in winter

b your house from getting hot in summer.

ISBN 978 1 4202 3735 1

6.3 Exploring heat

This section is different from other sections of the book. Instead of working through it page by page, you can choose to do one or more of the six activities on the following pages. You will need to apply what you have learnt in the first two sections and work things out for yourself. In some of the activities you will be designing your own experiments to solve a problem. If you need help with this, see Chapter 1.

1 Firewalking

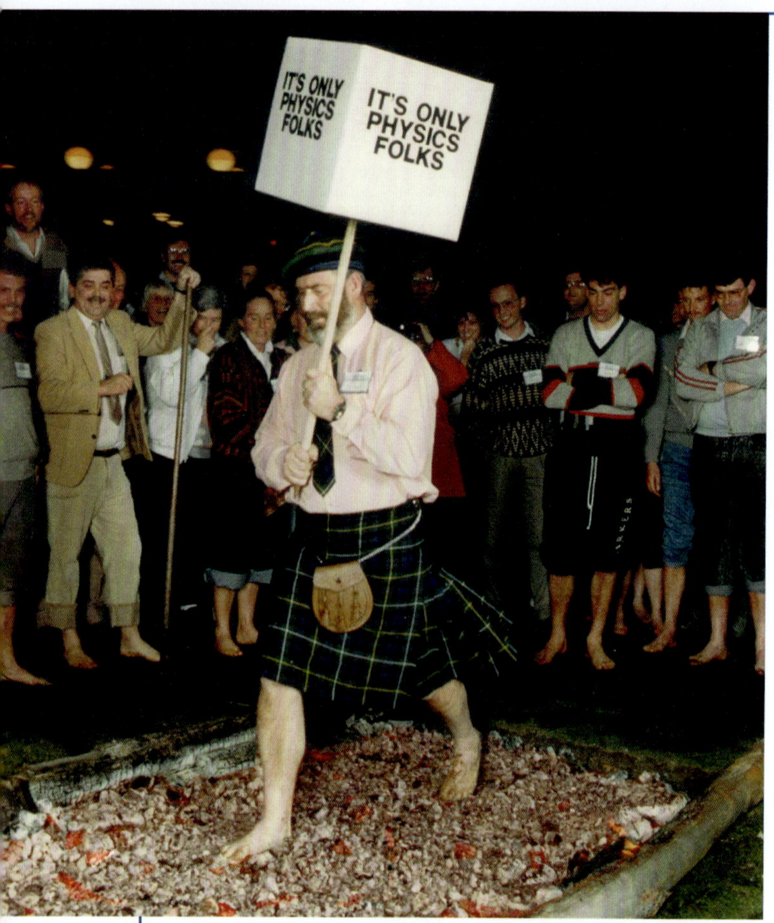

You may have seen firewalking on TV, where people walk barefoot across a pit of red-hot coals. Some people think that this shows how the mind can influence the body. But it can be explained in terms of heat transfer.

Even when you walk barefoot on a hot bitumen road, your feet can be burned as heat is transferred to them by conduction. So how can you walk on red-hot coals at about 800 °C?

The coals are charcoal—formed by the partial combustion of wood. Only the outer layer of each coal is actually burning. When a firewalker's foot touches a burning coal, a small amount of heat is transferred to the foot by conduction. This loss of heat is enough to temporarily reduce the surface temperature of the coal below ignition temperature, causing it to stop burning.

The secret to firewalking is that charcoal is a poor conductor of heat, and it takes about a second before enough heat is transferred through the dead outer layer of skin on the foot to the living tissue beneath, thereby causing a burn. So, provided that the foot is in contact with any one hot coal for less than a second, it will not be burned.

Despite all this, firewalking is still dangerous, and you should not try it yourself! Firstly, burns can occur where the skin is thinnest; for example, under the arch and between the toes. Secondly, if there is any burning wood mixed with the coals, it may produce hot gas jets capable of burning. Thirdly, small bits of coals can sometimes stick to the firewalker's feet. When this occurs the coal is in contact with the foot for longer than a second and a burn will result.

Exercises

1. What is the temperature of red-hot coals?
2. What are coals made of?
3. Are coals good conductors of heat or poor conductors?
4. What is the main way heat is transferred in firewalking?
5. What is meant by the term *ignition temperature*?
6. About how long does it take before living tissue beneath dead skin is burned?

ISBN 978 1 4202 3735 1

7 Why is it a problem if small bits of coal stick to the firewalker's feet?

8 Given the maximum contact time of one second, is it safe to walk across the coals at normal walking pace? Explain.

9 Suggest why the maximum contact time is slightly different for different people.

10 How does firewalking illustrate the difference between temperature and heat?

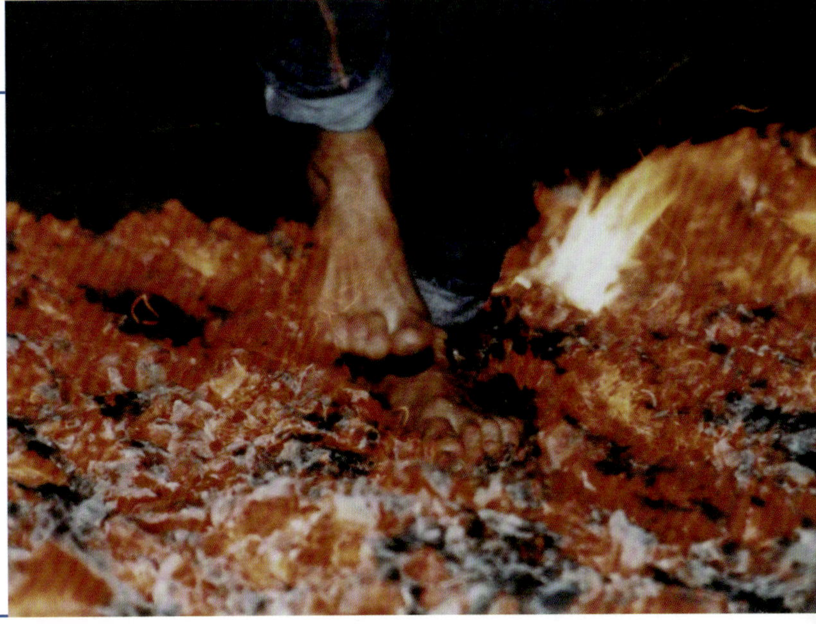

2 How does a thermos work?

Study the labelled diagram. Then write an explanation of how a thermos keeps liquids hot or cold. Make sure you explain how the various parts work to prevent heat flow by conduction, convection and radiation. Your teacher may be able to show you the inside of a thermos.

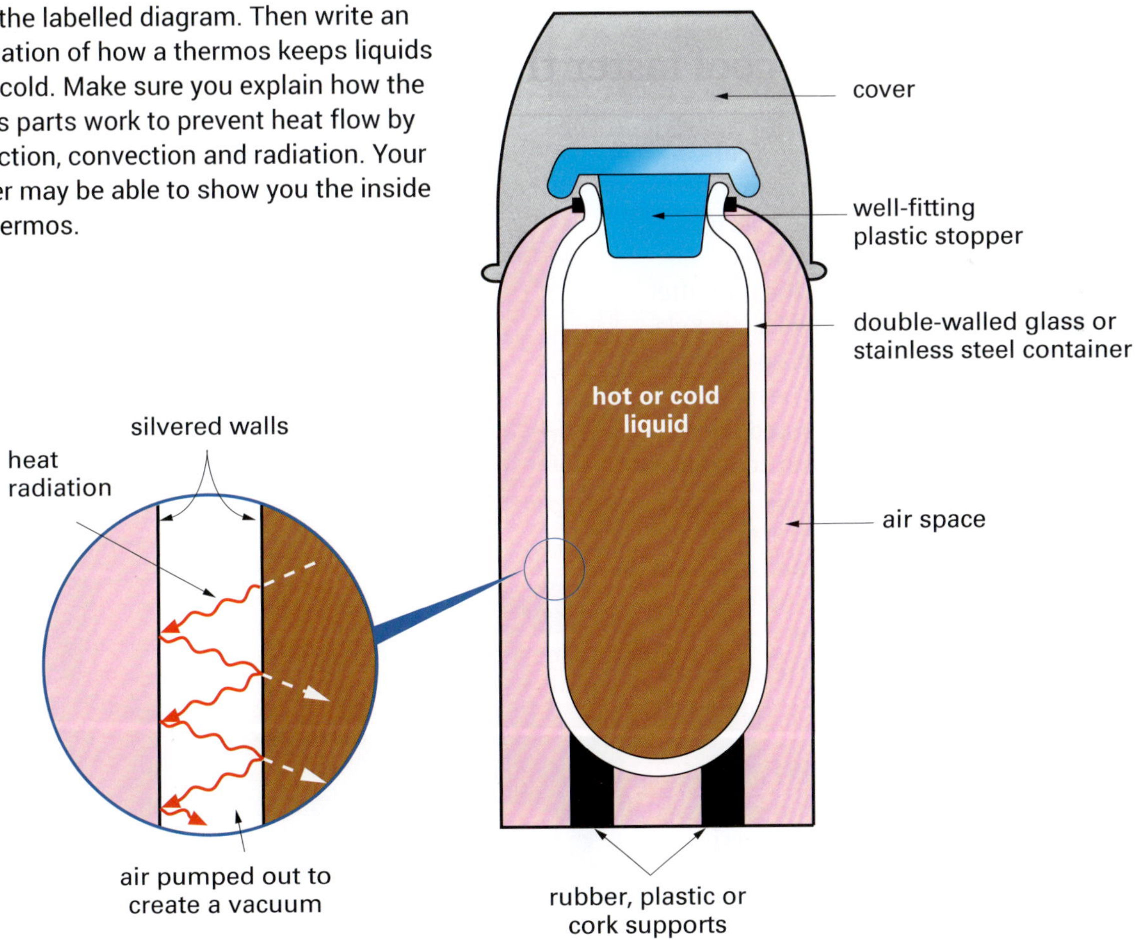

cover

well-fitting plastic stopper

double-walled glass or stainless steel container

hot or cold liquid

air space

silvered walls

heat radiation

air pumped out to create a vacuum

rubber, plastic or cork supports

ISBN 978 1 4202 3735 1

3 Which is the coolest colour to wear?

Which is the coolest colour to wear in summer?

Design an experiment to find out. You could use a method similar to that in Investigation 6.2 on page 144, or you could work out your own design. A datalogger with several temperature probes would be very useful here.

Write your report, giving your conclusion and commenting on the accuracy and reliability of your method and results. Include a recommendation to people wanting to keep cool in summer.

4 Does white coffee cool faster than black coffee?

One evening as Mahdi was making coffee for Kyle and herself the telephone rang. Mahdi was just about to add the milk to Kyle's coffee when he said, 'The coffee will probably stay hotter if you add the milk after I've finished on the phone.' Mahdi knew Kyle would be on the phone for ages, so she said, 'Wouldn't it be better if I added the milk now? I learnt at school that dark-coloured things like coffee give off more heat and cool faster than light-coloured things like milk.'

Who is right? Write a hypothesis about the cooling of coffee. Then design and carry out an experiment to test your hypothesis.

You will need to make careful measurements and record your results on a graph. (You may be able to use a datalogger with temperature probes.)

Is Mahdi's explanation correct? Is it to do with colour, or is it to do with the relative temperatures of the coffee and milk? How could you find out?

ISBN 978 1 4202 3735 1

5 Why use a lid?

You have been glancing through a book called *101 Ways to Save Energy in the Home*. One of the tips is to always put a lid on a saucepan when cooking. You wonder whether this is in fact true.

Based on what you have learnt in this chapter, write a hypothesis that you think is correct. Make sure that the hypothesis is written in such a way that you can test it. For example, you need to say what will be measured.

Now go ahead and test your hypothesis. If possible, repeat your experiment to make sure your conclusion is reliable.

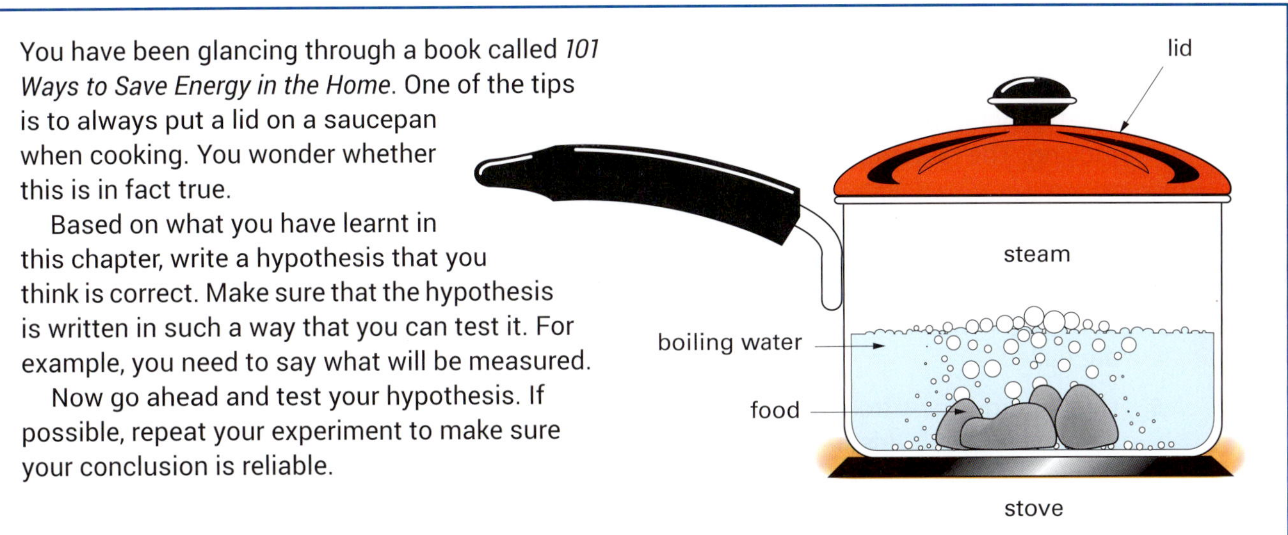

6 Designing a house

Your task is to design a house for your area that is cool in summer and warm in winter, using what you have learnt in this chapter about heat transfer.

Take into account how heat is gained and lost by an average house, as shown in the diagram. In your design you should consider:

- the position of the house
- the type of building materials used for the floor, walls and roof
- design features such as a flat or sloping roof, types of windows (e.g. single- or double-glazed) and ventilation
- the surrounds of the house, including the types of trees.

To find out more about energy-efficient house designs, follow the links to **Sustainable energy info** (fact sheets on building) and **Energy Smart house design**.

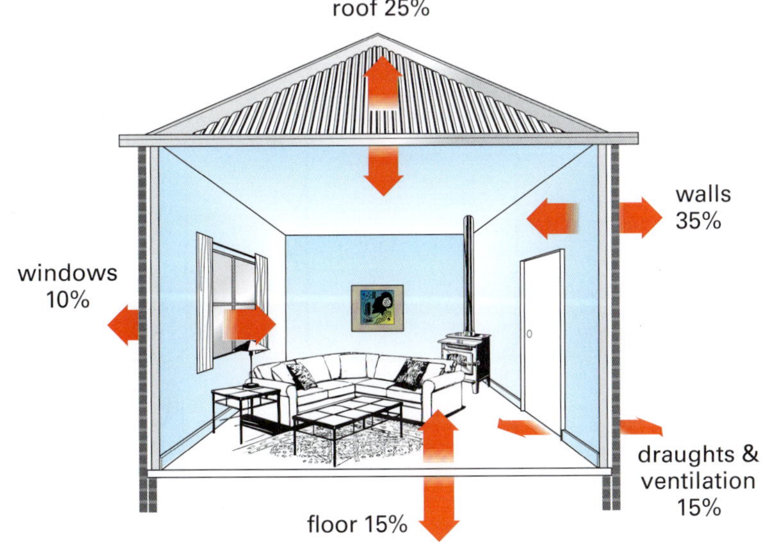

Figure 6.23 The percentages of the total heat transfer in various parts of an average house

ISBN 978 1 4202 3735 1

MAIN IDEAS

Copy and complete these statements to make a summary of this chapter. The missing words are on the right.

1 Heat is a form of ______, which can raise the ______ of an object.

2 The temperature of an object depends on how fast its ______ are moving. The faster they move, the higher the temperature.

3 The amount of heat gained or lost by an object depends on its ______, the temperature ______ and what it is made from.

4 ______ is the transfer of heat through a material by the collision of particles. Metals are the best conductors of heat. Poor conductors are called ______.

5 Heat energy flows from places where the temperature is ______ to where it is ______. Insulators are used to reduce the amount of heat ______.

6 ______ is the transfer of heat by circulating currents in liquids or gases.

7 Heat energy can be transferred across empty space by means of ______.

8 Dark-coloured surfaces ______ and emit radiation better than light-coloured surfaces.

energy
low
change
transfer
conduction
mass
high
particles
insulators
radiation
convection
absorb
temperature

CH•6 REVIEW

1 If one end of a copper rod is held in a burner flame, heat travels quickly along the rod to the other end. Substances such as copper that behave in this way are called good:

A absorbers.
B insulators.
C radiators.
D conductors.

2 A building is heated by running hot water through a number of radiators. The most efficient colour for these radiators would be:

A silver.
B white.
C black.
D red.

3 Copy and complete the diagrams below to show the convection currents you would expect to form in the water.

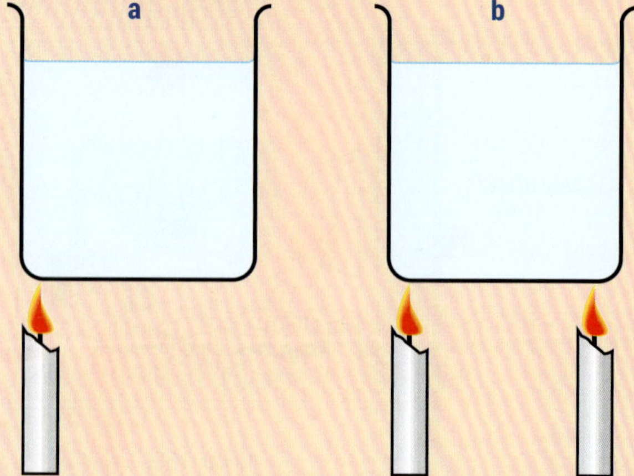

ISBN 978 1 4202 3735 1

4 Which of the following statements are true and which are false? Rewrite the false ones to make them correct.
   - **a** A cold object eventually heats up to the same temperature as its surroundings.
   - **b** Conduction is fast in insulators.
   - **c** Heat transfer by conduction is very slow in liquids and gases.
   - **d** The sun transfers heat energy to the Earth by the process of convection.
   - **e** The hotter an object is, the less radiation it emits.
   - **f** When an object absorbs radiation its temperature rises.
   - **g** Heat radiation travels at the speed of light.

5 If two objects are at different temperatures, what can you say about the movement of the particles in the hotter one?

6 Which has more heat energy—a teaspoon of water at 80 °C or a bucket of water at 80 °C? Explain.

7 If you hold your hand ***above*** a burning candle, you will burn yourself. Yet you can quite comfortably hold your hand *beside* the flame. Why is this?

8 Look at the diagram of a toasted cheese sandwich being cooked in a griller.
   - **a** How does heat travel from the heating element to the sandwich?
   - **b** Why can't the heat travel by conduction or convection?

9 Harry and Trent poured equal volumes of cold water into two identical styrofoam cups, then put identical thermometers in each. They put one cup in the sun and the other in the shade, and recorded the temperatures every 10 minutes. Here are their results.

Times (minutes)	Temperature (°C)	
	in sun	in shade
0	15	15
10	16.5	16
20	18	17
30	20	17.5
40	21	18.5
50	23	19.5
60	24	20.5

   - **a** Plot the results on a graph.
   - **b** What conclusion can you draw from the graph?
   - **c** What variables did Harry and Trent control in this experiment?
   - **d** Which method of heat transfer caused the increase in the temperature of the water in the cups?
   - **e** What would be the effect of painting the cup in the sun black?

10 A manufacturer claims that a certain insulating material is good 'to keep the cold out'. Is this expression accurate? Explain using a diagram.

11 Do sheep get colder when it is raining and their wool is wet? Design an experiment to find out, listing the steps you would need to take to make it a fair test.

Check your answers on page 297.

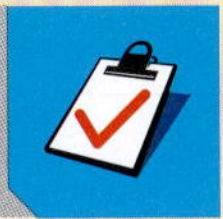

ISBN 978 1 4202 3735 1

IN THIS CHAPTER

Science Understanding

- examine specialised cells and tissues that make up the body's organs
- describe how organs function as part of a system
- describe how muscles that are attached to bones work in pairs to move your body
- describe the structure of each organ in the digestive system and relate its function to the overall function of the system
- investigate the structure of the circulatory, respiratory and excretory systems and explain their functions
- consider how technology has enabled organ transplants

Science Inquiry Skills

- use dissecting equipment safely to investigate the heart, lungs and kidneys
- work in a group and individually to test the food types in foods

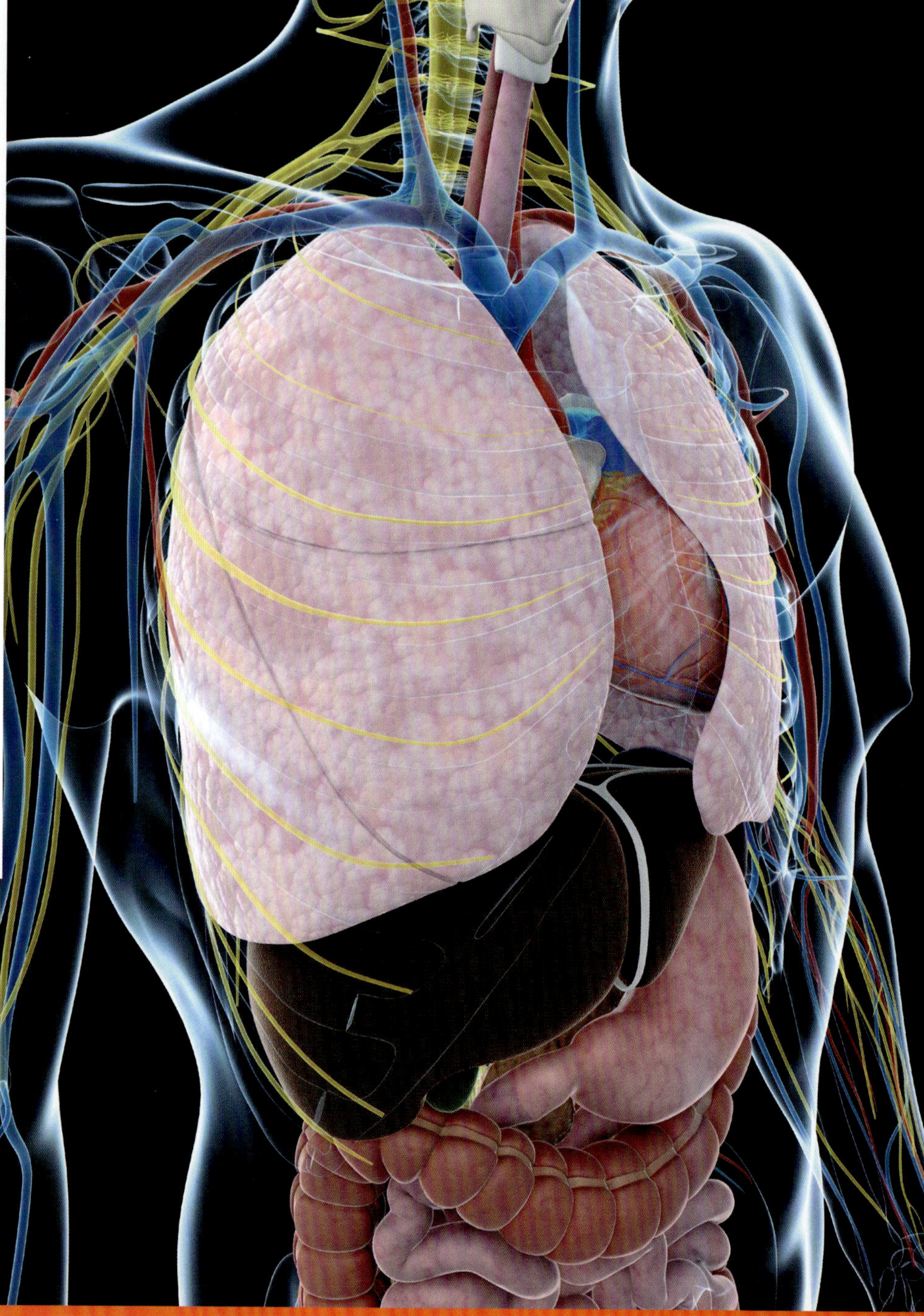

CH•7 The human body

ISBN 978 1 4202 3735 1

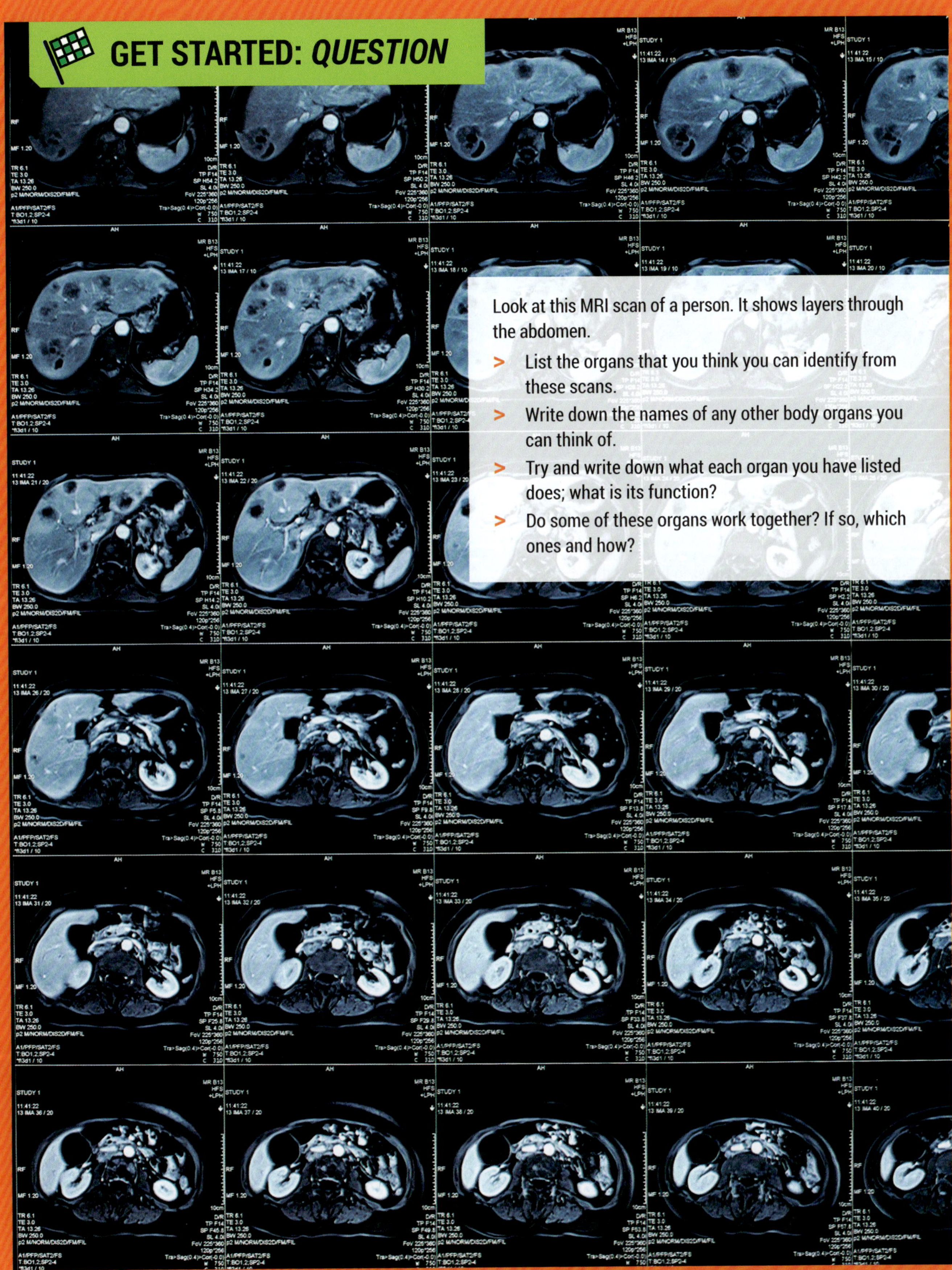

GET STARTED: *QUESTION*

Look at this MRI scan of a person. It shows layers through the abdomen.

> List the organs that you think you can identify from these scans.
> Write down the names of any other body organs you can think of.
> Try and write down what each organ you have listed does; what is its function?
> Do some of these organs work together? If so, which ones and how?

ISBN 978 1 4202 3735 1

7.1 How muscles work

Figure 7.1 Exercise causes changes in body systems.

What happens to your body when you start running fast? Your breathing and heart rate increase. You sweat a lot and your skin feels hot and often becomes red. Also, your muscles feel tired.

The human body is made up of many **systems**, which respond to changes. Your heart is a part of the circulatory system or blood transport system. Your lungs are part of the respiratory system or breathing system, and your muscles are part of your muscular system.

ACTIVITY

Group discussion—brainstorming

For this discussion, your group will be asked to come up with some ideas on *how the body reacts* in four situations.

Work in a group of 3 or 4. Elect a person to take notes. This person will record the ideas of the group.

In this group discussion, expect people to come up with a range of ideas, some of which may even seem a bit wacky! This is *brainstorming*. This technique often generates original, creative ideas.

Brainstorm each situation for a minute or two, and then discuss the group's ideas. Finally, select the best ideas.

Your teacher will organise a class discussion in which each group will put forward their ideas.

The four situations

1. You are walking along a dimly lit footpath at night. You suddenly see a shadow move in the bushes. You are scared.
2. You step onto a running machine at walking pace. You increase the speed until you are running fast.
3. You are dressed in light clothing (shorts and a t-shirt). The day becomes very cold and you have to wait 45 minutes at the bus stop for the next bus home.
4. You are very tired. You lie on your bed and fall into a deep sleep.

ISBN 978 1 4202 3735 1

Muscles

Muscles make bones move. Most of the 600 or so muscles in your body are connected to bones. Usually one of the ends of the muscle is joined to a bone that can move and the other end is joined to a bone that is fixed.

Muscles cannot push on bones. They can only work by pulling. When a muscle does this, it becomes shorter and fatter. This process is called **muscle contraction**. As the muscle contracts it pulls on the bone and the bone moves.

Muscles normally work in pairs. One muscle makes the bone move in one direction, and its partner makes the bone move in the other direction. For example, in Figure 7.3, when the biceps muscle contracts, the triceps muscle *relaxes*, and the arm bends upwards at the elbow. When the biceps muscle relaxes, the arm straightens.

If you bend your arm and feel the inside of your elbow, you will feel a hard 'cord'. This is a **tendon**. Tendons are hard, strong fibres that attach muscles to bones. There is a large tendon at the back of your ankle called the Achilles tendon. This attaches your calf muscle to the bones in your foot.

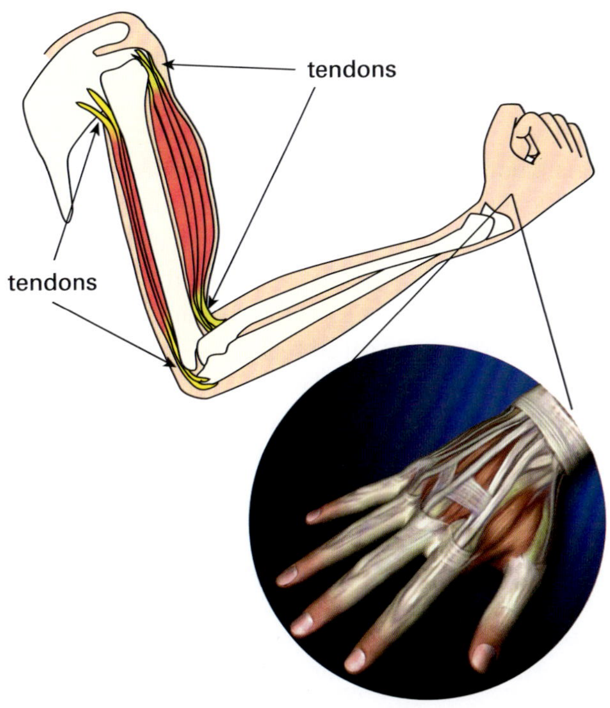

Figure 7.2 Tendons attach muscles to bones.

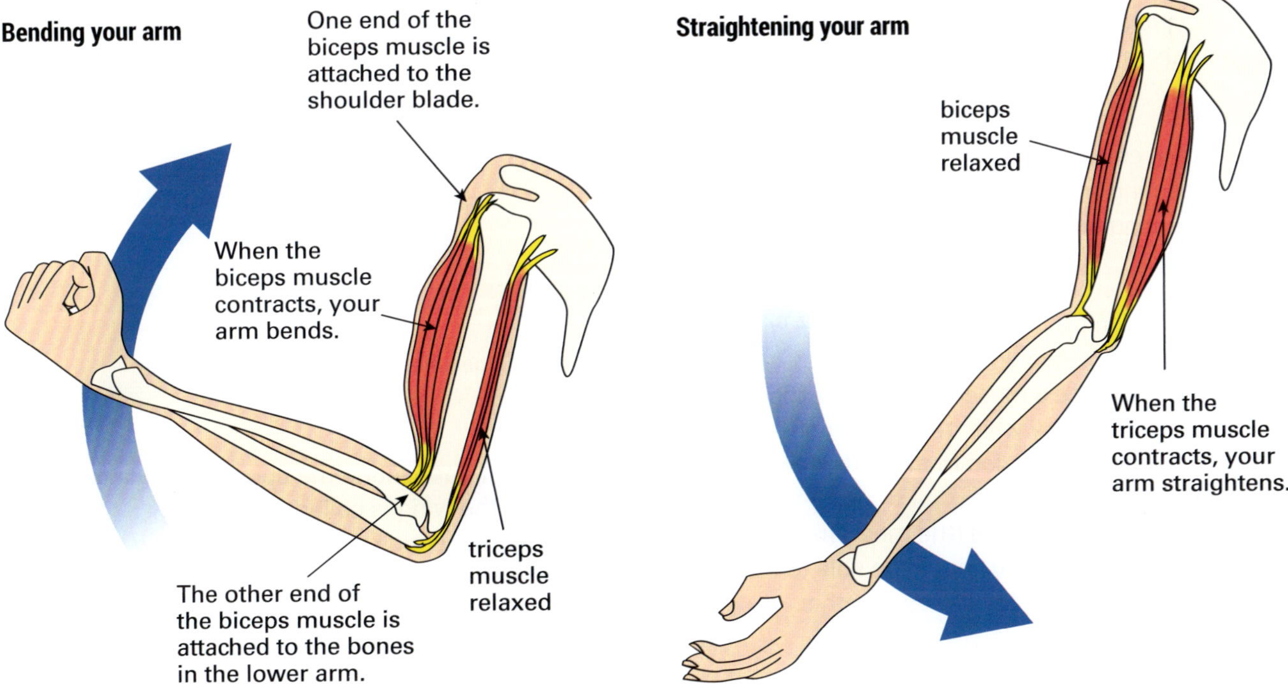

Figure 7.3 The biceps and triceps muscles work in pairs to bend and straighten your arm.

ISBN 978 1 4202 3735 1

You may have been wondering what holds bones together. You have learnt that muscles are connected to bones by tendons. But the muscles and tendons do not actually hold the bone joints together.

The bones in joints are held together by strong fibres called **ligaments**. In your knee joint, the ligaments allow the back-and-forth movement but stop the bones from coming apart. In dislocations, one bone is forced apart from the other and the force often tears the ligament. This can be extremely painful even after the bones have been returned to their correct positions.

At a joint, the ends of the bones are covered with cartilage. **Cartilage** is a soft but tough material that covers the bones at joints and prevents the bones from rubbing on each other. The whole joint is lubricated with fluid. This also stops the bones rubbing on each other, which can cause them to wear away.

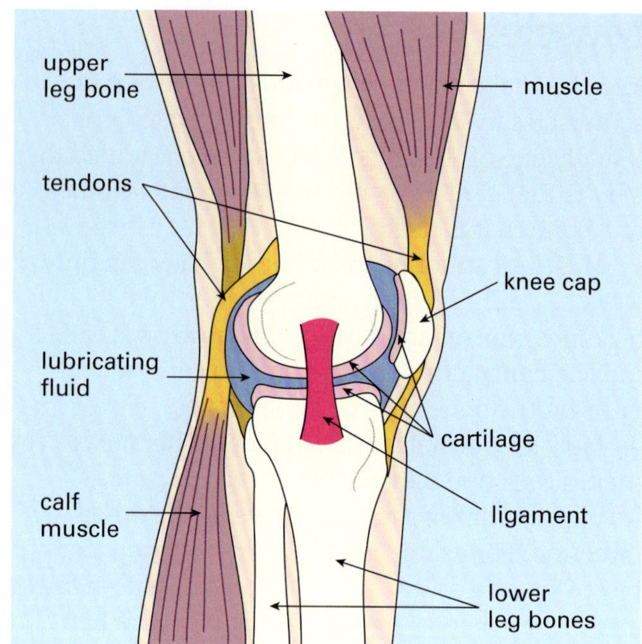

Figure 7.4 The side view of a knee joint

INVESTIGATION 7.1 Muscles in a chicken wing

Aim

To observe muscles, bones, tendons and ligaments in a chicken wing.

Materials

- a fresh chicken wing
- dissecting scissors and forceps
- dissecting board or newspapers

Method

1. Cut the skin from the upper part of the wing with scissors. You should now see the pink-coloured muscles. Trace a muscle down the wing until you see the shiny white tendon attached to it.
2. Try pulling on a tendon to move the lower part of the wing.
3. Remove all the muscle (the meat) from around the joint. You should be able to see the ligaments holding the bones together.

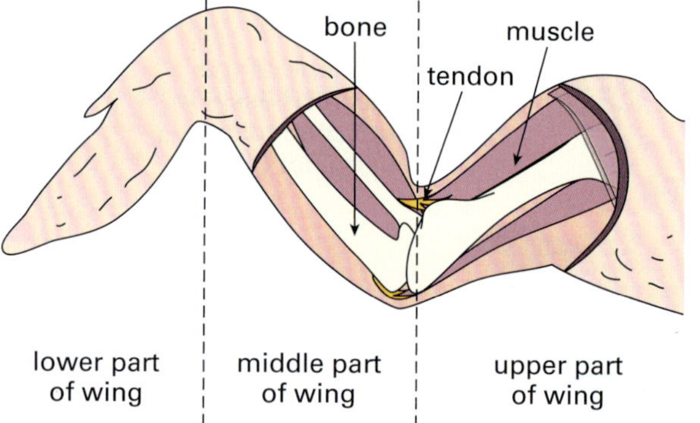

Figure 7.5 The different parts of a chicken wing

Discussion

1. The joint connecting the upper and middle part of the wing is called a hinge joint. In which direction does this joint allow movement?
2. Describe the differences in the appearance and the feel of a muscle and a tendon.

ISBN 978 1 4202 3735 1

Making muscles stronger

People who have been confined to bed for a long time find it difficult to do the physical activities they used to be able to do. This is because muscles actually shrink in size when they are not being used. When muscles are exercised, they grow in size. This is what happens when body builders work out at the gym.

Muscles that are exercised regularly are said to have *muscle tone*. Tone makes the muscles firm because they are never completely relaxed. This makes starting a contraction easier and helps muscles work quickly and efficiently without damage. If your muscles are in tone and you have to sprint 50 metres in an emergency, you can do it easily without feeling sore later.

Regular exercise also reduces the fat in muscles. Muscles that have lots of fat in them cannot contract efficiently.

ACTIVITY

You will need some bathroom scales and a partner to work with.

1 Put the scales on a table, sit on a chair and push down as hard as you can on the scales with both hands. This is testing the maximum force you can apply using your triceps. Ask your partner to read the scale.
Record your measurement (in kilograms) in a data table. To change it to newtons multiply by 10.

2 Place the scales upside down under the tabletop and push up as hard as you can. This will test your biceps strength. Ask your partner to read the scale.
Record your measurement.

Discuss how you would test the strength of other muscles such as your fingers or your thigh muscles. Try it.

Suggest why your biceps and triceps results were different.

Do you have equal strength in your left arm biceps and your right arm biceps? Try it.

Do you think that there is a connection between the size of a muscle and its strength? Explain your answer.

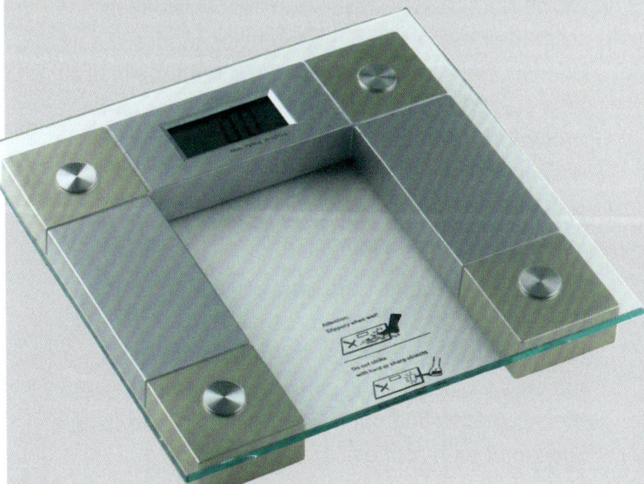

Figure 7.6 Testing the strength of your triceps. To calculate the force, multiply the reading in kilograms by 10.

ISBN 978 1 4202 3735 1

CHECK

1 Which of the following statements are true and which are false? Rewrite the false statements to make them correct.
   a Muscles pull on bones when they contract.
   b Tendons keep the bones together at a joint.
   c Muscles are connected directly to the bones.
   d Bones move when one muscle of the pair contracts and the other relaxes.
   e Regular exercise keeps muscles in tone.

2 How can bones move freely at a joint over a lifetime without wearing out?

3 Why do muscles have to be flexible and soft while tendons have to be firm and tough?

4 Figure 7.7 shows a drawing of a model arm. Copy the drawing in your notebook, and then answer the questions below.
   a Label the biceps, triceps, tendons, upper arm bone, lower arm bone and shoulder blade.
   b Beside each drawing, write a description of what is happening when the arm moves. Use the words contract, relax, biceps and triceps.

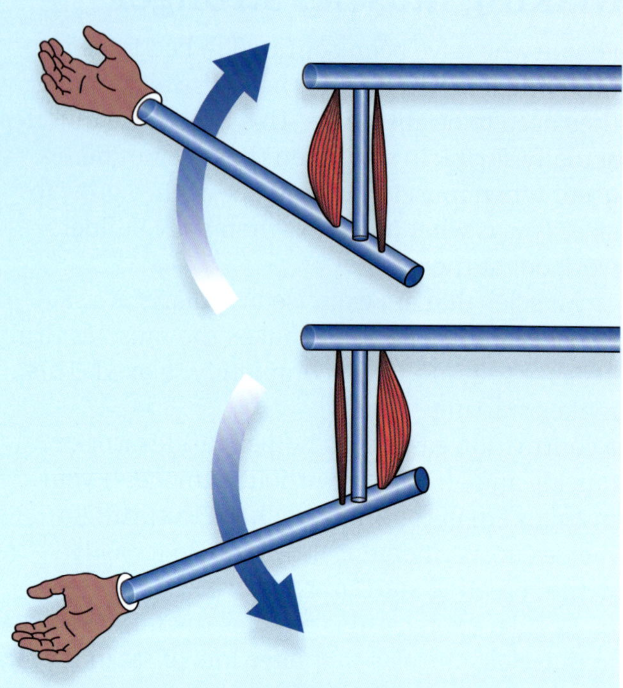

Figure 7.7 A model human arm

CHALLENGE

1 Muscles that work in pairs, like the biceps and triceps, are called *antagonistic* muscles. What do you think this word means? Check its meaning in a dictionary. Explain why you think it is an accurate name for these types of muscles.

2 The band of cartilage between the individual bones (vertebrae) of the backbone is quite thick. Suggest why this is so. You might like to check the vertebrae on a model human skeleton.

3 The bar graph below shows the results of a biceps and triceps strength test.

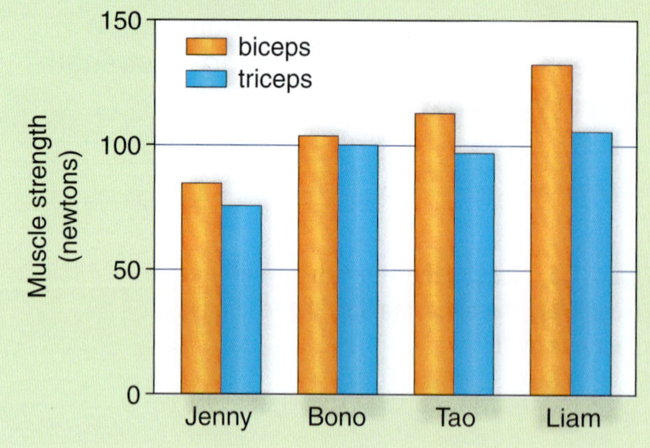

   a Use the graph to construct a data table listing the people in one column, their biceps strength in a second column, and their triceps strength in a third column.
   b Use the table to find the average strengths of the biceps compared to that of the triceps.
   c How much stronger, on average, is the biceps than the triceps? Express your answer as a percentage.

4 Some Roman slaves who constantly ran away from their owners had their Achilles tendon cut in one leg. Repeat offenders had the tendons in both legs cut. What was the purpose of this (cruel!) procedure? Demonstrate how the slaves would walk if they had this done to them.

5 Describe the features of a bone joint that makes it strong, flexible and long-lasting.

6 When people suffer from arthritis, the ends of the bones at the joint become swollen. Often the cartilage wears away and the ends of the bones rub on each other. This causes a lot of pain. Use the internet to find out how surgeons can replace a knee joint with artificial parts.

ISBN 978 1 4202 3735 1

7.2 Digestion

Food types

You need food for three reasons:

1 for energy
2 for growth and repair
3 to keep your body healthy and functioning correctly.

There are various substances in the food you eat. But the one thing that all food contains is water. For example, potatoes contain 77% water, lettuce 93% and eggs 75%, while peanuts contain only 5% water. The dry matter in foods is made of four main food types:

- **carbohydrates** (sugars, starch and fibre)
- **proteins**
- **fats**
- **vitamins and minerals.**

Carbohydrates

Carbohydrates include sugars, starch and fibre. Sugars and starch are used for energy. Sugars are found in fruits, honey and sweets. Starch is found in rice, potatoes, bread and pasta.

Fibre is a collective word that includes a number of plant substances such as cellulose. Fibre is found mostly in fruit, vegetables and cereals. It helps to keep the food moving in your gut and is important for the health of your gut.

Proteins

Proteins provide the material for the growth and repair of cells. They cannot be stored in the body, so some protein must be eaten regularly. Meat, fish, chicken, nuts, cereals, eggs and cheese are high in protein.

Fats

Fats are high in energy, producing about 2.5 times more energy per gram than carbohydrates. Fats are stored by your body as an energy reserve and to insulate your body from heat loss. Animal fats (e.g. butter) are usually solid at room temperature, while vegetable fats are usually liquid (e.g. olive oil).

Vitamins and minerals

These are found in very small quantities in foods, but are as important as the other food types. Vitamins are found in all fresh fruit, vegetables, beans, nuts and meats.

Figure 7.8 Can you identify which foods are high in fats, sugar, fibre, carbohydrates, proteins and vitamins? What do the proportions of each food group represent in this pie graph?

ISBN 978 1 4202 3735 1

SKILLBUILDER

Heating a liquid in a test tube

In Investigation 7.2 you will do a chemical test for glucose. In this test you use a Bunsen burner to heat a liquid in a test tube. This sounds simple, but it is a very difficult laboratory skill.

For this Skillbuilder, your teacher will give you the following equipment:

- test tube (in a test tube rack)
- test tube holder
- water
- Bunsen burner
- matches
- safety glasses

Method

1. Put on safety glasses.
2. Add a small amount of water to the test tube. The water should come up to about 1 cm in the tube.
3. Light the burner and turn the collar so that the flame is just off the yellow flame and turning blue.
4. Turn the gas down at the tap to give a small flame.
5. Hold the test tube with a test tube holder and heat the liquid gently as shown. Make sure you point the mouth of the test tube away from you and other people.
6. Move the test tube to and fro while you are heating. Don't heat the tube too strongly, otherwise the liquid will quickly boil and splash out of the tube.

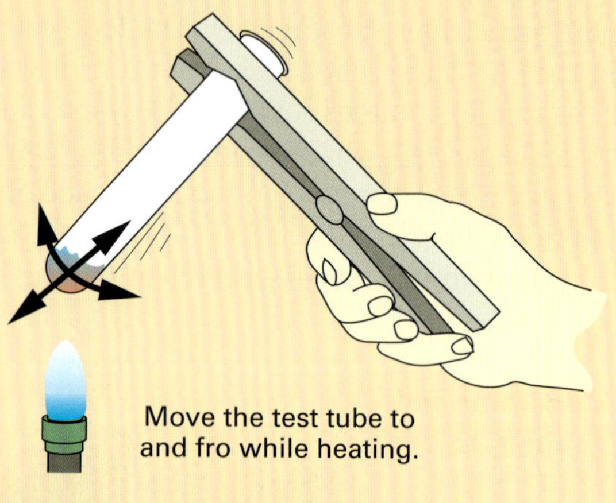

Move the test tube to and fro while heating.

INVESTIGATION 7.2 Testing foods

Aim

To test various foods for glucose and starch.

Materials

- glucose solution (10% glucose solution)
- starch suspension (20 g starch/L)
- Benedict's solution
- iodine solution (5 g I_2 in 100 mL of 10% KI)
- spotting tile
- test tube holder
- 3 test tubes, a stopper and a rack
- Bunsen burner and heatproof mat (or a boiling water bath for the class)
- small pieces of foods, e.g. cooked rice, fruit, bread, chicken, egg white

Risk assessment and planning

- Read each of the food tests in Part A very carefully.
- Using diagrams and labels only, describe what you have to do in each test.
- If you are going to heat a test tube containing Benedict's solution with a burner, you must first do the *Skillbuilder* on this page. Alternatively, your teacher may set up a boiling water bath for the test tubes.

ISBN 978 1 4202 3735 1

PART A Tests for food types

Method

Testing for glucose

1. Add 2 dropperfuls of glucose solution to a test tube and 2 dropperfuls of water to another test tube. (This is the *control tube*.)
2. Then add 2 dropperfuls of Benedict's solution to each test tube. Shake each tube to mix.
3. Use a test tube holder to heat each test tube very carefully over a small flame until it boils. Remember to constantly move the test tube to and fro while heating it.

   If your teacher has set up a boiling water bath for the class, place the test tubes in the water bath.
4. A red precipitate will form if glucose is present.

Testing for starch

Use a spotting tile for this test.

1. Add 5 drops of starch suspension to a spot on the tile. Add 5 drops of water to another spot.
2. Then add 2 drops of iodine solution to the starch and to the water.
3. A blue-black colour indicates starch is present.

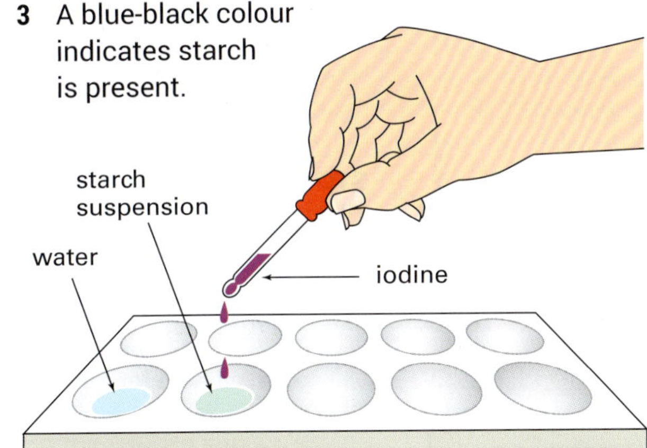

PART B Testing foods

Method

1. Wash and clean the two test tubes from Part A.
2. Select a piece of food and mash it up. Keep a little of it for the fat test, then add the rest to a clean test tube containing about 5 mL of warm water.
3. To dissolve as much of the food as possible, you need to add a stopper to the test tube and shake vigorously.

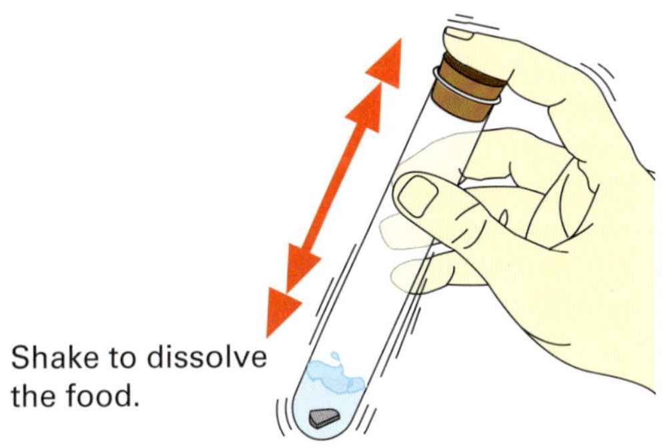

4. Pour equal amounts of the mixture into two clean test tubes.
5. Test for glucose and starch as you did in Part A.

   Record your results.
6. Select two other foods and repeat the above steps.

Discussion

1. Without looking at your book, briefly describe how you tested for glucose and starch.
2. The water test that you used for each food type in Part A is called an *experiment control*. What was the purpose of having an experiment control?
3. What food types were found in the foods you tested in Part B?

ISBN 978 1 4202 3735 1

The digestive system

When you take a bite out of a hamburger, you chew the mouthful of food a few times, and then swallow it. That is the last you see of the hamburger. How is the hamburger digested? The diagram of the **digestive system**, or *gut*, will help answer this question.

The job of the digestive system is to break down the food you eat into smaller particles, which are then able to pass from the small intestine into your blood. **Digestion** is both the physical breakdown of large lumps of food into smaller ones and the chemical breakdown that occurs with the help of substances called **enzymes** (EN-zimes). These substances speed up chemical reactions, which break down large insoluble food particles into small soluble ones.

Figure 7.9 The digestive system

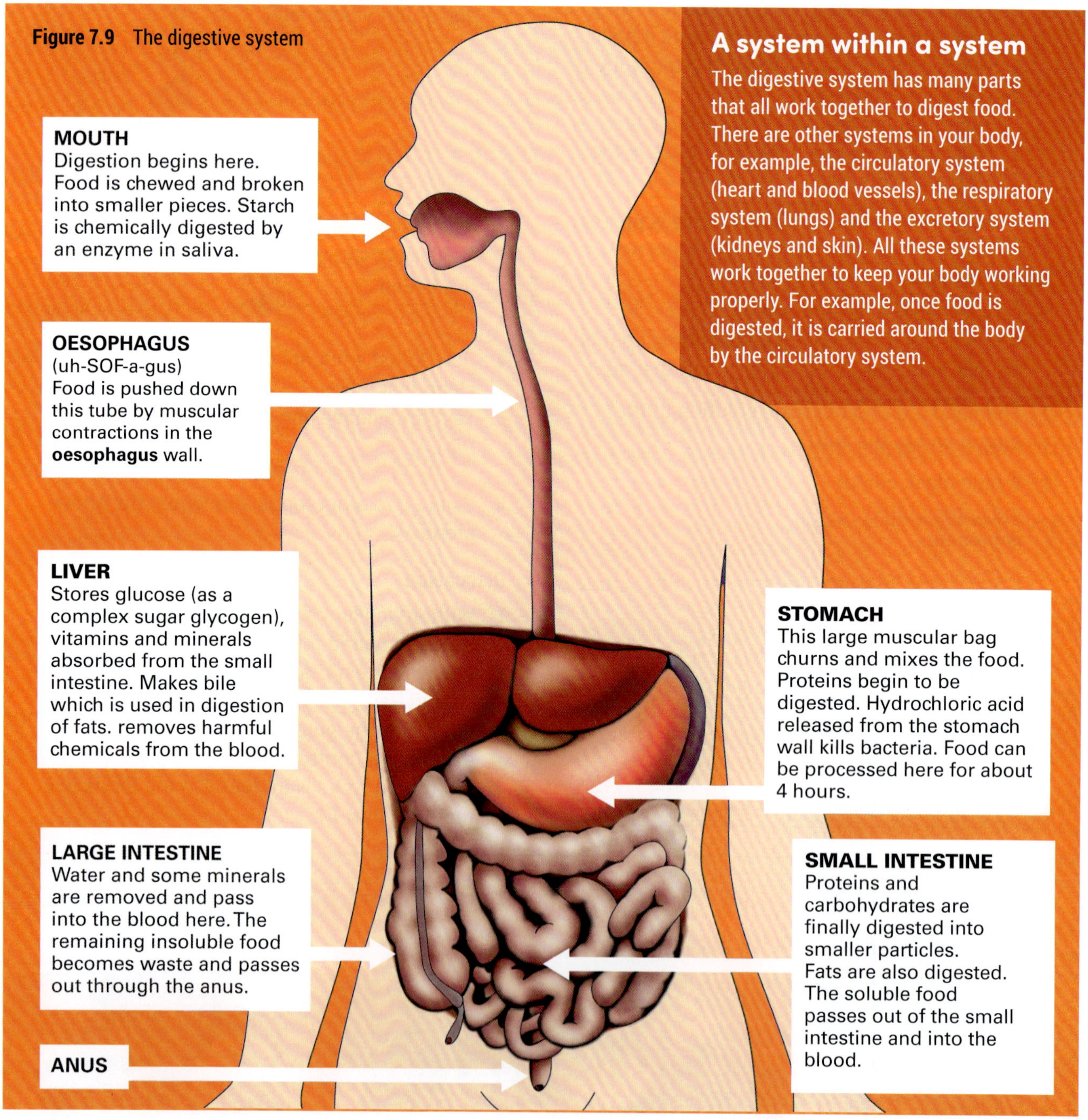

MOUTH
Digestion begins here. Food is chewed and broken into smaller pieces. Starch is chemically digested by an enzyme in saliva.

OESOPHAGUS
(uh-SOF-a-gus)
Food is pushed down this tube by muscular contractions in the **oesophagus** wall.

LIVER
Stores glucose (as a complex sugar glycogen), vitamins and minerals absorbed from the small intestine. Makes bile which is used in digestion of fats. removes harmful chemicals from the blood.

LARGE INTESTINE
Water and some minerals are removed and pass into the blood here. The remaining insoluble food becomes waste and passes out through the anus.

ANUS

STOMACH
This large muscular bag churns and mixes the food. Proteins begin to be digested. Hydrochloric acid released from the stomach wall kills bacteria. Food can be processed here for about 4 hours.

SMALL INTESTINE
Proteins and carbohydrates are finally digested into smaller particles. Fats are also digested. The soluble food passes out of the small intestine and into the blood.

A system within a system

The digestive system has many parts that all work together to digest food. There are other systems in your body, for example, the circulatory system (heart and blood vessels), the respiratory system (lungs) and the excretory system (kidneys and skin). All these systems work together to keep your body working properly. For example, once food is digested, it is carried around the body by the circulatory system.

ISBN 978 1 4202 3735 1

Food for the body

When it reaches the small intestine, the food is like thick creamy soup. The soluble food is made up of small particles that are able to pass through the small intestine wall. From here, they pass into the blood in the many blood vessels that surround the small intestine. This dissolved food travels to the liver and is then distributed to cells in all parts of the body. Here the soluble food particles leave the blood and pass into the cells.

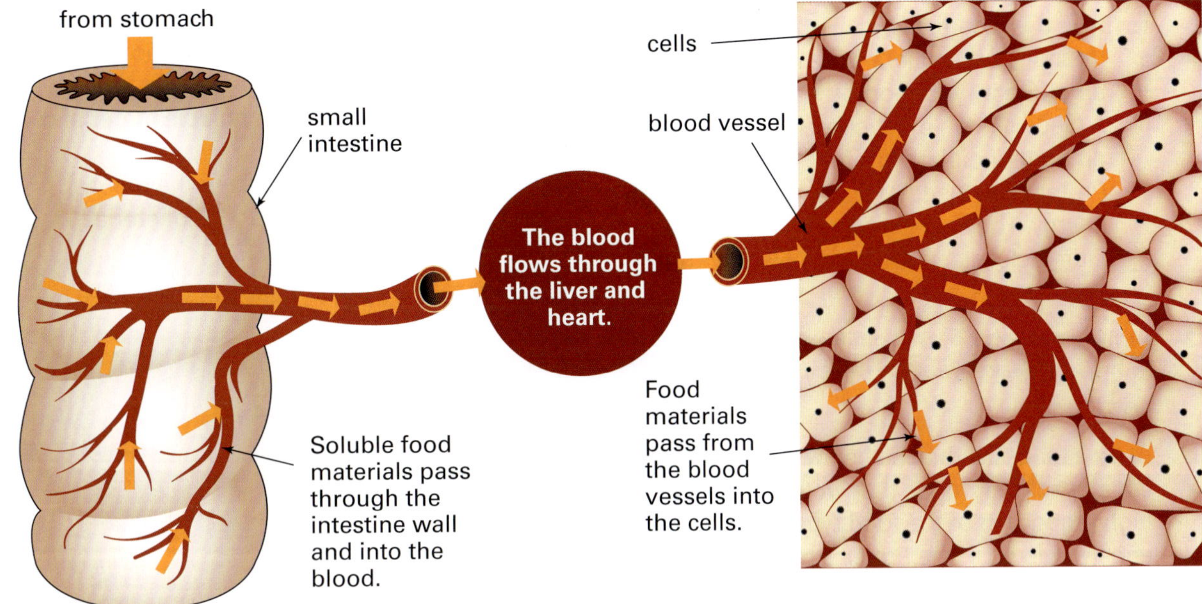

Figure 7.10 Blood carries food to cells around the body.

ACTIVITY

You are what you eat!

If you need food for energy, why can't you just eat fatty food, which is high in energy? The answer is that your body needs many different nutrients in foods to help it be healthy, in addition to its energy needs.

If you have a *balanced diet*, you are supplying your body's needs by eating food in the correct proportions. The table on the right shows an example of a balanced diet with the recommended amounts of foods to be eaten each day.

What's wrong with processed foods?

Processed foods are foods that are manufactured and they include biscuits, pies, chips and most fast foods. Processed foods often contain a high proportion of fats and very little protein. Always check the packaging for information about the fat and sugar content.

Foods	Daily amount
lean meat/chicken/fish/eggs	1–2 serves
dairy foods	2 serves
wholegrain bread/crispbread	2–3 serves
high-fibre cereal	1 serve
fresh fruit	2 serves
vegetables	1–2 cupfuls
fats and oil (added to food)	3 teaspoons

Questions

1. Use the table to plan your food intake for a day.
2. What foods have you eaten in the last few days that are not included in the table? Would you consider these foods to be high in fat or sugar?
3. Suggest how your food intake plan would change if you were an athlete.

ISBN 978 1 4202 3735 1

1 For each of the words below, write a sentence to show that you understand its meaning.

digestion	proteins
enzymes	carbohydrates

2 Which of the following statements are true and which are false? Rewrite the false statements to make them correct.
   - a Vitamins and minerals are needed in large amounts by your body.
   - b A piece of lean meat contains a large amount of carbohydrate.
   - c Proteins are used to supply your body with energy. They can be stored by the body.
   - d Fibre helps keep the food moving through your gut.

3 Write a paragraph to explain to someone a couple of years younger than you why we need food.

4 Draw a table like the one below. In each column list at least four foods that would contain a high proportion of the food type. For example, eggs would go in the *Proteins* column.

Carbohydrates	Proteins	Fats

5 Bronwyn and Leong sit down together to eat lunch. Bronwyn eats a roast chicken leg and two chocolate biscuits. Leong eats an apple, a salad sandwich on grain bread, and some sultanas.
   - a Whose lunch contains more protein?
   - b Whose lunch contains more fibre?
   - c What does *nutritious* mean? Who is eating the more nutritious lunch?

6 Copy and complete the following sentences.
   - a The speed of breakdown of foods into smaller particles is increased by _____.
   - b Digested food passes through the _____ _____ wall and into the blood.
   - c The oesophagus joins the _____ to the _____.
   - d In the large intestine _____ and _____ are removed and absorbed by the blood.

7 Look at the simple diagram of a human gut below. In which numbered part:
   - a does protein digestion first occur?
   - b is most of the digested food absorbed by the blood?
   - c do wastes and insoluble materials pass out of the body?
   - d does food enter from the mouth?
   - e is acid released to kill bacteria?
   - f are fats digested?

You will need to use some numbers more than once.

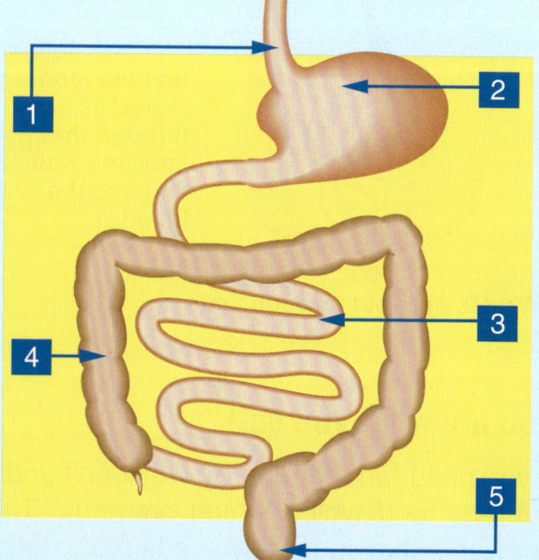

8 How does the function of the stomach differ from that of the small intestine?

9 Describe two functions of the mouth in digestion.

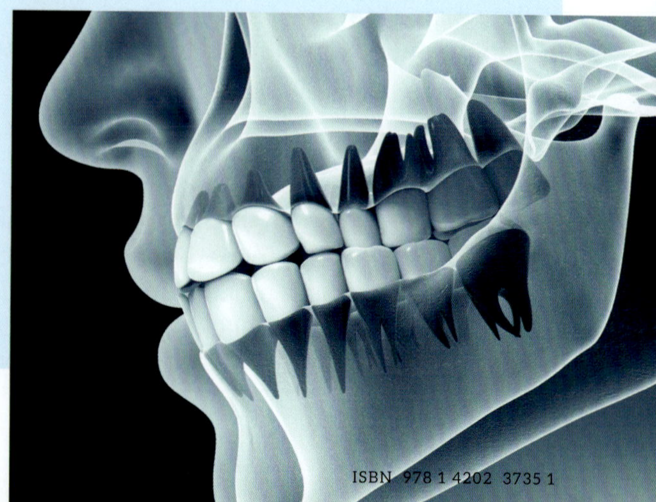

ISBN 978 1 4202 3735 1

CHALLENGE

1 In the mouth an enzyme breaks down the starch in foods into sugars (mainly glucose). In an experiment on starch and saliva, the equipment below was set up.

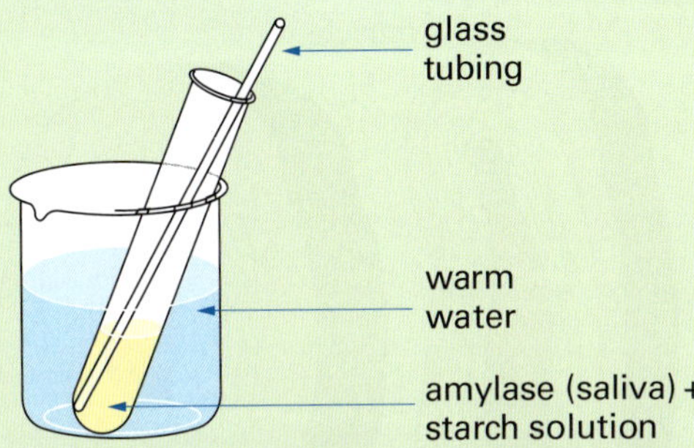

A drop was removed from the test tube using the glass tubing and placed on a spotting tile. Iodine was added to this drop. This procedure was repeated every 10 minutes for 1 hour.

a What results would you expect?

b What was the purpose of sampling at 10-minute intervals?

2 Suppose you were a ham sandwich. Write a fantasy story of what would happen to you if you were eaten and digested by a human.

3 If you take a mouthful of water and turn upside down, you can still swallow it. Explain why.

4 Lin was investigating the effect of temperature on the activity of the enzyme in saliva. She added 5 mL of starch, 1 mL of saliva and 3 drops of iodine to five test tubes. Then she put the test tubes in beakers of water at different temperatures. She recorded how long it took for the blue-black colour to disappear.

Test tube	Temperature of water (°C)	Time to change colour (min)
1	10	55
2	20	43
3	30	25
4	40	15
5	50	35

a Why did all the samples turn blue-black at the beginning? Why did they then change colour?

b Plot a line graph of the results. Then interpret your graph.

EXTRA FOR EXPERTS

The energy in food is measured in kilojoules (kJ). A kilojoule is quite a small amount of energy. It takes about 80 kJ of heat energy to boil a cup of water.

The amount of energy you use each day depends on three factors: how much you are growing, how active you are and how much you weigh.

The table below shows the approximate amounts of energy used per hour by a 60 kg person doing various activities.

Activity	Energy used (kJ per hour)
doing aerobics	7000
doing homework	500
housework	600
jogging	2500
lying still	300
playing ball games	2800
running fast	10 000
sitting	500
sleeping	250
standing	400
using a computer	350
walking, slow	600
walking, fast	2000
watching television	350

Tan Long weighs 60 kg. She works as a computer operator from 8 am to 4 pm. To get to and from work, she walks for 30 minutes to get to the train, has a 30-minute train ride, then walks for another 15 minutes. She does aerobics for an hour on the way home from work, does housework for an hour, eats dinner and watches television between 7.15 and 8.30 pm, reads until 10 pm, and then sleeps till 6 am. She sits and has breakfast until 6.45 am.

Use this information and the table above to estimate the amount of energy Tan Long uses each day.

ISBN 978 1 4202 3735 1

7.3 Body systems

Your body is like a factory. It takes in raw materials (food, water and air) and produces new products for the body to use. It also produces wastes (carbon dioxide and urine). Let's look at some systems that help keep your body functioning.

The heart and blood vessels

Your heart and blood vessels make up the body's **circulatory system**. The main part is the heart, a muscular organ that keeps pumping blood about 70 times a minute for the whole of your life.

The blood vessels that carry blood away from the heart are called **arteries**. **Veins** carry blood back to the heart. Arteries and veins have the same layers of elastic and muscular tissue, but the layers in the arteries are much thicker (see Figure 7.11). As the heart contracts, blood is forced through the arteries.

The large arteries and veins form many branches throughout the body. The narrowest arteries and veins branch into microscopic vessels called **capillaries**, which are very thin, usually only one cell thick. Food, oxygen and water pass through the capillaries to the cells, and wastes pass back, as shown in the diagram below.

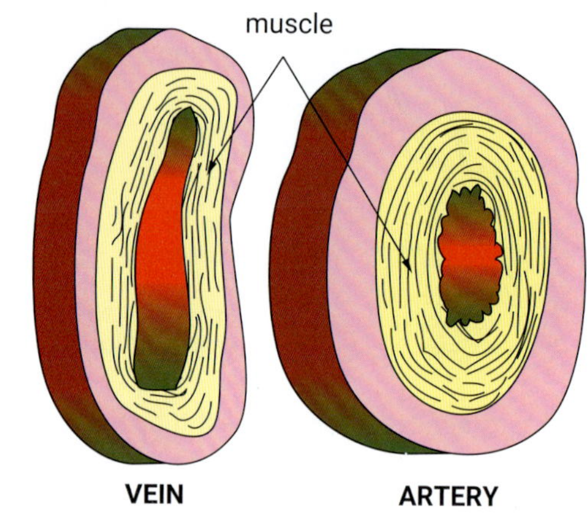

Figure 7.11 Arteries have thicker walls than veins.

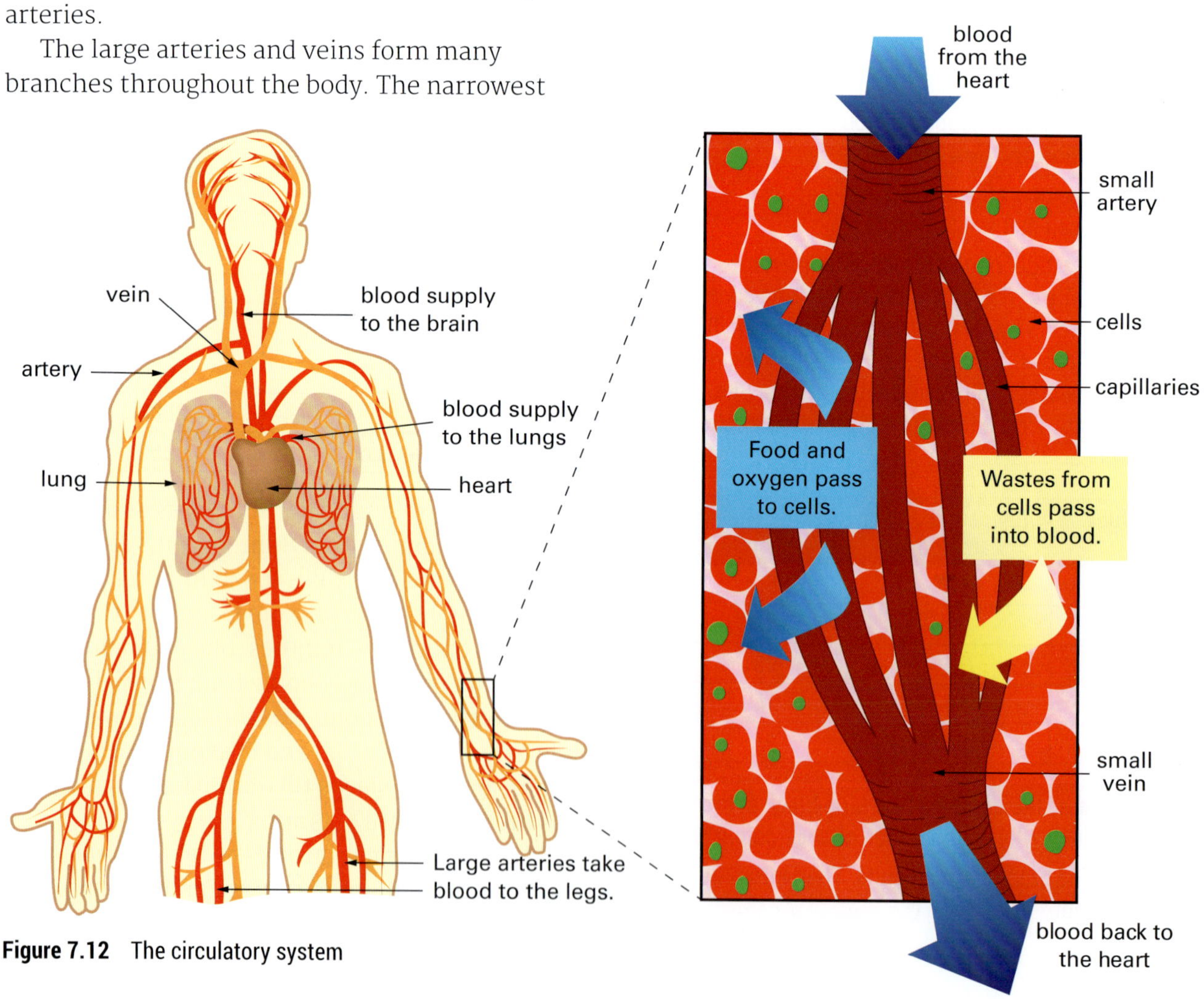

Figure 7.12 The circulatory system

ISBN 978 1 4202 3735 1

The blood system

Aim

To investigate your pulse and observe the blood capillaries in a fish's tail.

Materials

- a watch with a second hand, or digital watch
- small aquarium fish, e.g. guppy
- microscope and microscope slide
- cotton wool
- aquarium or pond water

Risk assessment and planning

- Read through Part A and decide who is going to do what sort of exercise. Design a data table for the results.
- Make a list of all the precautions you will take to make sure the fish in Part B is not harmed in any way.

PART A Measuring pulse

1. Use your index finger to find your partner's pulse in the artery in their wrist.
   Record the number of beats per minute and call this the resting pulse rate.
2. Have your partner exercise (e.g. by standing and sitting rapidly) for 2 minutes. Immediately after the exercise, take their new pulse rate.
   Record your results.
3. Record how long it takes for the pulse rate to return to the resting rate.

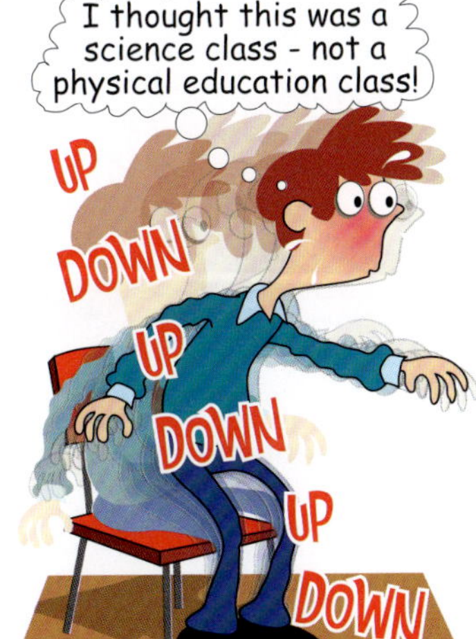

PART B Capillaries

Your teacher will do this part of the investigation as a class demonstration. Note: You can view capillaries on a computer or TV monitor via a video camera fitted to a microscope. Take care of the fish and return it to the aquarium immediately after use.

1. Soak some cotton wool in pond water, squeeze out most of the water and lay it on a microscope slide.
2. Carefully lay the fish on the cotton wool and place some more wet cotton wool on top of the fish. This will hold the fish in place and stop it from drying out.
3. Make sure the tail is sticking out of the cotton wool, as shown.

4. Look at the tail through low power on a microscope. Then switch to higher power to observe the capillaries and blood cells.
   Use a diagram to record your observations.

Discussion

1. How does your heart (pulse) respond to a change in activity in your body?
2. Suggest why there is a change in the pulse rate with exercise. Include the needs of the body cells in your explanation.
3. Why is it necessary for your heart to continue beating when you are asleep?
4. Does a fish have a pulse? Suggest reasons for your answer.

ISBN 978 1 4202 3735 1

How the heart works

The heart of a mammal has four chambers: two smaller, thin-walled ones on the top of the heart called **atria** (singular **atrium**); and two larger, more muscular chambers called **ventricles**.

Look at the cut-away diagram of the heart in Figure 7.13. Notice that the *left* ventricle is on the *right-hand* side of the diagram. This is not a mistake. Figure 7.13 shows the heart as it would be in a person facing you.

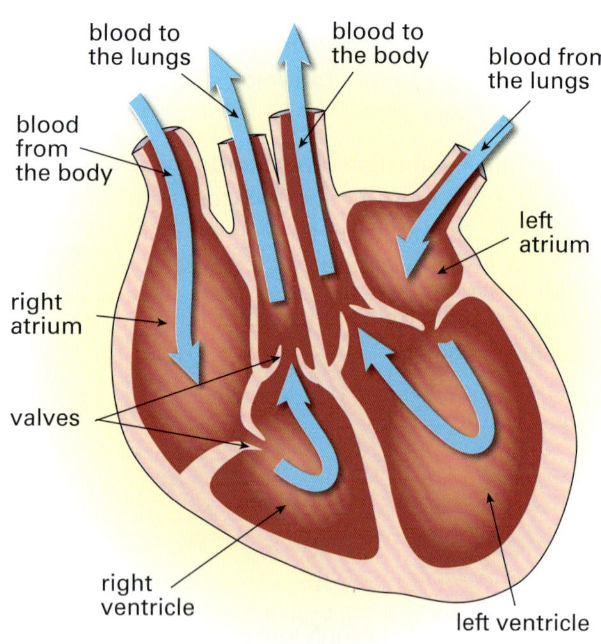

Figure 7.13 A cut-away diagram of the heart

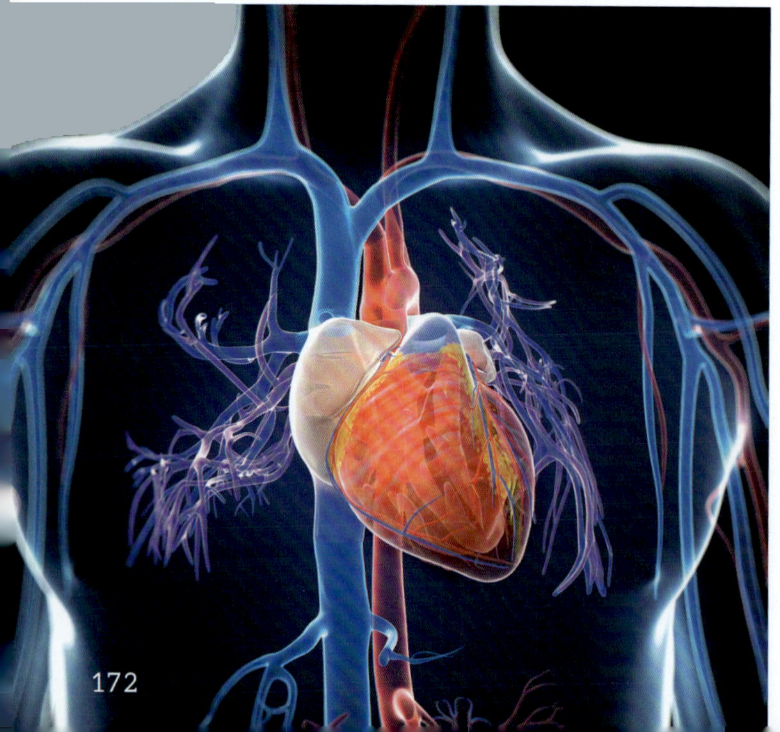

The heart pump

The blood in your body moves within a one-way, closed circuit. The heart pumps blood out through the arteries to the capillaries and back through veins to the heart again. However, the heart is a double-action pump. At each beat, the right-hand side pumps blood to the lungs at the same time as the left-hand side pumps blood to the body. Look at Figures 7.14 and 7.15.

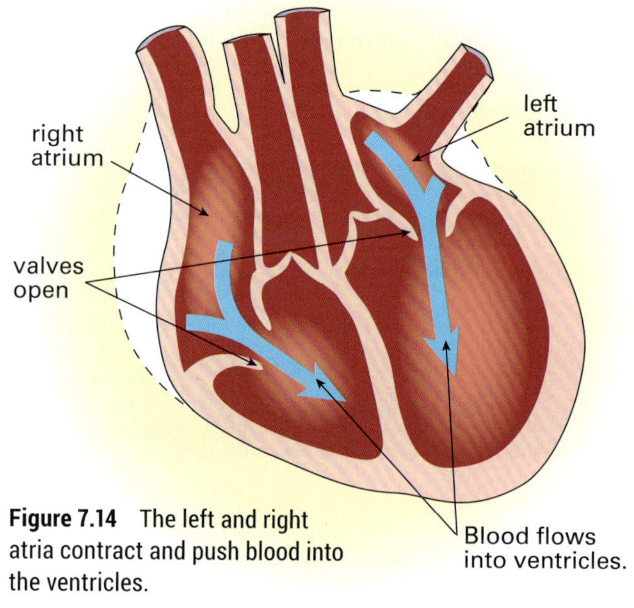

Figure 7.14 The left and right atria contract and push blood into the ventricles.

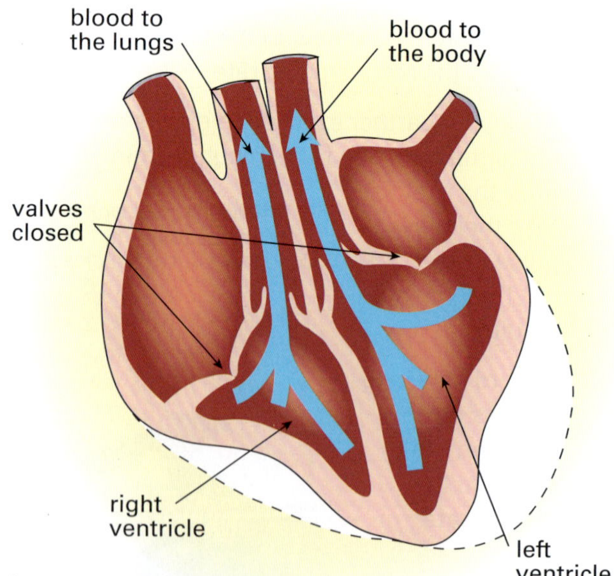

Figure 7.15 Both ventricles contract immediately after the atria contract. The right ventricle pushes blood to the lungs, and the more muscular left ventricle pushes blood to the body.

ISBN 978 1 4202 3735 1

INVESTIGATION 7.4 A heart dissection

Aim

To observe the structure of a mammalian heart.

Materials

- bullock's or sheep's heart
- dissecting board
- disposable gloves
- scalpel (or single-edge razor blade), scissors and probe
- paper towel

Risk assessment and planning

- Read through the steps in the Method very carefully. You will be using very sharp dissecting instruments, a scalpel and scissors. Take care with these instruments.
- Make a list of all the precautions you will need to take when dissecting the heart.

Method

1 When handling the heart, you *must* wear rubber gloves
2 Locate the right side of the heart. This side feels softer than the left side because it has thinner walls.
3 Use a scalpel to make the first cut to open the right atrium and ventricle.

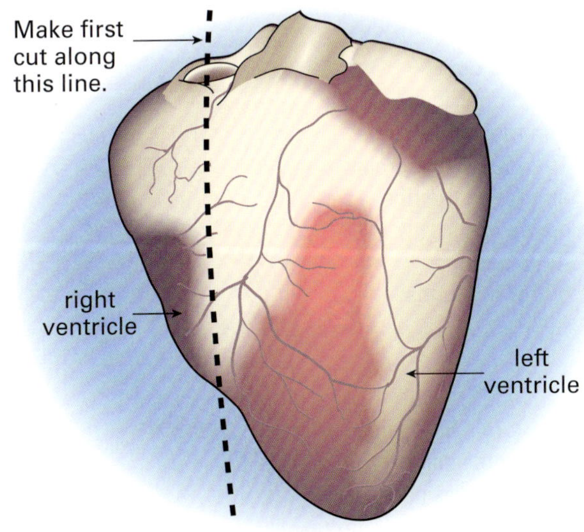

Look for the heart valves between the chambers. Notice the tendons holding the valves to the inside of the heart.

4 Now cut open the left atrium and ventricle.

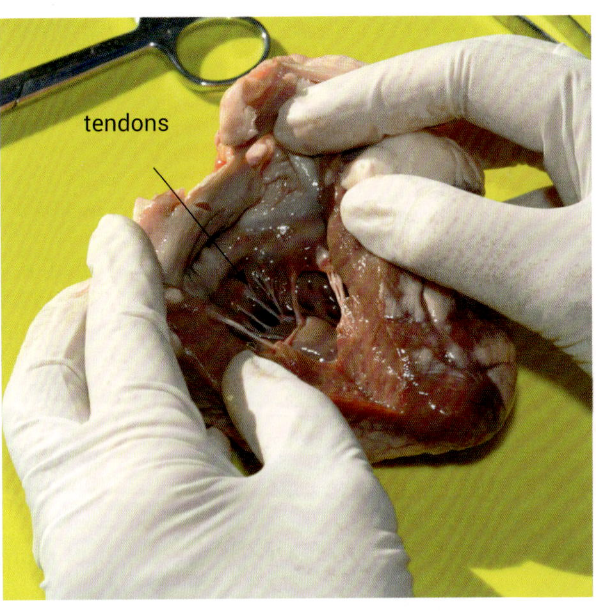

5 Use the probe to lift up the white-coloured tendons of the heart valves.
Compare the thickness of the muscular walls of the two ventricles.
6 Look for the valves in the large artery that leaves the left ventricle. Push the probe up the large artery.

Note: Your teacher will tell you how to dispose of the remains of the heart.

Discussion

1 Write compare and contrast sentences to summarise your observations of the left and right ventricles.
2 What is the function of the atria? Suggest why the muscular walls are much thinner than the ventricle walls.
3 The large artery that carries blood away from the left ventricle is called the aorta. Suggest why there is a large valve at the start of the aorta where it joins the left ventricle.

ISBN 978 1 4202 3735 1

Transporting oxygen

The two lungs are part of your **respiratory system**. They are large pink organs found inside the chest cavity. The lungs appear solid but are soft and sponge-like. The pink colour is due to the many blood capillaries in the lung tissue.

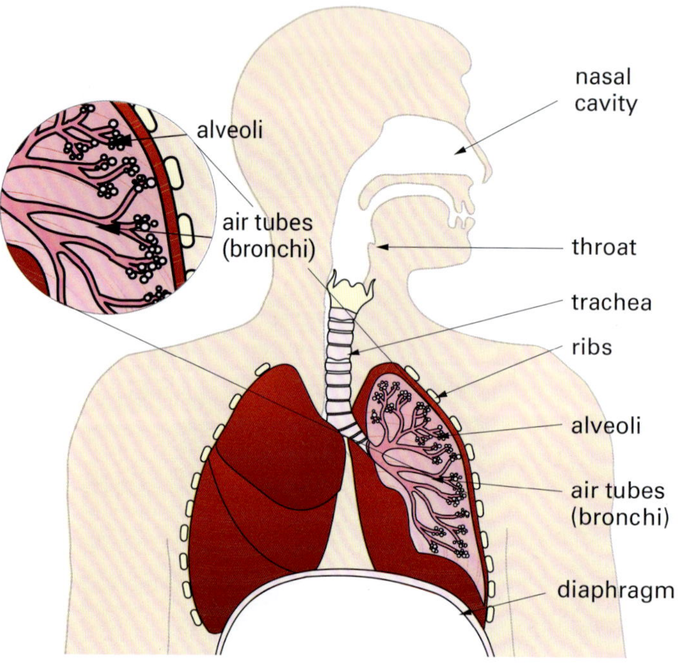

Figure 7.16 Oxygen from the air passes into the blood in the lungs, and waste carbon dioxide passes from the blood and is breathed out.

Air enters the lungs from the nose or mouth and then the **trachea** (track-EE-a) or windpipe. The air is moved in and out of the lungs by the movements of the muscles around the ribs and the large muscular diaphragm (see Figure 7.18). The air moves through the trachea and into smaller air tubes called bronchi (BRONK-ee), which end in minute air sacs called **alveoli** (AL-vee-OH-lee). The total surface area of the alveoli in the lungs is enormous—about 80 m^2, or about half the size of a tennis court.

The oxygen in the air breathed in passes through the thin walls of the alveoli and into the blood in the capillaries. From here, the blood is pumped throughout the body. The blood coming into the lungs from the body contains a lot of carbon dioxide. This passes from the blood into the alveoli and is breathed out.

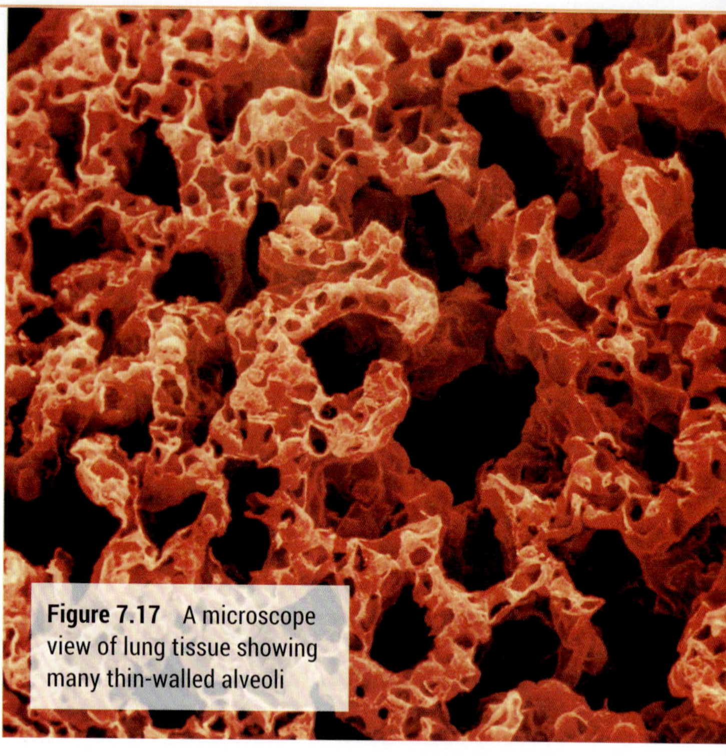

Figure 7.17 A microscope view of lung tissue showing many thin-walled alveoli

Breathing in

1. The diaphragm contracts and moves downwards.
2. The rib muscles contract and move the ribs upwards and out.
3. Air is drawn into the lungs.

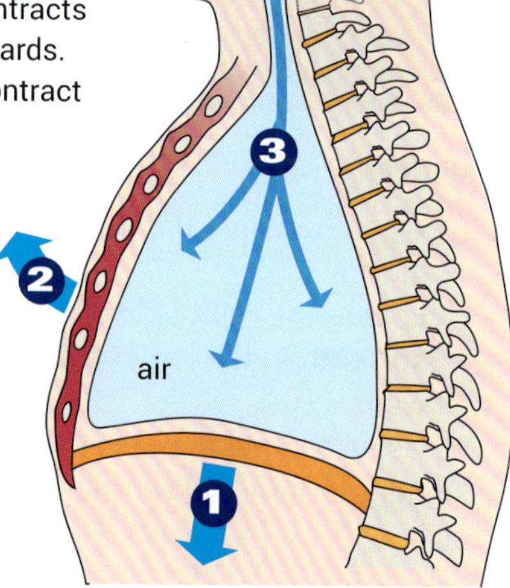

Breathing out

1. The diaphragm relaxes and moves upwards.
2. The rib muscles relax and move the ribs downwards.
3. Air is pushed out of the lungs.

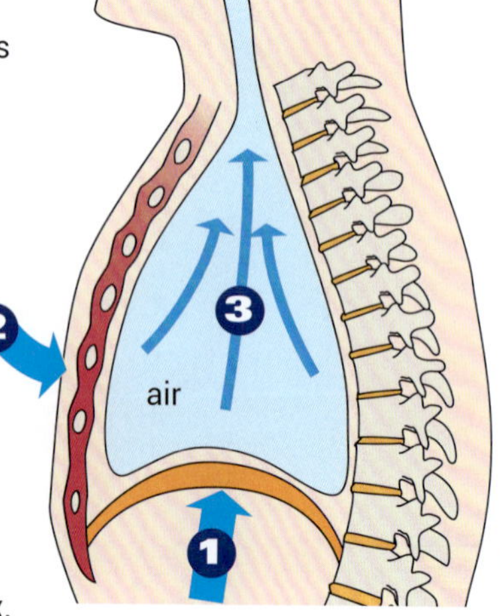

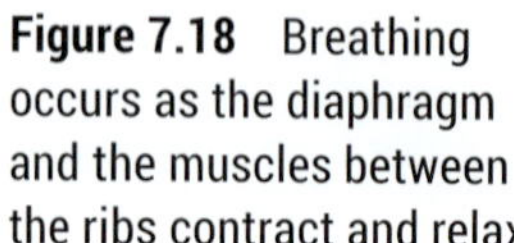

Figure 7.18 Breathing occurs as the diaphragm and the muscles between the ribs contract and relax.

ISBN 978 1 4202 3735 1

Getting rid of wastes

As mentioned earlier, your body is like a factory. It takes in raw materials (food, water and air) and produces new products. It uses energy in these processes and it produces wastes. The wastes are gases, liquids and solids.

Gaseous wastes—carbon dioxide

The most important reaction in your body is *respiration*, which produces carbon dioxide and water. The body reuses much of the water, but carbon dioxide is not used and has to be removed through your **lungs**.

ACTIVITY

Testing for carbon dioxide

Pour some fresh limewater into a test tube. Blow gently through a straw into the limewater.

- What do you observe?
- What can you infer from this observation?

Liquid wastes—urine

Most of the wastes produced by your body cells are soluble in water and are therefore able to be transported away by the blood. Many of these waste products are taken to the **liver** for processing.

The liver is a very important organ in the body. It not only stores and distributes digested food, but it also breaks down many harmful substances, such as alcohol.

Urea is one of the substances produced by the liver. Urea is soluble and so is carried in the blood from the liver to the **kidneys**, where it is then removed.

Blood is supplied to each of the two kidneys by a large artery called the renal artery (*renal* means *of the kidney*). About 1 litre of blood passes through the kidneys each minute. This blood is filtered and the wastes and some water pass out of the kidney to the bladder. The liquid waste is called *urine*.

The removal of wastes from the body is called **excretion** (ex-KREE-shun). The kidneys and liver are part of the **excretory system**.

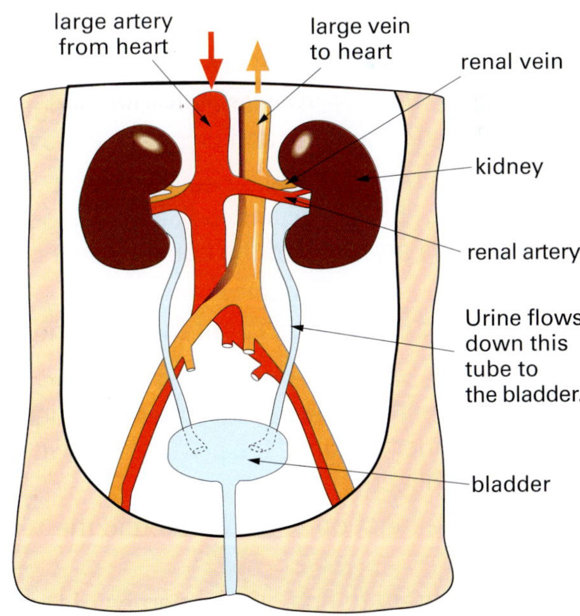

Figure 7.19 Kidneys are the main organs of excretion—the removal of wastes through dissolving in water.

ISBN 978 1 4202 3735 1

Sweat on your skin also removes salts and other soluble substances. But the skin is not considered part of the excretory system because the main purpose of sweat is to regulate your body temperature. When the water in sweat evaporates, it takes energy from your skin and cools it. This process helps lower your temperature.

Solid wastes—faeces

The solid wastes are called **faeces** (FEE-seas) and consist of leftover material from the food you eat (mainly fibre), as well as bacteria (about 30% of the mass), water and other products of cell reactions. The faeces pass out of your body through the anus. The brown colour of faeces is due to substances produced in the liver when blood is broken down.

Bird poo

Birds excrete urine and faeces from one opening in their body. Usually urine and faeces come out together as a dropping. Some of the soluble wastes in the urine give it a whitish colour. The dark, almost black, specks in the droppings are the faeces.

ACTIVITY

Part A Looking at lungs

Your teacher will show you a pair of sheep's lungs attached to the trachea.

Observe the colour and texture of the lungs and the trachea.

Infer the function of the bands of cartilage in the trachea.

Observe what happens when the lungs are inflated with air.

Part B Looking at kidneys

You will need a sheep's kidney, a single-edge razor blade (or scalpel) and disposable gloves.

1. Put on the gloves and peel off the fat around the kidney and look for the blood vessels attached to the concave side of the kidney.
2. Use the razor blade to cut the kidney in half.

The outer pale pink region is called the *cortex* and is where the wastes are filtered. The light-coloured inner region is the *pelvis* and is where the urine collects.

Infer the function of the fat around the kidney.

Use a library to find out the names of the various parts of the kidney, and how the kidney filters the blood.

Your teacher will set up a microscope to view a thin section of kidney.

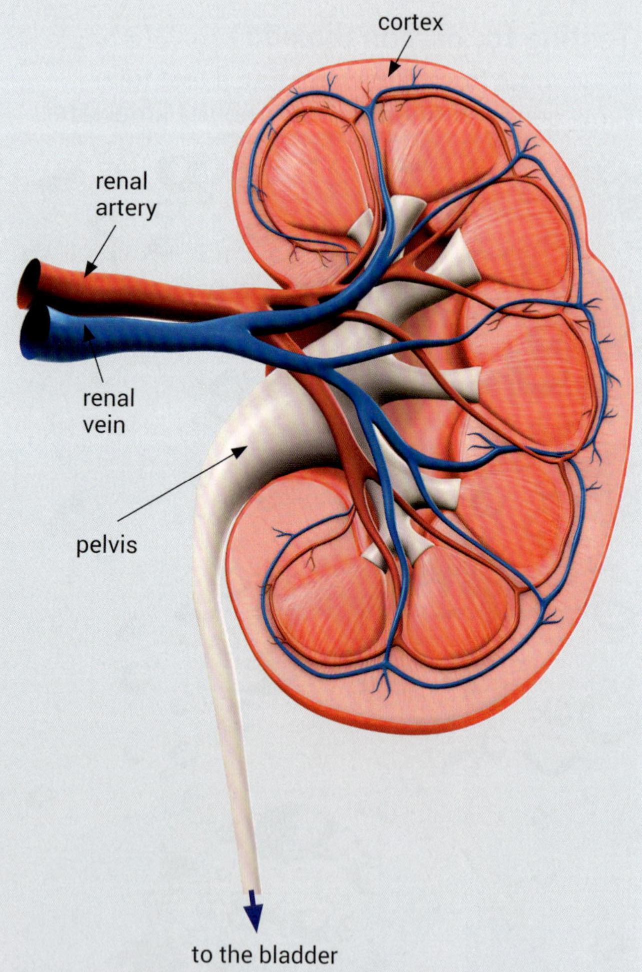

Note: Your teacher will tell you how to clean up and prepare the remains of the lungs and kidneys for disposal.

ISBN 978 1 4202 3735 1

SCIENCE AS A HUMAN ENDEAVOUR

Organ transplantation

Organ transplantation is moving one person's organ to another person's body, to replace a damaged or faulty organ. Many organs can now be transplanted, including skin, heart, lungs, kidney and liver. Most organs are donated by people when they die, although some people donate one of their kidneys while still alive. Other body tissues such as the cornea of the eye can also be transplanted to help people with non life-threatening diseases.

Before an organ can be transplanted, there must be a match between the donor and patient. If the organs do not match, they can be rejected by the patient's immune system. The immune system will think the organ does not belong and will attack it, just like your body attacks bacteria that enter your body. To avoid this, the patient usually has to take medication that stops organ rejection for the rest of their life, even with a good organ match.

Most of the time about 1600 people are waiting for organ transplants in Australia. For these people it is often a matter of life and death; if they do not receive a transplant in time they will die. These patients wait for a matching organ to become available when someone dies, e.g. in an accident or from a disease. One donor can save about 10 lives by providing various organs. In 2015, 435 organ donors gave 1241 Australians a new chance in life, so there is a shortage of donor organs.

Each person must decide whether they will be an organ donor; a doctor cannot just take your organs after you die. You must give permission, or your parents can do it for you if you are under 18.

EXPLORE ONLINE

1. Undertake a simulated organ transplant operation. Many simulations are available online.
2. Find out about how to become an organ donor. Discuss whether you think organ donation should be compulsory. Think about the reasons why some people may agree or disagree, such as religious reasons.
3. Find out how many transplants were done in Australia last year, and what organs were transplanted.
4. Investigate some of the technologies that have made organ transplants more successful.

Figure 7.20 Australia needs more organ donors.

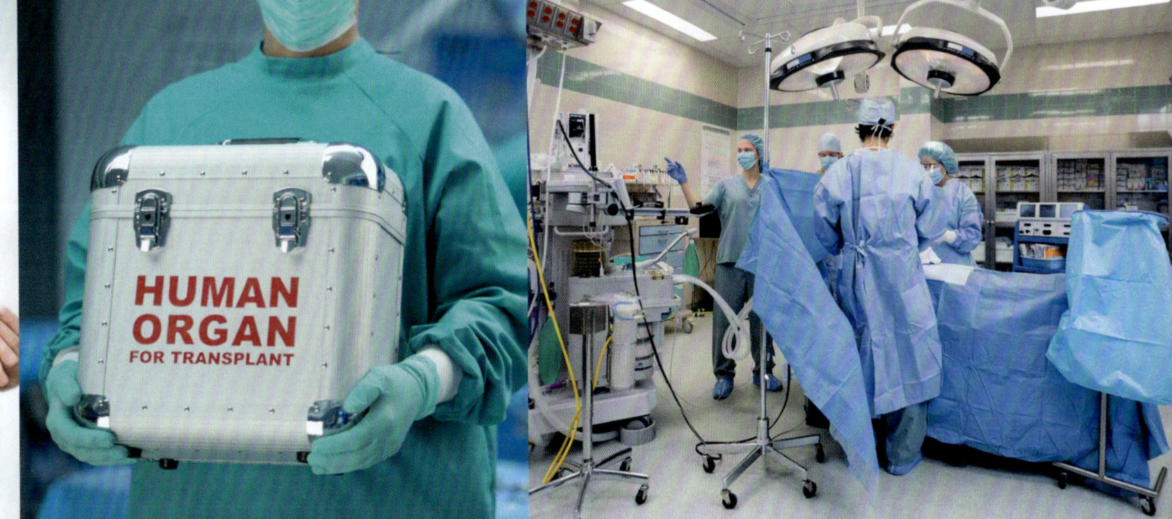

Figure 7.21 Organs must be obtained quickly after death to be used for a transplant, and are transported to the operating theatre on ice.

Figure 7.22 Technology and medical advances have made transplants more successful.

ISBN 978 1 4202 3735 1

CHECK

1 Which of the following statements are true and which are false? Rewrite the false statements to make them correct.
   a Blood consists of blood cells suspended in plasma.
   b Lung tissue has very few blood vessels.
   c During exercise, the amount of blood flowing to the body cells decreases.
   d Arteries have the same structure but much thicker walls than veins.
   e Urine is produced by the liver and is collected in the bladder.

2 What is a pulse? Why does the pulse rate vary? What actions do you have to take to lower your pulse rate?

3 Suppose you analysed the blood in an artery in your arm. You then did the same to the blood in a nearby vein.
   a Which substances would you find more of in the artery than in the vein?
   b Which substances would you find more of in the vein than in the artery?

4 What is urine? Where is it made and what happens to it in the body?

5 The diagram below shows simplified blood vessels.
   a Which parts are veins, arteries and capillaries?
   b In which direction does the blood flow? How do you know?

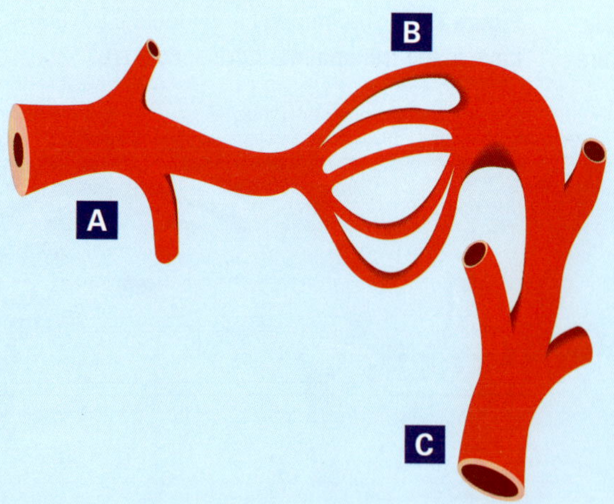

6 Why does the air you breathe out contain less oxygen and more carbon dioxide than the air you breathe in?

7 Suggest why your pulse rate increases when you see signs that you are in danger.

8 During exercise your heart rate increases. Suggest why your breathing rate also increases during exercise.

9 Samples of blood from two blood vessels in a person's arm were analysed and the results recorded.

Blood vessel	Amount of oxygen	Amount of carbon dioxide	Blood pressure
A	high	low	high
B	low	high	low

Which vessel is the artery? Give *two* reasons for your answer.

10 Your heart pumps about 70 mL of blood with each beat. Estimate the volume of blood it would pump in 24 hours. What assumptions have you made in your calculations?

11 Food and oxygen that are carried by the blood pass out of the capillaries into the cells. However, food and oxygen do not pass out of the arteries or the veins. Suggest a reason for this.

ISBN 978 1 4202 3735 1

CHALLENGE

1 Your body contains about 5 litres of blood.
 a If the kidneys filter 1 litre of blood in 1 minute, how much blood is filtered in a day?
 b How many times are the 5 litres of blood filtered in a day?
 c You produce about 1500 mL of urine each day. Express the amount of urine produced as a percentage of the total amount of blood filtered in a day.

2 The veins in your body have valves that allow blood to flow in one direction only.

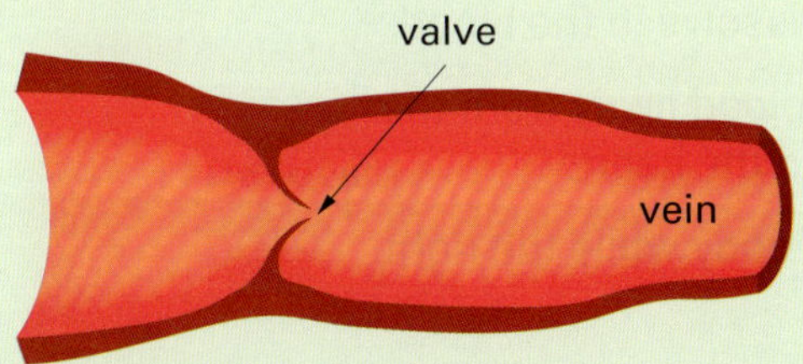

 a In which direction would the blood flow in the diagram above? How do you know?
 b Suggest why arteries do not have valves.

3 The pie charts below show the amounts of gases in air breathed in and out.
 a What percentage of air does the human body use?
 b What percentage of oxygen in the air is used?
 c Which gas is produced as waste?
 d Is nitrogen used by the body? Are you sure of your answer from this data? Would you need any other information before you could be sure?

4 The windpipe, or trachea, connects your throat to your lungs. It has little rings of cartilage along its length. You can feel these rings of cartilage by moving your fingers up and down your neck near your voice box. Suggest a reason for the rings of cartilage.

5 Arteries are located much deeper inside the body than veins are. For example, the veins in the arm can be seen just below the skin, whereas the arteries cannot be seen. What are the advantages of this arrangement?

6 A mouse's heart beats about 800 times a minute. It lives for about 2 years. A cat's heart beats 200 times a minute and it lives for about 15 years. A dog's heart beats 120 times per minute and lives for about 14 years. An elephant lives for 70 years and has a heartbeat of 25 beats per minute.
 a Record this information in a data table.
 b Which one of these animals has the most heartbeats per lifetime?

7 A fit person generally has a lower heart rate than a person who is unfit. Use the internet to find out why.

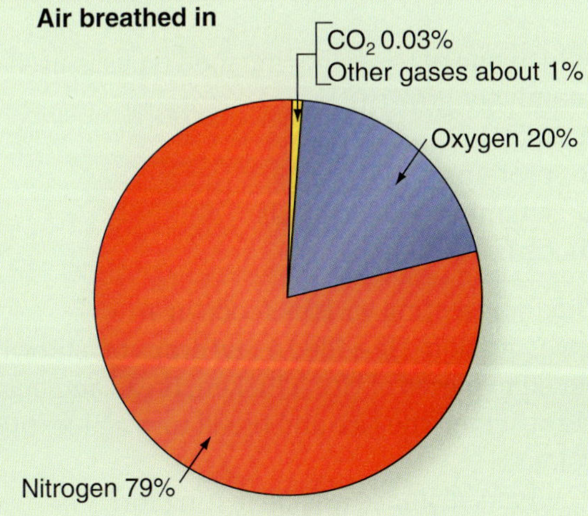

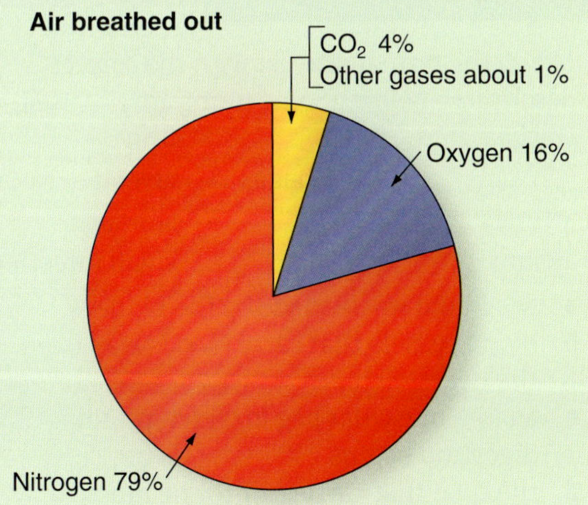

ISBN 978 1 4202 3735 1

MAIN IDEAS

Copy and complete these statements to make a summary of this chapter. The missing words are on the right.

1 Muscles that move your body are attached to ______. Most of these muscles work in pairs: one ______ and the other relaxes.

2 All organisms need food for ______, for ______ and to keep their bodies healthy and functioning correctly.

3 Carbohydrates include ______ and are used for energy. Fats are also an energy source, while proteins provide materials for growth and ______ of the body.

4 ______ is a process that breaks down large lumps of food into soluble materials containing small particles that can dissolve in the blood.

5 ______ are very thick-walled vessels that carry blood away from the heart. Thinner-walled ______ carry blood back to the heart.

6 In the heart, the left ______ pumps blood to the body while the right ventricle pumps blood to the ______.

7 You breathe air into your lungs when the ______ and muscles between the ribs contract.

8 ______ is removed from the blood by the lungs, dissolved wastes are filtered from the blood in the ______, and solid wastes pass out of the body through the anus.

repair
bones
arteries
ventricle
contracts
energy
diaphragm
sugars and starch
growth
lungs
kidneys
digestion
veins
carbon dioxide

CH•7 REVIEW

1 Match the term in the list with the correct description below. (Some terms are not used.)

trachea	liver	urine
ventricle	oesophagus	alveoli
arteries	enzyme	stomach
atrium	faeces	small intestine

a the liquid waste made by the kidneys
b blood vessels that carry blood away from the heart
c a large muscular bag in which protein digestion begins
d the tube that carries air to the lungs
e where carbohydrates, proteins and fats are digested to small particles
f the larger, more muscular chamber of the heart

2 Which one of the following food types is used mainly for growth?
A fats
B proteins
C vitamins and minerals
D carbohydrates

3 Brad was testing various foods in an investigation. He added a few drops of a brown liquid to pieces of rice, chicken, bread and butter. He observed the rice and bread turn a blue-black colour.
What substance was he testing for?
A sugar
B protein
C fat
D starch

ISBN 978 1 4202 3735 1

4 The following questions refer to the diagram below.

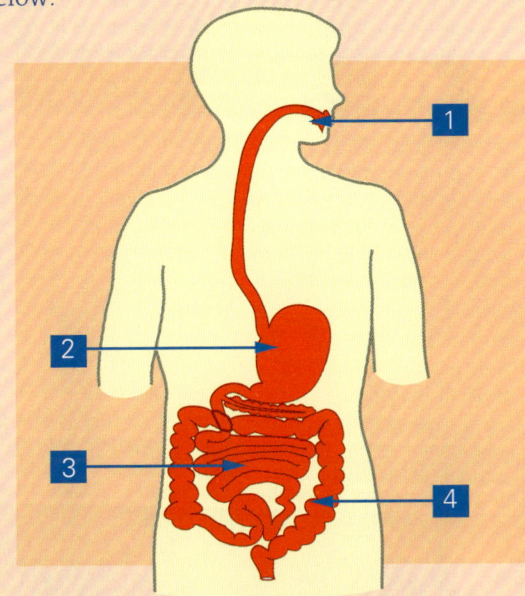

a Where are most substances absorbed into the blood?
b Where is food first acted on by enzymes?
c Where is food stored for short periods of time?
d Where are carbohydrates first digested?
e Where are water and some minerals absorbed into the blood?

5 Which organs are responsible for the removal of solid, liquid and gaseous wastes from the body.

6 The diagram below shows a simple model of a human heart.
a Which blood vessel, A or B, would have thicker walls? Explain your answer.

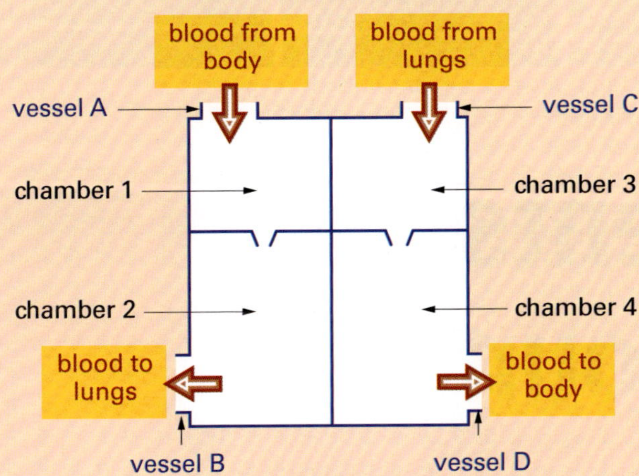

b Does the blood in chamber 1 contain more or less oxygen than the blood in chamber 3? Explain your answer.
c Write a paragraph describing the flow of blood through the four chambers and four blood vessels of the heart.

7 The amount of carbohydrate, protein, fat and water was measured in 100 grams of each of the following four foods: oranges, wholemeal bread, eggs and fried beef sausages.

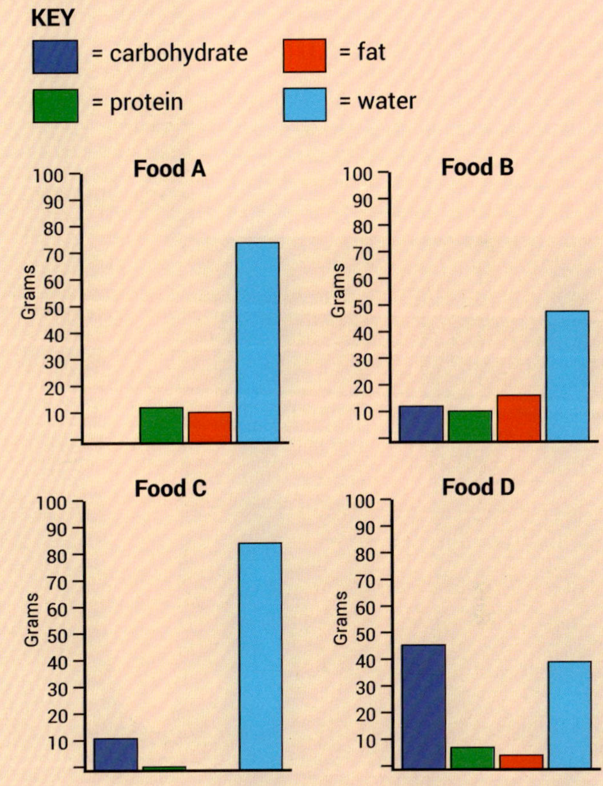

a Which of the foods (A, B, C, D) had the smallest amount of protein? The largest amount of water?
b Match the four foods to the graphs. Give reasons for your choices.
c Why was the same mass of food (100 grams) used in each of the tests?

Check your answers on page 299.

ISBN 978 1 4202 3735 1

IN THIS CHAPTER

Science Understanding

> model the arrangement of particles in elements and compounds
> represent elements and compounds by symbols and formulas
> investigate the reactions involved in the making and breaking of compounds
> locate elements in the periodic table
> assess the progress being made with hydrogen cars to overcome the environmental problems with petrol cars

Science Inquiry Skills

> use molecular models to represent the ways atoms join together to make molecules

CH•8 Elements and compounds

ISBN 978 1 4202 3735 1

GET STARTED: *IMAGINE*

Meet Super-Sci. She can make herself smaller and smaller until she is not much bigger than the tiny invisible particles in all matter. These are called atoms and molecules.

Super-Sci travels through the air where she sees nitrogen and oxygen molecules. These are 'double atoms', made of two atoms stuck together. They are whizzing past at about 1800 km/h, all moving in different directions. Occasionally they collide with each other.

When Super-Sci dives into the harbour, she finds herself surrounded by water molecules. Each molecule consists of three atoms. The molecules are much closer together than they are in the air. They often touch each other and are constantly moving past one another, continually changing their positions.

Finally Super-Sci tries to push her way into the steel in the bridge, but the ball-like iron atoms are so close together she can't crawl through. The iron atoms stay in their places, but they are constantly vibrating. Super-Sci counts the atoms and calculates that about eight million of them placed side by side would fit across the head of a pin!

Imagine you are a TV news or current affairs presenter. Prepare a news item on Super-Sci's fantastic voyage.

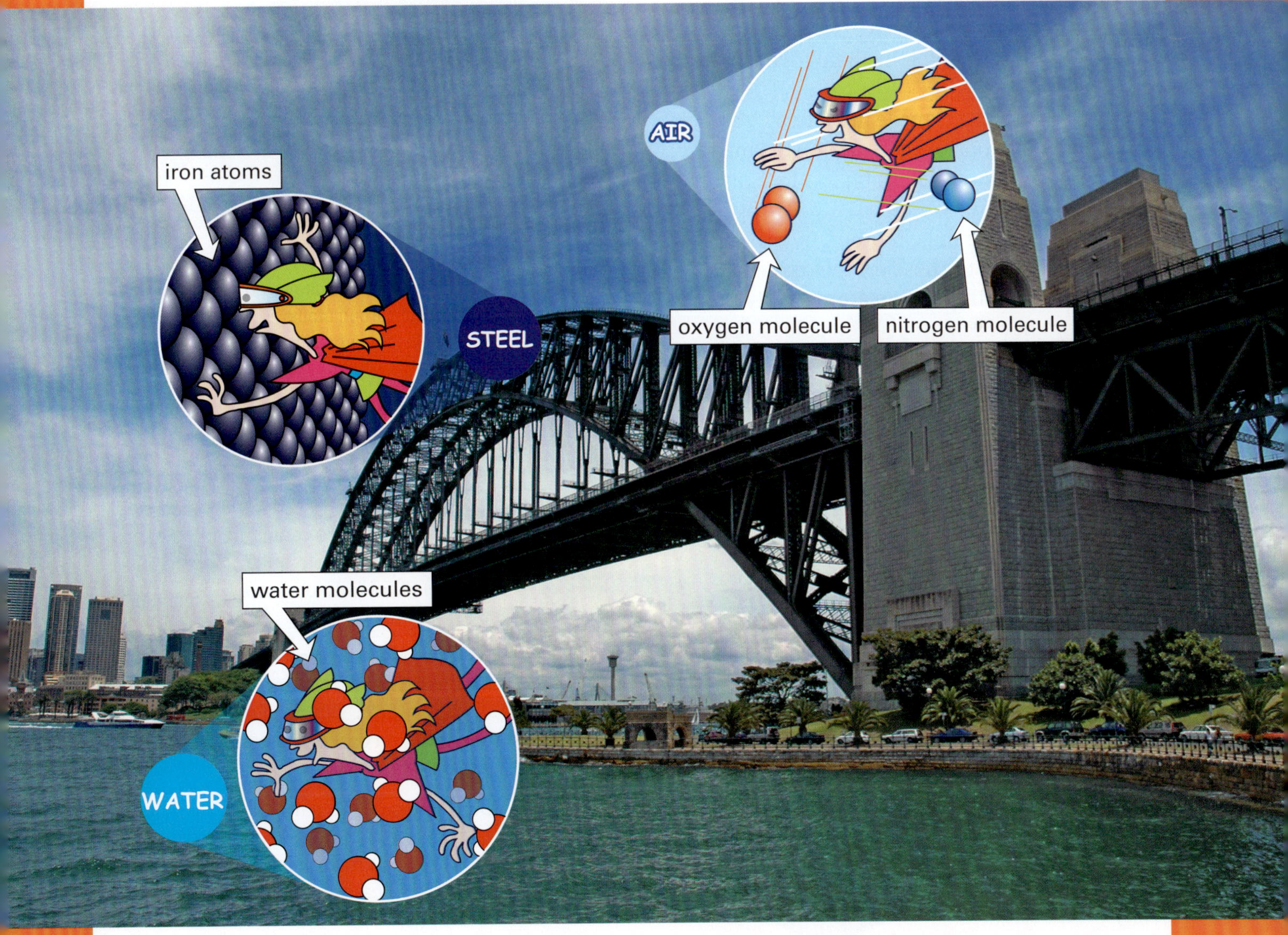

AIR
iron atoms
STEEL
oxygen molecule
nitrogen molecule
water molecules
WATER

ISBN 978 1 4202 3735 1

8.1 Atoms and molecules

In Chapter 2 you learnt about the tiny particles that make up all matter. For example, if you could break a piece of gold into smaller and smaller bits you would eventually end up with a single atom of gold.

Atoms are the basic building blocks of all matter—both living and non-living. They are incredibly small. To give you some idea of their size, the dot at the end of this sentence contains millions of them.

Atoms are not usually found on their own. Two or more atoms joined together is called a **molecule**. For example, an oxygen molecule consists of two oxygen atoms held together by a chemical bond.

A water molecule is made up of two hydrogen atoms bonded to one oxygen atom. This means water contains two different types of atoms.

Molecules vary in size from tiny hydrogen molecules up to the huge protein molecules in your body. Each of these protein molecules contains about half a million atoms. Only in recent years have scientists been able to use special microscopes to 'see' atoms and molecules.

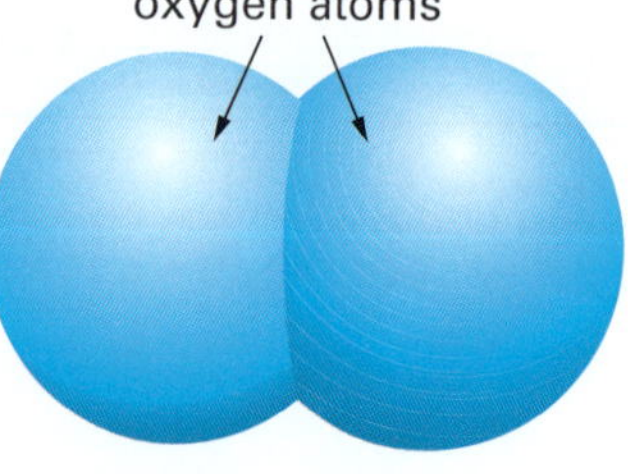

Figure 8.1 An oxygen molecule is made up of two oxygen atoms bonded together.

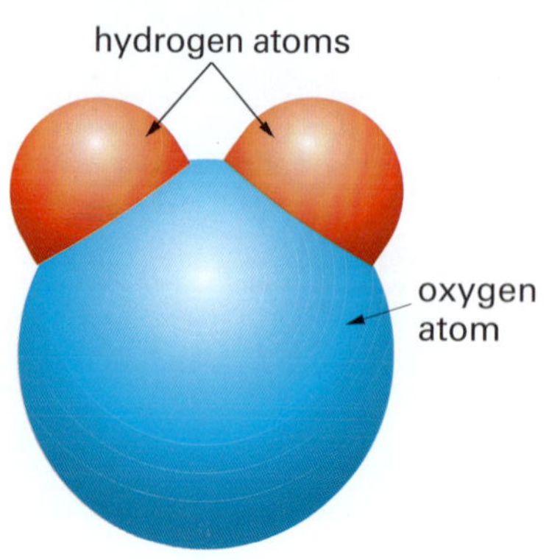

Figure 8.2 A water molecule is made up of an oxygen atom combined with two hydrogen atoms, one on each side, a bit like Mickey Mouse's ears.

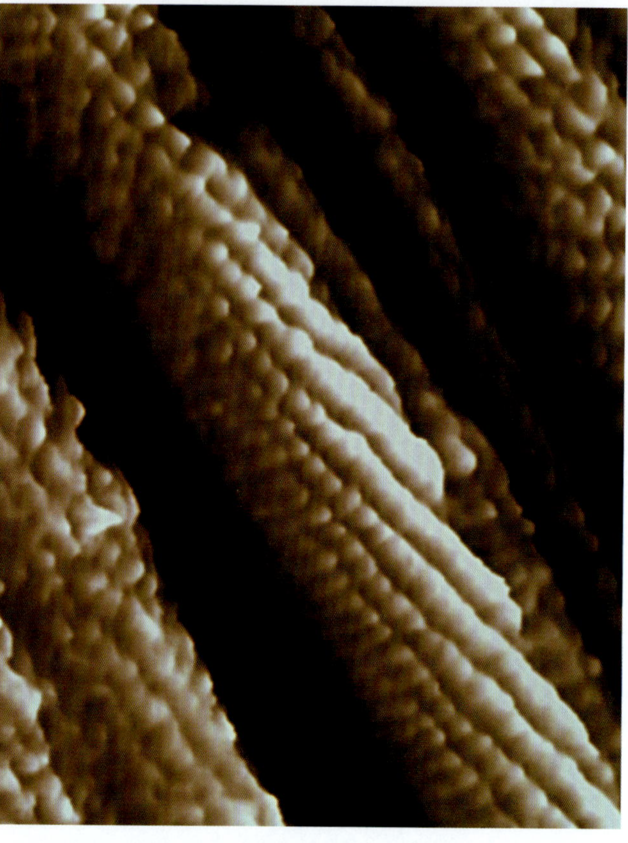

Figure 8.3 Carbon nanotubes are large molecules made of carbon atoms joined together. The bumps on the surface of these tubes are individual carbon atoms. This image was made using a scanning tunnelling microscope.

CHECK

1. a Describe what an atom is.
   b Give an example of an atom.
2. How are atoms usually found in nature?
3. a Explain what a molecule is.
   b Give an example of a molecule.
   c What holds two atoms together in a molecule?
4. An oxygen molecule is shown in Figure 8.1. An ozone molecule has three oxygen atoms. Draw a diagram of what you think an ozone molecule would look like.
5. Carbon dioxide is a molecule made up of one carbon atom between two oxygen atoms—all in a straight line. Draw what you think a carbon dioxide molecule looks like.

ISBN 978 1 4202 3735 1

SCIENCE AS A HUMAN ENDEAVOUR

John Dalton (1766–1844)

John Dalton was born in 1766 and spent his childhood in a small English town. He soon became interested in mathematics and science, and when he was 12 he started a school of his own. This school seems to have been quite a success, despite the difficulty he had keeping the other children in order, especially those who were older than he was. Dalton continued teaching and lecturing throughout his life. He never married, and he said this was because his head was 'too full of triangles, chemical processes, and electrical experiments to think much of marriage'.

Dalton was a Quaker, and Quaker men and women had to dress in dark clothes. He was also colour-blind. The story is told that he once bought his mother a pair of bright scarlet stockings. He thought they were 'bluish-drab and Quakerish', and was very upset when his mother said she could not wear them because they were too bright. She had to call in a neighbour to convince her son that the stockings were bright scarlet and not bluish-drab.

Dalton made over 200 000 recorded weather observations during his life. However, his greatest achievement was his atomic theory. He did a series of experiments and hypothesised that the atoms of any one element are identical to each other but different from those of all other elements. He also suggested that chemical reactions take place through rearrangements of atoms.

Dalton imagined his atoms to be like pool balls, and he devised symbols for the different atoms. Some of his ideas later proved to be incorrect. For example, he inferred that a water molecule is made up of one oxygen atom and one hydrogen atom, instead of *two* hydrogen atoms. However, the atomic theory used today is basically the same as the theory Dalton proposed 200 years ago.

Questions

1. Which nationality was John Dalton?
2. What did he do for a living?
3. In your own words, explain why Dalton never married.
4. What was Dalton's atomic theory?
5. How did he explain chemical reactions?
6. Suggest why Quakers wore drab clothing.
7. How does a hypothesis like Dalton's become a theory?
8. Is Dalton's atomic theory the same as the particle theory you learnt about in Chapter 2? Explain.

Figure 8.4 Dalton listed various atoms and some other chemicals, giving them names and symbols.

ISBN 978 1 4202 3735 1

8.2 Elements and compounds

Elements

If you could look inside a piece of iron like Super-Sci did on page 183, you would find that it is made of millions and millions of tiny iron atoms—all the same. Similarly, a piece of copper is made of copper atoms only. But the piece of copper is different from the iron, because the copper atoms are different from iron atoms. Pure substances, such as iron and copper, whose atoms are all the same are called **elements**.

Figure 8.5 below shows a child building with Lego bricks. Thousands of different models can be built from a small number of different types of blocks. In a similar way, everything around us is made from just over one hundred different elements.

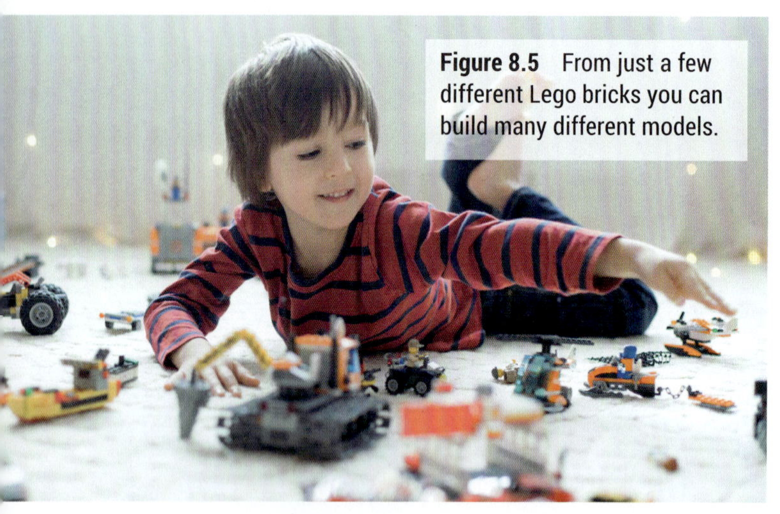

Figure 8.5 From just a few different Lego bricks you can build many different models.

Common elements

The first elements discovered were the metals gold, tin, copper and iron. Over the years more and more elements were discovered. In total, 90 elements have been found in the Earth's rocks, soil, air and water. Another 20 or so elements, which do not occur naturally, have been made by nuclear scientists. Radioactive plutonium is one of these synthetic elements. Some elements, such as gold and silver, are very rare. Other elements are very common. For example, oxygen makes up about 21% of the air.

Some common elements are listed in the table on page 188. They can be classified into two main groups—metals and non-metals. (Metals conduct electricity; most non-metals do not.) The elements can also be classified according to whether they are solids, liquids or gases at room temperature (20 °C).

Each element is represented by a **symbol**. This is a shorthand way of writing the name of the element. Sometimes the symbol is the first letter of the English name of the element; for example, carbon **C**. However, some elements have the same first letter; for example, carbon and calcium. In these cases a second letter is used: calcium **Ca**. Note that the first letter is a capital, but the second letter is not. In some cases the symbol comes from a Greek or Latin name. For example, the symbol for gold is **Au**. This comes from the Latin word aurum, which means 'shining dawn'. Some elements are named after famous people or places; for example, einsteinium and francium.

Figure 8.6 Some common elements you may know

sulfur

gold

carbon (graphite)

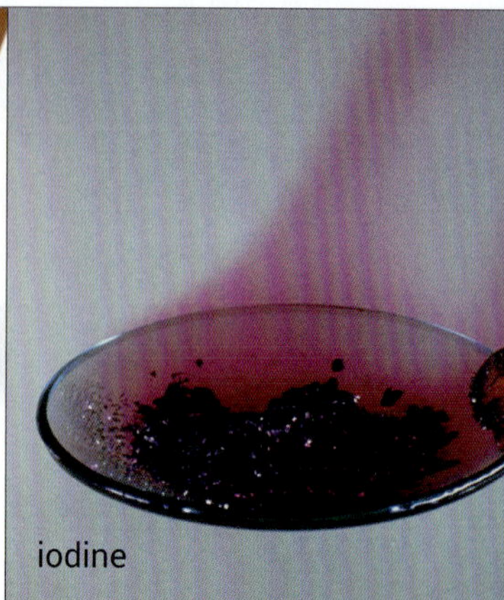

iodine

ISBN 978 1 4202 3735 1

SCIENCE AS A HUMAN ENDEAVOUR

Marie Curie (1867–1934)

Marie Curie was born in Poland in 1867. At school she was always top of her class, and she went to university in Paris. Marie and her husband Pierre, who was also a scientist, became interested in pitchblende, an ore of uranium that was radioactive. It gave off a strange new radiation, including the newly discovered X-rays. They found that it was even more radioactive than pure uranium. So what else could be in the ore that gave out radiation? Marie thought she was on the track of a new element.

Marie bought a tonne of pitchblende and had it dumped outside the shed where she worked in Paris. She and her husband ground the heap of ore to a powder, 20 kilograms at a time. They dissolved each lot of powder in acid, and evaporated the solution to form crystals. After four years of backbreaking work, Marie and Pierre had a tiny pile of white crystals a little bigger than the head of a pin. These crystals contained a new element called *radium*. In the dark it glowed with a bluish light.

Whenever Marie worked with radioactive radium, her hands became covered with sores, burns and blisters. This led to the discovery that radium can be used to kill diseased cells in cancer tumours. Even though the gram of radium she had extracted was worth millions of dollars, she gave it to her university. During World War I, Marie organised mobile X-ray vans so that pieces of shells in wounded soldiers could be found and removed.

In 1934 Marie Curie died of leukaemia, a disease probably caused by the radioactive materials she had worked with. She was the first woman to receive a Nobel Prize—one in physics and one in chemistry. During her life she had discovered two new elements—radium and polonium. In 1946 American scientists discovered another radioactive element. It was called *curium* in honour of Marie and Pierre Curie.

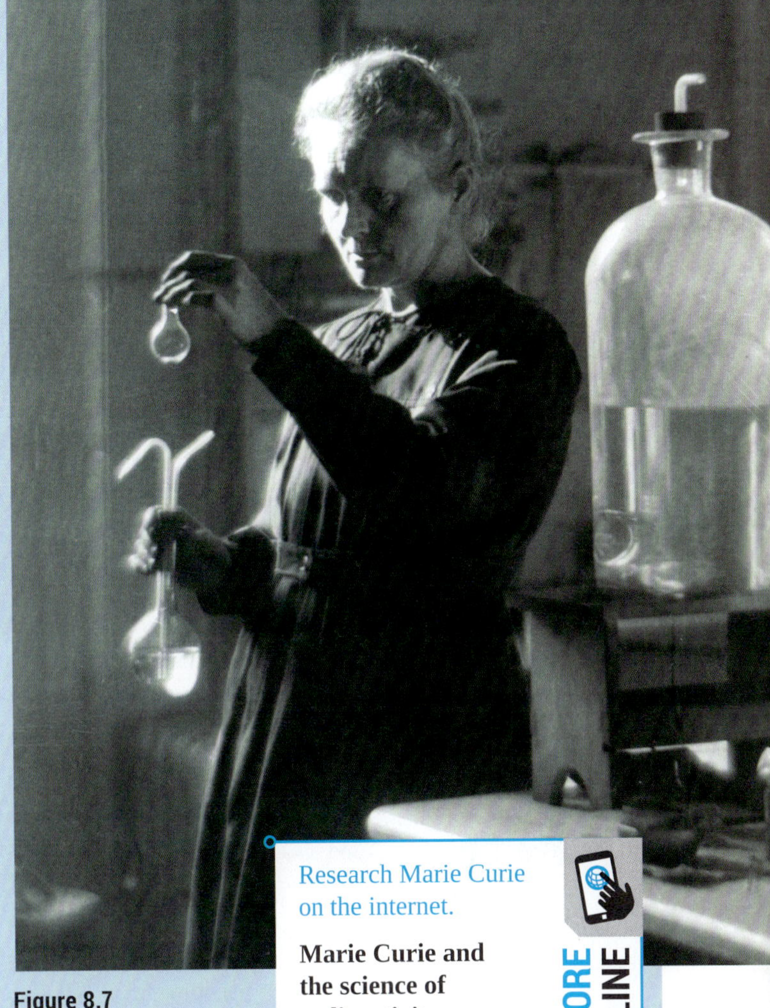

Figure 8.7 Marie Curie in her laboratory

EXPLORE ONLINE

Research Marie Curie on the internet.

Marie Curie and the science of radioactivity.

Her life in detail, well-illustrated.

Questions

1. When and where was Marie Curie born?
2. What was radium used for?
3. What was the name of the ore from which she obtained radium?
4. Suggest why Marie named one of the elements she discovered *polonium*.
5. Which new element was named in honour of Marie and Pierre Curie after their deaths?
6. Suggest how Marie could have protected herself from radiation.

ISBN 978 1 4202 3735 1

ACTIVITY

Use the table below to answer these questions.

1 Write down the symbols for the following elements:

calcium	iron	nitrogen
carbon	lead	oxygen
hydrogen	magnesium	sodium

2 Which elements have the following symbols?

Al	Au	Br	Cl	Cu
Hg	Fe	P	S	Zn

3 Which one of the elements in the table has the highest melting point?

4 Which is the most recently discovered element in the table? When was it discovered?

5 Which of the elements are solids, which are liquids, and which are gases?

solids: melting point and boiling point above 20 °C (room temperature)

liquids: melting point below 20 °C, but boiling point above 20 °C

gases: melting point and boiling point below 20 °C

Put your answers in a table.

6 Are metals usually solids, liquids or gases at room temperature?

7 Which is the lightest gas?

8 Find the 22 elements listed on this page in the periodic table opposite. Name three that are in the same vertical group.

Element	Symbol	Metal or non-metal	Melting point (°C)	Boiling point (°C)	Density (g/cm³)	Date of discovery
aluminium	Al	metal	660	2060	2.7	1825
argon	Ar	non-metal	−189	−188	0.0017	1894
bromine	Br	non-metal	−7	58	3.1	1826
calcium	Ca	metal	850	1440	1.6	1808
carbon	C	non-metal	3500	4200	2.2	ancient
chlorine	Cl	non-metal	−101	−35	0.003	1774
copper	Cu	metal	1080	2500	9.0	ancient
gold	Au	metal	1060	2700	19.3	ancient
hydrogen	H	non-metal	−259	−253	0.00008	1766
iodine	I	non-metal	114	183	4.9	1811
iron	Fe	metal	1540	3000	7.9	ancient
lead	Pb	metal	327	1744	11.3	ancient
magnesium	Mg	metal	650	1110	1.7	1808
mercury	Hg	metal	−39	357	13.6	ancient
nitrogen	N	non-metal	−210	−196	0.00117	1772
oxygen	O	non-metal	−219	−183	0.00132	1774
phosphorus	P	non-metal	44	280	1.8	1669
plutonium	Pu	metal	640	3230	19.8	1940
silver	Ag	metal	961	2200	10.5	ancient
sodium	Na	metal	98	890	0.97	1807
sulfur	S	non-metal	119	444	2.1	ancient
zinc	Zn	metal	419	910	7.1	1700

ISBN 978 1 4202 3735 1

Periodic table of the elements

Elements in the same vertical column have similar properties

Key

6 C Carbon

- atomic number (6)
- symbol (C)
- name of element (Carbon)

The colour of the name for each element indicates its state at room temperature:

- black—solid
- blue—liquid
- pink—gas
- red—synthetic

Chemical families

- Alkali metals
- Alkaline earth metals
- Transition metals
- Rare earth metals
- Other metals
- Metalloids
- Other non-metals
- Halogens
- Noble gases

1	2	3	4	5	6	7	8	9	10	11	12	13	14	15	16	17	18
1 H Hydrogen																	2 He Helium
3 Li Lithium	4 Be Beryllium											5 B Boron	6 C Carbon	7 N Nitrogen	8 O Oxygen	9 F Fluorine	10 Ne Neon
11 Na Sodium	12 Mg Magnesium											13 Al Aluminium	14 Si Silicon	15 P Phosphorus	16 S Sulfur	17 Cl Chlorine	18 Ar Argon
19 K Potassium	20 Ca Calcium	21 Sc Scandium	22 Ti Titanium	23 V Vanadium	24 Cr Chromium	25 Mn Manganese	26 Fe Iron	27 Co Cobalt	28 Ni Nickel	29 Cu Copper	30 Zn Zinc	31 Ga Gallium	32 Ge Germanium	33 As Arsenic	34 Se Selenium	35 Br Bromine	36 Kr Krypton
37 Rb Rubidium	38 Sr Strontium	39 Y Yttrium	40 Zr Zirconium	41 Nb Niobium	42 Mo Molybdenum	43 Tc Technetium	44 Ru Ruthenium	45 Rh Rhodium	46 Pd Palladium	47 Ag Silver	48 Cd Cadmium	49 In Indium	50 Sn Tin	51 Sb Antimony	52 Te Tellurium	53 I Iodine	54 Xe Xenon
55 Cs Caesium	56 Ba Barium		72 Hf Hafnium	73 Ta Tantalum	74 W Tungsten	75 Re Rhenium	76 Os Osmium	77 Ir Iridium	78 Pt Platinum	79 Au Gold	80 Hg Mercury	81 Tl Thallium	82 Pb Lead	83 Bi Bismuth	84 Po Polonium	85 At Astatine	86 Rn Radon
87 Fr Francium	88 Ra Radium		104 Rf Rutherfordium	105 Db Dubnium	106 Sg Seaborgium	107 Bh Bohrium	108 Hs Hassium	109 Mt Meitnerium	110 Ds Darmstadtium	111 Rg Roentgenium	112 Cn Copernicium	113 Nh Nihonium	114 Fl Flerovium	115 Mc Moscovium	116 Lv Livermorium	117 Ts Tennessine	118 Og Oganesson

57 La Lanthanum	58 Ce Cerium	59 Pr Praseodymium	60 Nd Neodymium	61 Pm Promethium	62 Sm Samarium	63 Eu Europium	64 Gd Gadolinium	65 Tb Terbium	66 Dy Dysprosium	67 Ho Holmium	68 Er Erbium	69 Tm Thulium	70 Yb Ytterbium	71 Lu Lutetium
89 Ac Actinium	90 Th Thorium	91 Pa Protactinium	92 U Uranium	93 Np Neptunium	94 Pu Plutonium	95 Am Americium	96 Cm Curium	97 Bk Berkelium	98 Cf Californium	99 Es Einsteinium	100 Fm Fermium	101 Md Mendelevium	102 No Nobelium	103 Lr Lawrencium

ISBN 978 1 4202 3735 1

INVESTIGATION 8.1 Flame tests

Aim

Different metals produce different colours in a flame. The aim of this investigation is to identify various metallic elements using flame tests.

Materials

- Bunsen burner
- small atomiser bottles containing 0.5 M solutions of the following or other soluble metal salts:

barium chloride	potassium chloride
calcium chloride	sodium chloride
copper sulfate	strontium chloride

- unknown metal solution (step 4)

Risk assessment and planning

- Draw up a suitable data table to record your results.
- What safety precautions will be necessary?

Method

1 Light the burner and adjust it to the blue flame.

2 Hold the first atomiser bottle just below the top of the burner, about 20 cm away from the flame. Spray the mist so that it goes up into the flame and observe the flash of colour as the solution vaporises. (You may need to repeat this if you did not see the colour clearly.)

3 Repeat the procedure with the atomisers of the other solutions.

Record the flame colours for the different metals in your data table.

4 Now that you know the colour each metal produces in a flame, your teacher will give you an unknown metal salt. Test it and infer which metal it contains.

Particles in elements

Metals, such as gold, are composed of collections of single atoms. In non-metals the atoms are bonded together. For example, diamond consists of carbon atoms, each linked to four other carbon atoms, as shown in Figure 8.8.

Some gaseous elements contain separate molecules. For example, the molecules hydrogen (H_2), nitrogen (N_2), oxygen (O_2) and chlorine (Cl_2) each contain a pair of atoms. The gas ozone (O_3), which protects us from UV radiation from the sun, has a molecule containing *three* atoms of oxygen.

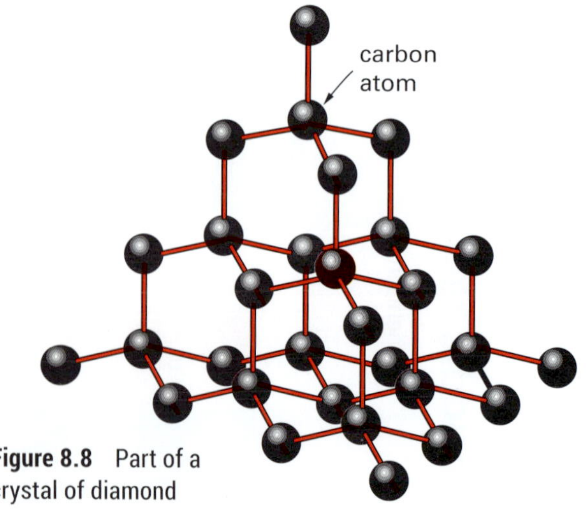

Figure 8.8 Part of a crystal of diamond

ISBN 978 1 4202 3735 1

SCIENCE AS A HUMAN ENDEAVOUR

Fireworks

The Chinese were probably the first to use fireworks when they discovered how to make the black powder we call gunpowder in about 850 CE. They wrapped the black powder in bamboo or paper tubes to make crude missiles and flares that could be used to frighten away potential invaders. This was the beginning of pyrotechnics, which means 'the art of making fireworks'.

The Italians were probably the first to experiment with coloured fireworks in the early 1700s. The white colours of fireworks are due to the metals aluminium and magnesium burning at about 3000 °C. The gold colours are due to iron and charcoal burning at a lower temperature. The other colours are produced by adding small amounts of other metals. For example, barium gives you a green colour, copper gives you blue and strontium gives you red.

A fireworks shell is a cardboard sphere filled with hundreds of little black balls called stars, which contain the colour-producing elements. The stars are surrounded by the bursting charge. Multiple-burst shells are designed with several separate compartments. At the bottom of the shell is a compartment that contains the black powder lift charge.

To set off the fireworks, pyrotechnicians place the shell in a plastic cylinder and light the main fuse. This in turn lights the side fuse that ignites the lift charge at the bottom. This propels the shell high into the sky, where the time-delay fuse ignites the bursting charges to propel the stars out of the shell.

Andrew and Christian Howard are brothers and the directors of Howard and Sons Pyrotechnics, who light up our skies with spectacular fireworks displays.

Andrew's interest in fireworks was sparked at the age of seven, when his father took over the business from his grandfather. When asked how he got started, Andrew said 'There are TAFE courses that specialise in high explosives, but not in fireworks. There are no textbooks covering our trade, so basically we learn from the dos and don'ts. We abide by some fairly strict safety regulations, particularly to ensure no spectators or staff are injured.' Christian has arranged the special effects in movies such as *The Matrix* and live shows for bands such as Metallica and Pink Floyd.

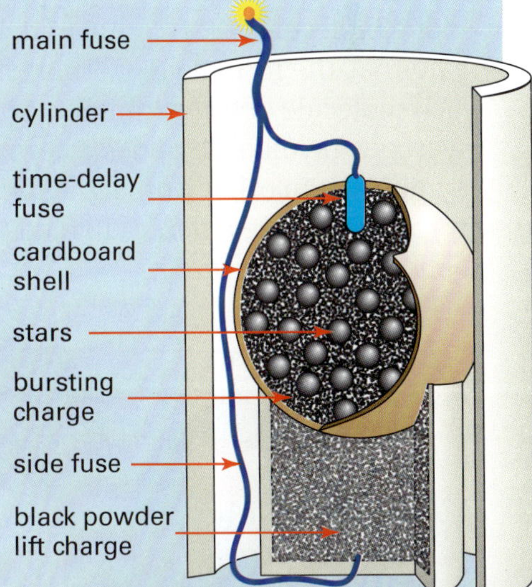

Figure 8.9 The design of a fireworks shell

For great fireworks photos, go to **Howard and Sons Pyrotechnics**.

ISBN 978 1 4202 3735 1

Compounds

If you look at Figure 8.2 on page 184 you will see that water molecules contain two different kinds of atoms—hydrogen and oxygen. Similarly, molecules of the poisonous gas carbon monoxide contain one carbon atom combined with one oxygen atom. And molecules of carbon dioxide (which plants use in photosynthesis) contain one carbon atom combined with two oxygen atoms. Substances that are made of two or more different kinds of atoms are called **compounds**.

Chemical formulae

A **chemical formula** is a shorthand way of showing which elements are in a compound. It also tells you how many atoms of each element are present in one molecule of the compound. For example, water has the formula H_2O. This tells you that each molecule of water contains two atoms of hydrogen (symbol H) and one atom of oxygen (symbol O). In other words, the hydrogen and oxygen are in the *ratio* 2:1. To read the formula aloud you say 'H two O'.

Sodium chloride (common salt) has the formula NaCl. To read such a formula you say the letters in order: N-a-C-l. Sodium chloride is a solid compound and has the structure shown below, with sodium and chloride particles packed together tightly. There are no separate molecules, but the formula tells you that there are equal numbers of sodium and chlorine atoms.

Iron oxide (rust) has the formula Fe_2O_3 (F-e-two-O-three). It has a similar structure to sodium chloride, but there are two particles of iron for every three particles of oxygen. (The iron and oxygen are in the ratio 2:3.)

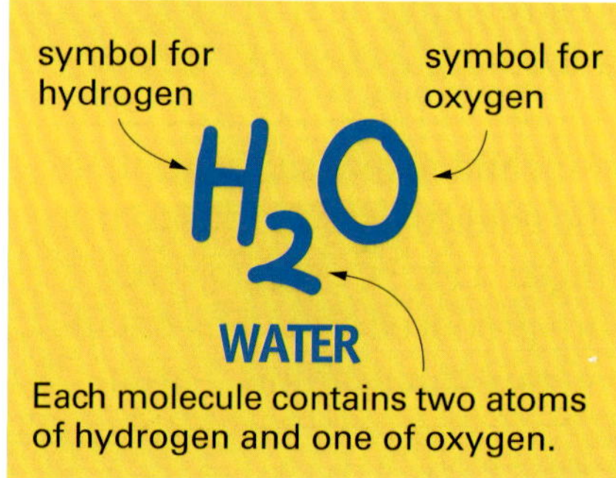

Figure 8.10 Chemical formula for water

Figure 8.11 Some compounds you know.

ISBN 978 1 4202 3735 1

ACTIVITY

Molecular models

Because atoms and molecules are so small, we use models to represent them. There are two main types of molecular models. In both types the atoms are represented by coloured balls of different sizes. Different colours represent different atoms.

In ball-and-stick models (see Figure 8.8 on page 190) the balls are joined by sticks to form molecules. There are no such sticks connecting atoms—the sticks merely represent the bonds between the atoms. When you use these models you will notice that the bonds between atoms are at definite angles.

The other type of model is the space-filling type (Figure 8.2 on page 184), in which the atoms fit together at the correct angles.

Make models to represent these molecules:

hydrogen (H_2)	ammonia (NH_3)
oxygen (O_2)	methane (CH_4)
water (H_2O)	carbon dioxide (CO_2)

Draw a diagram of each molecule, labelling the different atoms.

Examine the models to see how many bonds each type of atom can form. Record your results in a table.

Living and non-living

All things, whether living or non-living, are made up of elements and compounds of these elements. Common non-living things such as salt and sugar are usually compounds. For example, salt (sodium chloride) is a compound of sodium and chlorine. Sugar is a compound of carbon, hydrogen and oxygen. Many substances, such as petrol, are a mixture of a number of different compounds.

Living things contain a large number of different compounds, some very simple (for example water), and others very complex (for example proteins, fats and carbohydrates). These compounds are made up of about twenty essential elements, as shown in Figure 8.12.

About 65% of your body mass is water (hydrogen and oxygen). The proteins, fats and carbohydrates all contain carbon. Hence the high proportion of oxygen, carbon and hydrogen in your body.

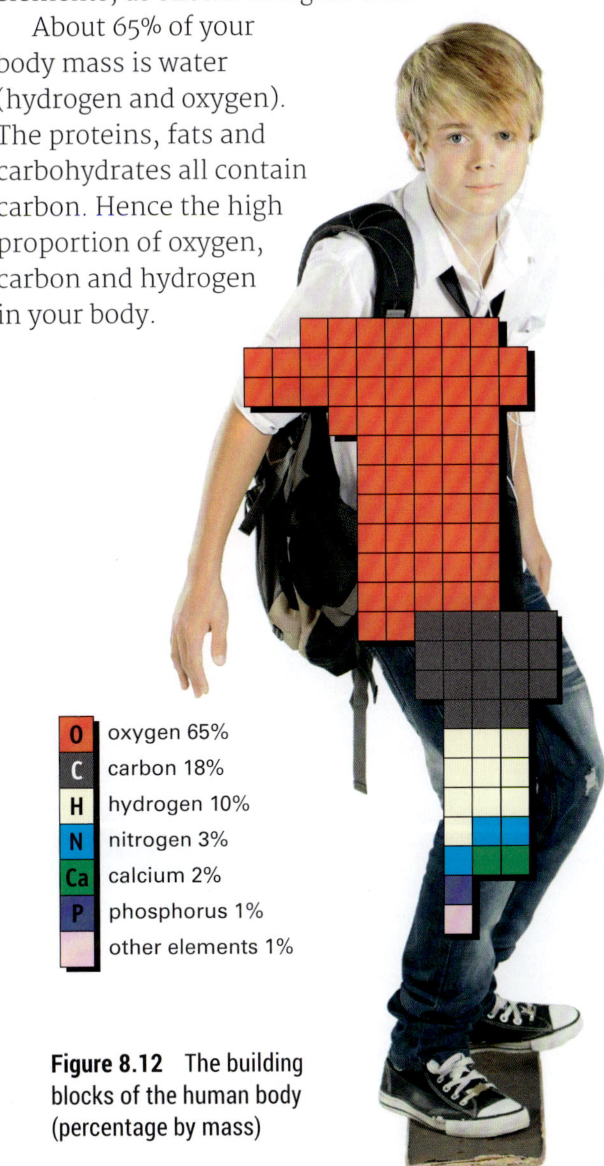

Figure 8.12 The building blocks of the human body (percentage by mass)

ISBN 978 1 4202 3735 1

DNA—a complex compound

The basis of life is a compound called DNA, which determines what you are like. It contains only the elements carbon, hydrogen, oxygen, nitrogen and phosphorus. However, DNA is very complex, and the various atoms can be combined in millions of different ways. The result is that there are millions of different types of DNA. What makes you different from everybody else is the way in which the atoms in the DNA in your body are put together.

What is life?

All matter can be divided into living and non-living things. Cells are the building blocks for living things. But cells and all non-living things are made of elements and compounds, which in turn are made of atoms and molecules. The salt in your body is the same as the salt on the kitchen table. And the calcium carbonate in an eggshell and in your bones is the same as the calcium carbonate in limestone.

Just what gives a living thing life is not well understood. In the 19th century scientists said that living things contained a mysterious 'vital force'. However, it is now known that living things contain very complex compounds. Somehow life is associated with these complex compounds.

Scientists have been able to work out the structures of living substances. In fact, they have even been able to make quite complex substances in the laboratory. One such substance is insulin, one of the smallest proteins. Some day scientists may be able to create life itself!

Some common compounds

Compound	Formula	Notes
methane	CH_4	in natural gas used for cooking
silicon dioxide	SiO_2	commonly known as sand
sugar (glucose)	$C_{12}H_{22}O_{11}$	produced by plants during photosynthesis
hydrochloric acid	HCl	used for cleaning concrete, and many industrial processes
hydrogen sulfide	H_2S	'rotten egg' gas
sodium bicarbonate	$NaHCO_3$	bicarbonate soda or baking soda, used in cooking
hydrogen peroxide	H_2O_2	used in hair bleach
octane	C_8H_{18}	the main component of unleaded petrol
ammonium hydroxide	NH_4OH	commonly known as ammonia, used as a glass cleaner
dinitrogen monoxide (nitrous oxide)	N_2O	laughing gas, used as an anaesthetic

Figure 8.13 A model of a small section of the complex DNA molecule: oxygen—red, carbon—black, hydrogen—white, nitrogen—blue, phosphorus—purple

ISBN 978 1 4202 3735 1

CHECK

1 Ask someone to check your spelling of these words:

carbon dioxide	hydrogen
compound	molecule
element	sodium chloride
formula	symbol

2 What is the difference between:
  a an atom and a molecule?
  b an atom and an element?

3 Which of the following are elements? aluminium, carbon monoxide, copper, iron oxide, kerosene, mercury, phosphorus, sand, sugar, water

4 Which element is present in all of the following compounds?

SO_2

H_2S

H_2SO_4

$CuSO_4$

5 Suppose you represent the atoms in three different elements by ●, ■ and ▲. How many different molecules could you form by linking these atoms together.
  a two at a time?
  b three at a time?

CHALLENGE

1 Use Figure 8.12 on page 193 to draw a bar graph and a pie chart of the elements in the human body.

2 Why are there so many more compounds than elements?

3 Draw up a table with two columns headed 'Elements' and 'Compounds'. Put each of the following into the correct column:

Al	SiO_2
CO_2	Cu
N_2	NH_3
H_2SO_4	O_3
A_u	S_8
$AlPO_4$	HCl

4 Write the formula for each of the following molecules:
  a nitrogen dioxide contains one nitrogen atom and two oxygen atoms
  b propane contains three carbon atoms and eight hydrogen atoms
  c glucose contains six carbon atoms, twelve hydrogen atoms and six oxygen atoms.

5 A tiny crystal of magnesium chloride contains 2 billion magnesium atoms and 4 billion chlorine atoms. What is the formula for the compound?

EXPLORE ONLINE

Select an element and use library resources to find out what you can about it. Here are some things you might look up:

1 name of discoverer, date of discovery, how the element was named

2 properties and uses of the element.

Here are some useful links:

Web Elements

Select an element for a range of information including photos, cartoons and audio descriptions.

CHEM4KIDS

Simple information on the first 36 elements, with puzzles and help with pronouncing their names.

The Visual Elements Periodic Table

Very colourful and interactive

It's elemental

Below the table you can click on a number of online games based on the elements.

The Periodic Table of Comic Books

Click on an element to see a list of comic book pages involving that element.

ISBN 978 1 4202 3735 1

8.3 Chemical reactions

Over the years scientists have experimented with substances—mixing them, heating them and passing electricity through them. For example, a chemical reaction occurs when sugar is heated. It splits into two simpler substances—water and carbon (which is black). Scientists discovered that carbon and other elements cannot be split into anything simpler by chemical reactions. They cannot be split because they contain only one sort of atom.

Using chemical reactions, scientists are also able to make many new compounds. For example, when carbon is heated it reacts with the oxygen in the air to form the compound carbon dioxide.

Most substances are not pure, but are mixtures of two or more different substances (elements or compounds) that are not chemically combined. Air is a good example of a mixture. It contains many different elements and compounds whose proportions are not always the same.

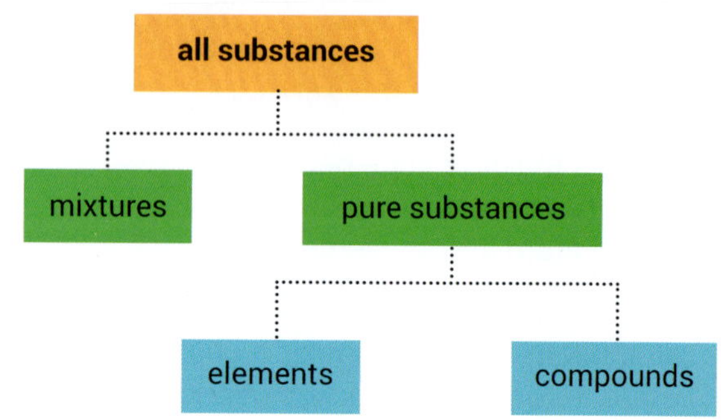

In Investigation 8.2 you can use a chemical reaction to make a compound. Then in Investigation 8.3 you can break a compound down into its elements.

INVESTIGATION 8.2 Making a compound

Aim

To make a compound from the elements iron and sulfur.

Risk assessment and planning

- Read through the experiment and note the places where safety precautions will be necessary.
- Discuss the experiment with your teacher. Only one group at a time can use the fume cupboard, and your teacher may prefer to demonstrate all or part of the experiment.

Materials

- powdered sulfur
- iron powder
- dilute hydrochloric acid (1 M)
- spatula
- bar magnet
- 3 small test tubes
- Bunsen burner
- tripod and heatproof mat
- aluminium foil
- crucible
- pipeclay triangle

PART A Testing iron and sulfur

Method

1 Place a small amount of iron powder in a test tube. Use a magnet as shown to test whether you can pull the iron powder up the side of the test tube. If you can, then the iron powder is magnetic.

Test some sulfur in the same way.

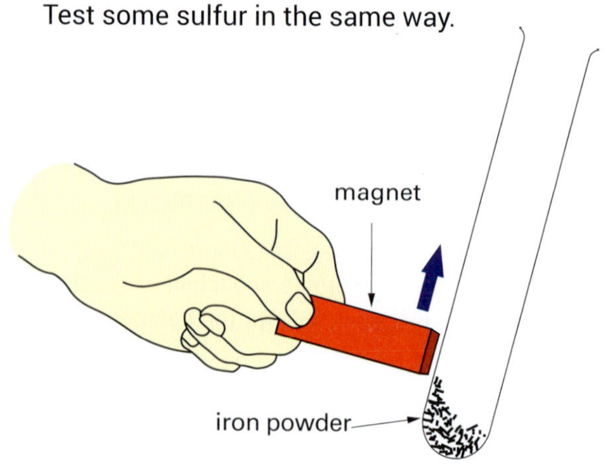

ISBN 978 1 4202 3735 1

2 Add a few drops of dilute hydrochloric acid to some iron powder in a test tube.

Observe what happens.

Do the same with the sulfur.

3 Put two spatulas of sulfur in another test tube, then two spatulas of iron powder. Mix them well by shaking the test tube.

Test the mixture with the magnet. What do you notice?

Discussion

1 Did the properties of the iron and sulfur change when you mixed them?

2 Was there a chemical reaction when you mixed them?

3 Have the iron and sulfur formed a mixture or a compound? Explain your answer.

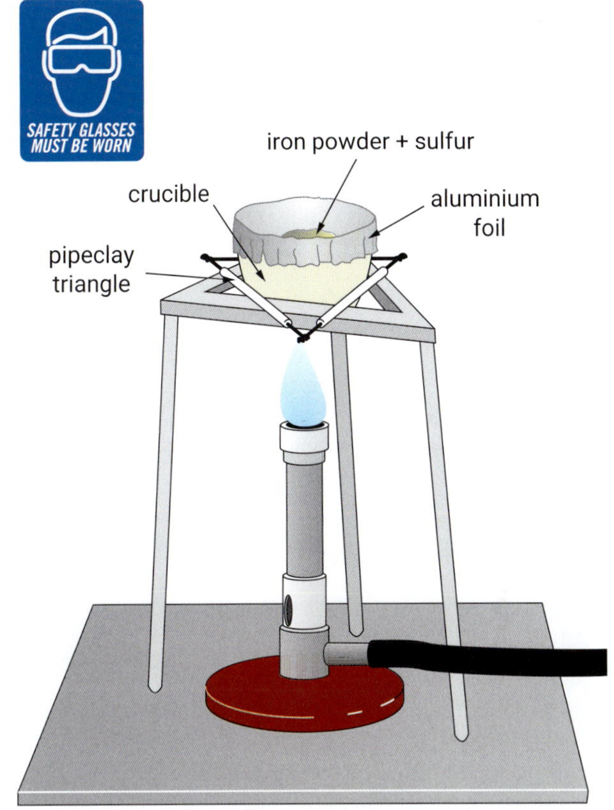

PART B Making iron sulfide

Method

1 Line the crucible with some aluminium foil and pour in the mixture of iron and sulfur. (The aluminium foil doesn't react—it simply stops the hot mixture sticking to the crucible.)

Caution: The fumes from burning sulfur are poisonous. It is essential to use a fume cupboard so that you don't breathe in any of the fumes.

2 Put the crucible in a pipeclay triangle on a tripod, as shown. Heat it with a Bunsen burner. As soon as the mixture begins to glow, turn off the burner.

3 When the crucible has cooled examine the new substance that has formed.

Describe the properties of the new substance.

Is the substance magnetic?

Can you separate the iron and sulfur?

4 Add some dilute hydrochloric acid to a small piece of the substance in the crucible. (This should also be done in a fume cupboard as the 'rotten egg' gas given off is poisonous.)

Record your observations.

Is this the same gas that was formed when you added hydrochloric acid to iron powder? How could you tell?

Discussion

1 Did the properties of the iron and sulfur change when you heated the mixture? Explain.

2 Was there a chemical reaction? How do you know?

3 What was needed to make the reaction occur?

4 Have the iron and sulfur formed a mixture or a compound? Explain your answer.

5 The compound you have made is called *iron sulfide*. What are the elements in it?

ISBN 978 1 4202 3735 1

Breaking a compound

Aim

To find out what substances are produced when water is decomposed (split up) by passing electricity through it.

Materials

Corrosive

- dilute **sulfuric acid** (1 M)
- voltameter (vol-TAM-e-ter)
- 2 pyrex test tubes
- wooden splint, e.g. paddle-pop stick
- distilled water
- power pack
- taper

Risk assessment and planning

Read through the investigation. A voltameter is an expensive piece of equipment and the school probably has only one. So your teacher will probably set up the equipment for you.

Method

1 Set up a voltameter as shown.

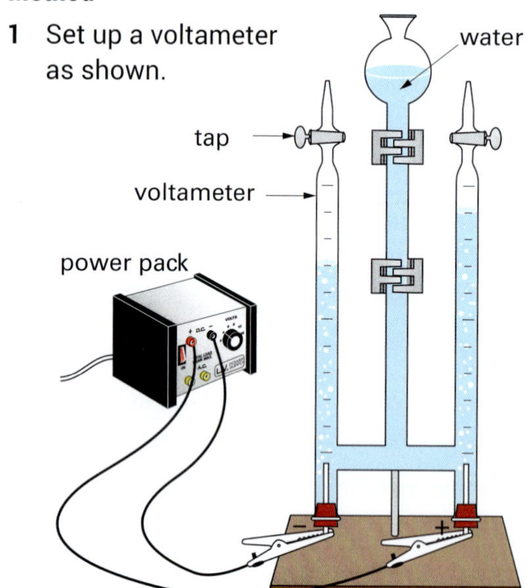

2 Open the taps at the top and add water containing a few millilitres of dilute sulfuric acid to the middle tube. (The acid makes the water conduct electricity more easily.) When the side tubes are full, close the taps.

3 Connect the voltameter to a power pack set on 6 volts DC. Turn it on.

4 Allow the current to flow for about 15 minutes, and observe the gases that collect in the tubes.

Compare the volumes of the gases in the two tubes.

5 Invert a dry test tube over the tube with the most gas in it. Then open the tap and collect the gas.

Light a taper, tilt the test tube upwards, and put the burning taper near its mouth.

A 'pop' indicates the gas is hydrogen.

After the 'pop', look for water droplets inside the tube. Infer where they came from.

6 Collect a test tube full of gas from the other voltameter tube. Light the wooden splint, then blow it out so that it has a glowing tip. Put the glowing splint into the test tube. If it bursts into flame this indicates the gas is oxygen.

Discussion

1 What were the two gases produced when electricity was passed through water?

2 Copy and complete this sentence. Water is a compound of the elements _____ and _____.

3 When hydrogen burns it combines with the oxygen in the air. Infer what substance is formed. (See step 5 above.)

4 The volume of hydrogen produced was twice the volume of oxygen. Suggest a reason for this.

ISBN 978 1 4202 3735 1

Chemical equations

When you mixed iron and sulfur and heated the mixture, a chemical reaction occurred. The iron and sulfur were the reactants—the substances you started with. The iron sulfide was the product of the reaction. The equation for the reaction is:

iron	+	sulfur	→	iron sulfide
Fe	+	S	→	FeS

The reactants and products in a reaction can be solids, liquids or gases. For example, in Investigation 8.3 liquid water decomposed to produce hydrogen gas and oxygen gas.

water	→	hydrogen	+	oxygen
$2H_2O$	→	$2H_2$	+	O_2

During this reaction, molecules of water break apart and form molecules of hydrogen and oxygen.

Water molecules always contain two atoms of hydrogen bonded to every atom of oxygen. So when water decomposes, two molecules of hydrogen are formed for every molecule of oxygen, as Figure 8.14 shows. This is why the volume of hydrogen gas produced in Investigation 25 was exactly twice the volume of oxygen gas produced. Similarly, when a compound is made, exact quantities of the different elements react.

The atoms in some molecules are more tightly bonded together than in other molecules. As a result, when molecules are rearranged during a chemical reaction, energy may be needed or energy may be released. The reaction between iron and sulfur needed heat to make it go, and the decomposition of water needed electricity. On the other hand, burning produces heat, and the reactions that occur inside batteries produce electricity.

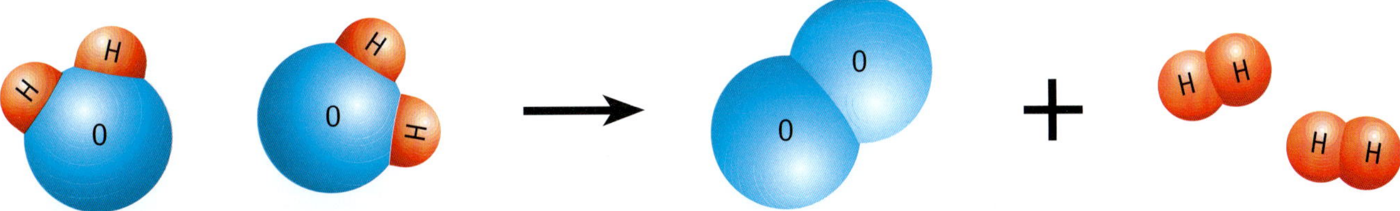

Figure 8.14 The compound water decomposes to form the elements hydrogen and oxygen.

SCIENCE AS A HUMAN ENDEAVOUR

Hydrogen—fuel of the future

The problem of cars

There are two major problems with our use of cars. First, they use petrol which is made from oil—a fossil fuel. Oil is a non-renewable resource and it won't be long before it is all used up. Second, when petrol is burnt in the engine of a car, the exhaust contains carbon dioxide—a greenhouse gas that many scientists infer is responsible for global warming. Electric cars solve the second problem—they don't produce carbon dioxide. But they need electricity to charge their batteries, and most electricity is produced by burning coal, which produces carbon dioxide.

Hydrogen cars

One solution may be hydrogen cars. In Investigation 8.3 you learnt that hydrogen can easily be produced by passing electricity through water. This process is called **electrolysis** (ee-lek-TROL-e-sis), which is reversible.

water	⇌	hydrogen	+	oxygen
$2H_2O$	⇌	$2H_2$	+	O_2

ISBN 978 1 4202 3735 1

A **fuel cell** makes this reaction go *backwards*—it combines hydrogen and oxygen to produce water and electricity. A hydrogen car has a fuel tank containing hydrogen, which reacts in the fuel cell with oxygen from the air to produce electricity. The electricity then powers an electric motor. The big advantage of a hydrogen car is that the exhaust contains only water, which doesn't cause pollution. So why aren't we using hydrogen cars? The answer is that there are many problems to solve before hydrogen cars become a reality on our roads.

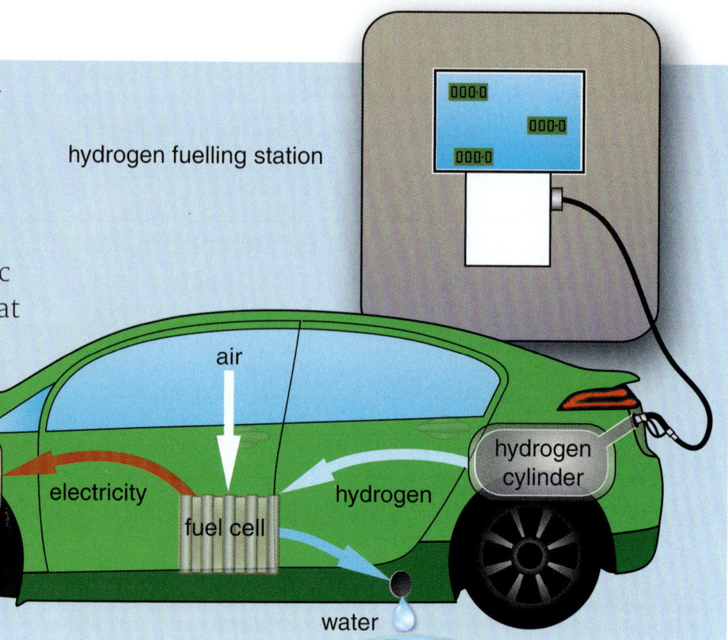

Figure 8.15 A hydrogen car

Making hydrogen fuel

There is no shortage of hydrogen around us. It is by far the most common element in the universe. Our sun and all the other stars are about three-quarters hydrogen. So are the outer planets in our solar system. The only problem is that hydrogen gas does not occur naturally on Earth. It is such a reactive element that it is always combined with other elements. For example, water (H_2O) is a compound of hydrogen and oxygen. Methane (CH_4), in natural gas, is a compound of hydrogen and carbon. However, scientists have worked out how to use chemical reactions to manufacture hydrogen. For example, if you react natural gas with steam at high temperature (900 °C) you end up with a mixture of hydrogen and carbon dioxide. This is how most hydrogen is made now. However, this process still uses a fossil fuel, and still produces carbon dioxide.

Figure 8.16 A solar-powered hydrogen fuelling station

ISBN 978 1 4202 3735 1

A much better way to make hydrogen is to use electrolysis, but this uses a lot of electricity. This isn't such a problem, though, if the electricity has been generated using renewable resources. Figure 8.16 shows a hydrogen fuelling station in Germany where solar power is used to produce liquid hydrogen. At the University of Queensland researchers are using microscopic algae to produce hydrogen. The University of New South Wales is investigating the use of sunlight to split water into hydrogen and oxygen.

Storing hydrogen

Once the hydrogen has been generated it has to be stored. Because hydrogen is a gas at normal temperatures it must be compressed at high pressure in a car's fuel tank, otherwise refuelling is needed every few kilometres. Also, the tanks needed to store compressed hydrogen are heavy and expensive. The hydrogen-powered marathon car at the 2000 Sydney Olympics ran on *liquid* hydrogen. To make liquid hydrogen you need to cool the gas to −253 °C and keep it at this temperature. Liquid hydrogen, however, provides only a quarter of the energy of the same volume of petrol. This means you need a tank four times the size of a normal petrol tank to drive the same distance. To overcome this problem scientists are experimenting with squashing the hydrogen into the spaces between atoms in metal compounds called *hydrides*. Heat is then needed to release the hydrogen. They are also experimenting with storing the hydrogen inside carbon nanotubes. These are long cylindrical molecules made only of carbon atoms (see Figure 8.17).

Hydrogen can be transported using special tankers, or by pipelines, but installing a network of pipelines is expensive—$300 000 to $600 000 per kilometre. In some parts of the world they have started building 'hydrogen highways' with hydrogen filling stations along major highways.

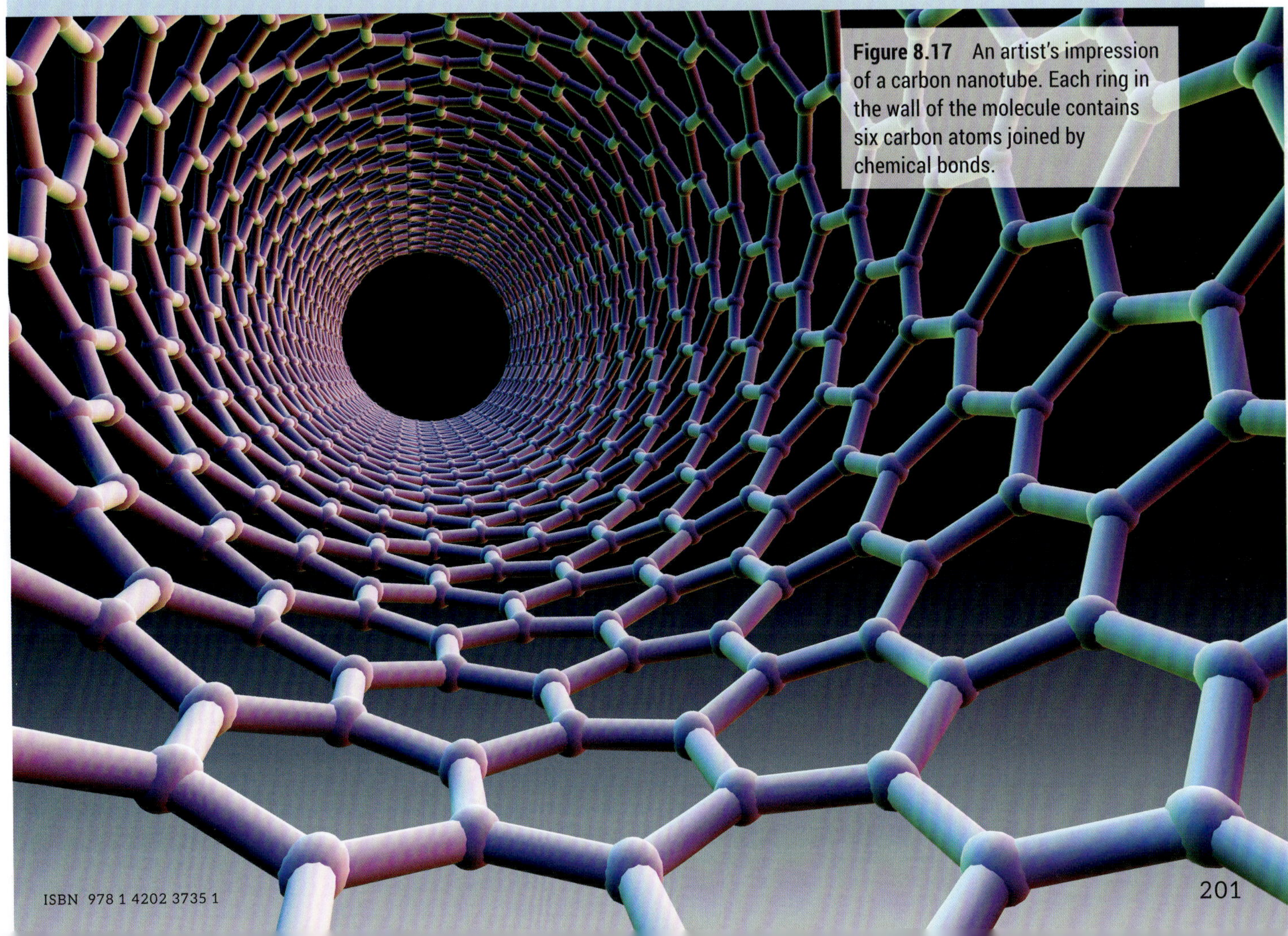

Figure 8.17 An artist's impression of a carbon nanotube. Each ring in the wall of the molecule contains six carbon atoms joined by chemical bonds.

ISBN 978 1 4202 3735 1

Questions

To make sure you understand what you have read you will answer questions at three levels—each level becoming more and more difficult.

Level 1—Comprehension

For these questions you can find the answers in the text on the last three pages.

1 What is produced when you react hydrogen with oxygen?

2 How is most hydrogen made now?

3 What is electrolysis?

4 What are 'hydrogen highways'?

Level 2—Application

Here you are asked what you think having read the text.

1 Why are hydrogen cars better for the environment than petrol-fuelled cars?

2 What is the main problem with using liquid hydrogen in cars?

3 Hydrogen does not exist naturally as an element—only as compounds. List two of these compounds.

4 Is hydrogen safe to use in cars? Explain.

Level 3—Analysis

Here you have to use your understanding of the text and your own knowledge in a wider context.

1 What are the reactants in a hydrogen fuel cell? What are the products?

2 What is the chemical formula for water? What does this mean?

3 Explain how hydrogen can be made using solar power.

4 List three problems that must be overcome before hydrogen cars are common on our roads. Which do you think is the biggest problem? Why?

Figure 8.18 A hydrogen fuelling station in California

CHECK

1 State whether each of the following statements is true or false. Rewrite those that are false to make them correct.
   a New substances are produced in a chemical reaction.
   b The rusting of iron is a chemical reaction.
   c Hydrogen is another name for water.
   d Hydrogen sulfide contains two elements—hydrogen and sulfur.
   e In a mixture, the parts can be separated only by a chemical reaction.
   f Compounds cannot be broken down into simpler substances.
   g The same elements can combine to form many different compounds.

2 Classify the following substances as elements, compounds or mixtures.

air	protein
copper	pure water
hydrogen	rust
iron oxide	soft drink
mercury	sulfur dioxide

3 Sodium is a soft silvery metal that reacts violently with water, and chlorine is a poisonous green gas. Sodium chloride (common salt) contains these two elements, but is safe to eat. How can you explain this?

4 Pure substance X is a green powder. When heated, it gives off a gas and changes to a black powder. Is substance X an element or a compound? Give reasons.

ISBN 978 1 4202 3735 1

CHALLENGE

1 Diagrams A to D below represent the particles in different substances. Which represents:

a an element?
b a compound?
c a mixture of elements?
d a mixture of compounds?

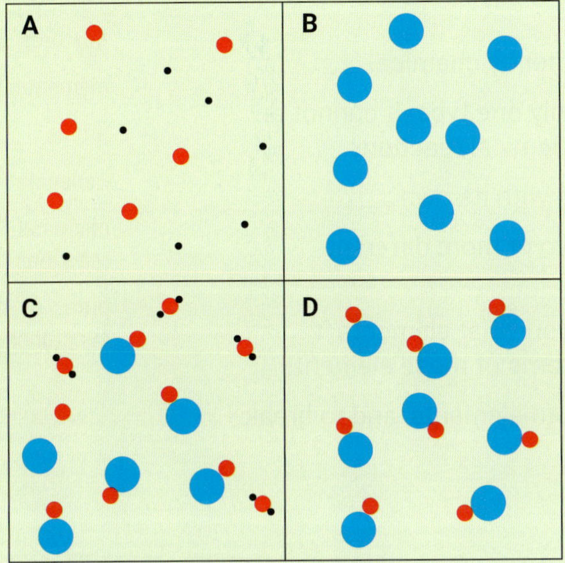

2 a What substance do you predict will be formed when hydrogen and oxygen react together? Explain your prediction.

b Write a word equation for the reaction.

3 Tamara heated a white powder, and two different gases were given off. One was a poisonous brown gas called nitrogen dioxide, and the other was oxygen. A red substance was left behind in the test tube.

a Is the white powder an element or a compound?
b Which elements can Tamara be sure are in the white powder?
c When Tamara continued to heat the red substance, she was left with a silvery liquid called mercury. More oxygen was also produced. What are all the elements in the white powder?

4 Hydrogen peroxide is a compound of hydrogen and oxygen. However, it is quite different from water. It is used in bleaches and antiseptics, and in rocket fuel. If water and hydrogen peroxide are both made up of hydrogen and oxygen, why are they so different? Write an explanation.

5 The diagram below illustrates how ammonia gas is made from nitrogen and hydrogen gases.

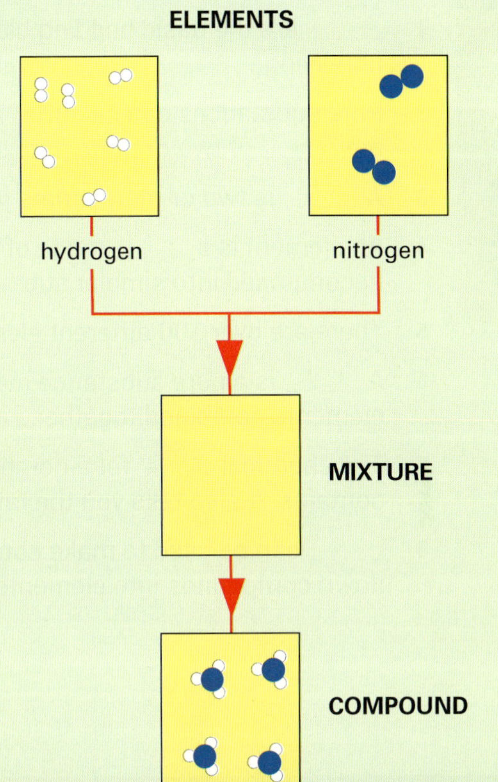

a What is the ratio of nitrogen atoms to hydrogen atoms in the ammonia molecule?
b In what ratio do the nitrogen and hydrogen react? Is this the same ratio as in the ammonia product?
c What is the total number of atoms in the product? Is this the same as the total number of atoms in the reactants?
d Copy the diagram and draw the molecules in the box labelled MIXTURE.
e Write a word equation for the reaction.

ISBN 978 1 4202 3735 1

MAIN IDEAS

Copy and complete these statements to make a summary of this chapter. The missing words are on the right.

1 _____ are the basic building blocks of all _____, both living and non-living.

2 Pure substances can be either _____ or compounds. Most substances are _____ of two or more pure substances.

3 A _____ is two or more atoms joined together by chemical _____.

4 An element is a _____ made of atoms of only one type. It cannot be decomposed into simpler substances by chemical reactions.

5 There are over 100 different elements, each with its own _____.

6 A _____ is a pure substance made up of two or more different elements combined together.

7 The chemical _____ for a compound tells you what elements it contains. It also tells you the ratio of the atoms of these elements.

8 _____ can be used to make compounds from elements, and to break down compounds into elements.

matter
symbol
formula
bonds
compound
mixtures
elements
atoms
molecule
chemical reactions
pure substance

CH•8 REVIEW

1 Which one of the following can you normally see without a microscope?
- **A** cells
- **B** elements
- **C** molecules
- **D** atoms

2 Copper, iron and chlorine are all:
- **A** compounds.
- **B** mixtures.
- **C** elements.
- **D** metals

3 How many naturally occurring substances are there that cannot be broken down into simpler substances by chemical reactions?
- **A** ninety
- **B** hundreds
- **C** many thousands
- **D** not known

4 Which one of the following is a compound?
- **A** sodium
- **B** chlorine
- **C** sugar
- **D** hydrogen

5 Nitrogen dioxide is a compound that contains nitrogen and oxygen in the ratio of one atom of nitrogen to two atoms of oxygen. Its formula would be:
- **A** NO
- **B** N_2O
- **C** NO_2
- **D** N_2O_2

ISBN 978 1 4202 3735 1

6 If ● and ● represent two different atoms, which one of the following would best represent:

a an element?

b a compound

c a mixture

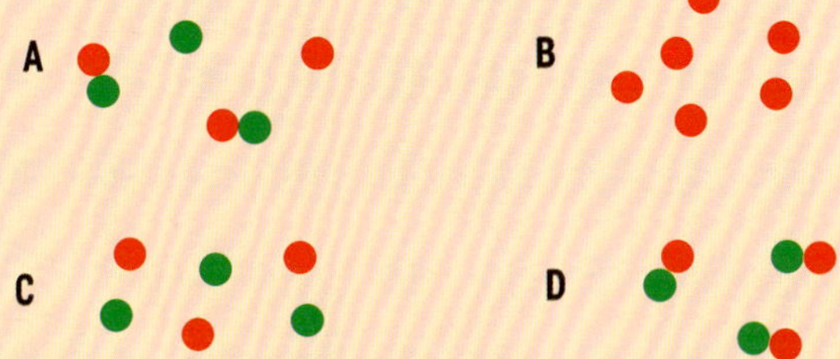

7 The formula for ammonia is NH_3. How many atoms are there in a molecule of ammonia?

8 Nicholas knows that compounds containing sodium (a metal) produce a golden-yellow colour in a flame. He also knows that compounds containing iodine produce a purple gas when heated with concentrated sulfuric acid. He tests four chemicals and records his results.

Substance	Yellow flame	Purple gas
1	✓	✗
2	✗	✓
3	✗	✗
4	✓	✓

a Which of these chemicals contain the element sodium?

b Which contain the element iodine?

c Which is likely to be the compound sodium iodide?

9 Jake pours some acid onto the element zinc. Hydrogen gas is formed and a colourless solution is left. He tests the colourless solution and finds that it contains only two different elements—zinc and chlorine. On the basis of these tests Jake can conclude that the acid contains:

A hydrogen only.

B hydrogen and chlorine only.

C chlorine only.

D zinc and chlorine only.

10 Write several complete sentences to explain the differences between an atom, a compound, an element and a molecule.

11 Stephanie passes an electric current through water in a voltameter, as in Investigation 8.3. She finds that the water slowly disappears and she is left with two gases—hydrogen and oxygen. She then mixes the two gases and puts a match to them. An explosion occurs and water is formed again.

Knowing the structure of a water molecule, how can you explain Stephanie's results?

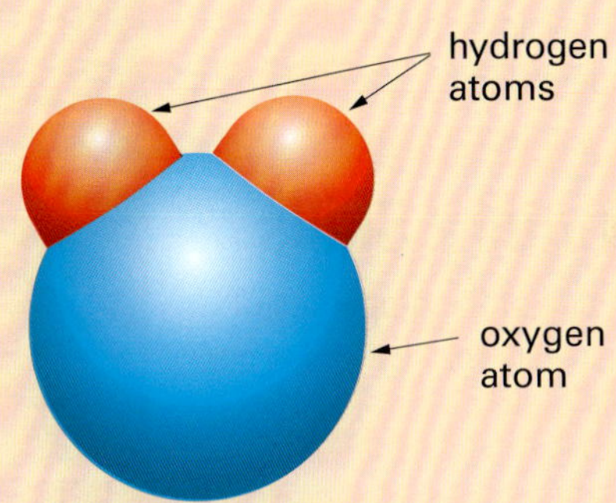

12 A scientist has two different compounds. She knows that the molecules in one can be represented by ●■ and the molecules in the other can be represented by ●■●. However she does not know which is which. How could she use chemical reactions to find out?

Check your answers on page 299.

ISBN 978 1 4202 3735 1

IN THIS CHAPTER

Science Understanding

- investigate the reflection of light from mirrors and its refraction through various media, for example air, water and glass
- describe corrective eye technologies and how the human eye receives light
- compare and contrast the way in which energy is transferred as light waves and sound waves
- investigate the properties of sound
- consider how technologies have been developed using optical fibres and polarising filters
- use the invention of the microscope to illustrate how advances in scientific understanding often rely on developments in technology

Science Inquiry Skills

- use a knowledge of the properties of light to explain why the sky is blue and how rainbows form

CH•9 Light and sound

ISBN 978 1 4202 3735 1

GET STARTED: *EXPLORE*

Work in a small group to discuss each of the following. Keep your answers for later on in this chapter.

- You are playing pool and you have to pocket the yellow ball by hitting it with the white ball. How does a knowledge of reflection help you pocket the yellow ball? Which pocket will you aim for? Why?
- When the sun shines onto a large crystal hanging in your bedroom window, you sometimes get a rainbow image on your wall. How is the rainbow of colours formed?
- How would you write the word SELF on paper so that when you hold it in front of a mirror you can see the reflection of the word written correctly?
- Two actors stand on a stage. One of them wears a white costume. When a spotlight with a coloured filter is turned on, one actor's costume looks red and the other's looks black. What colour is the filter and what colour is the other costume?

ISBN 978 1 4202 3735 1

This activity will help you see what you already know. Form a group and discuss each of the following questions. Be prepared to discuss your answers with the class.

1 Light and sound are forms of energy. All forms of energy have a starting point or source. Name two sources of light and two sources of sound.

2 Which of the following shapes can be seen as the letter K in a mirror?

What does this tell you about the images formed in a mirror?

3 When you hit a metal cymbal with a drumstick, it rings. If you put your hand on the cymbal and hit it again, it does not ring. Explain why this happens.

4 Three rays of light shine onto a glass lens as in the diagram below. What will happen to the light rays when they pass through the lens?

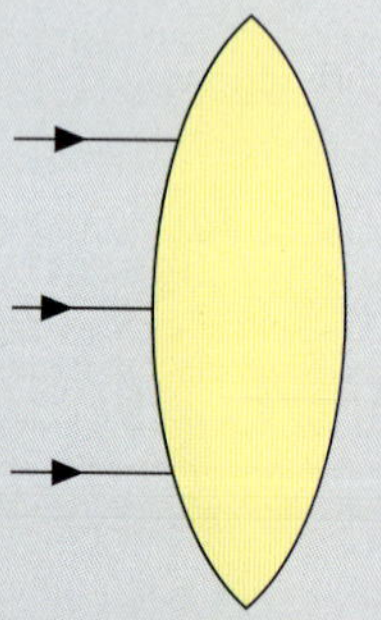

5 The diagram below shows four light rays reflecting from a plane mirror. Each light ray is coloured. There are five errors in the diagram. Can you find them?

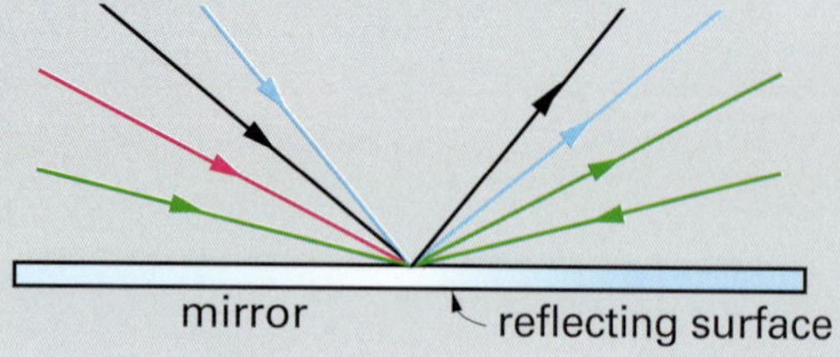

9.1 Properties of light and sound

In the first Get started problem in which you had to pocket the yellow ball, you used the fact that the angle at which the ball strikes the cushion is equal to the angle at which it leaves the cushion. This is the same principle as the reflection of light. Reflection is one of the *properties* of light and sound.

Both light and sound are forms of energy and both can be transformed into other sorts of energy. For example, light can be transformed into chemical energy in a leaf during photosynthesis, or into electrical energy in a solar cell. Sound can be transformed into kinetic energy in a radio speaker. The transformation of energy is another property of both light and sound.

Another property of light is that it can travel in straight lines. Surveyors rely on this property when they use their instruments to find boundary lines or take measurements for new roads.

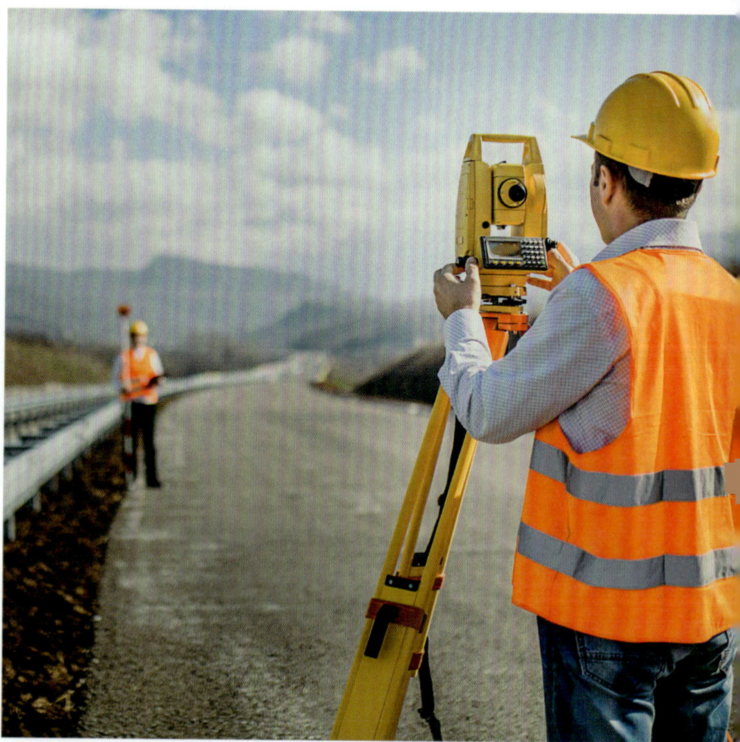

Figure 9.1 Surveyors use the property that light travels in straight lines to measure distances and heights.

ISBN 978 1 4202 3735 1

The law of reflection

When light strikes a mirror, the reflected light ray bounces off the mirror at the same angle as it strikes the mirror. This is the **law of reflection**.

When doing experiments on the law of reflection, scientists measure the angle between the light ray and an imaginary line, called the *normal*. This line is at right angles to the surface. The light ray coming towards the surface is called the *incident ray*, and the outgoing one is called the *reflected ray*. The angles formed between the rays and the normal are called the *angle of incidence* and the *angle of reflection*. These two angles are always equal no matter how the light rays strike the surface. This is the law of reflection.

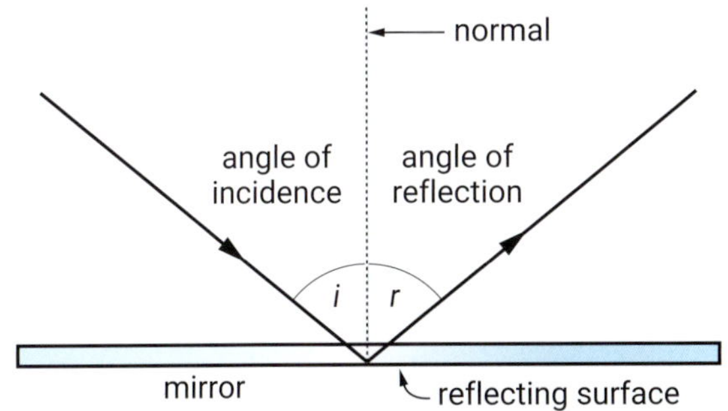

Figure 9.2 The law of reflection

Reflection from curved mirrors

The law of reflection applies to curved mirrors as well as plane (flat) mirrors. In the diagram below, two parallel light rays hit a concave mirror (concave means curved inwards like a cave). These rays reflect off the mirror and meet at a point called the **focus**. The focal length of the mirror is the distance of the focus from the mirror's reflecting surface.

Notice that the light rays reflect off the curved mirror surface and obey the law of reflection—the angle of incidence (i) is equal to the angle of reflection (r).

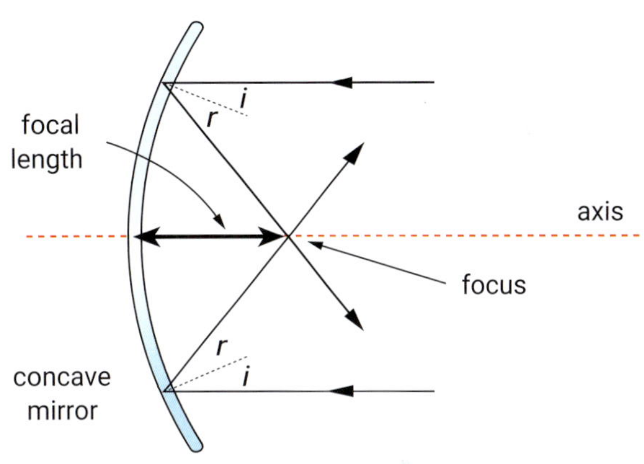

Figure 9.3 Parallel light rays reflect from a curved mirror and meet at a point called the focus.

The history of mirrors

Around 600 BCE, the early Etruscans and Greeks used polished discs of thin bronze as mirrors. During Roman Christian times, small metallic mirrors made of highly polished silver or steel were worn by fashion-conscious men and women.

Mirrors made of glass with a very thin layer of metal were first used in the 1300s. However, it wasn't until 1564, when the mirror-makers of Venice formed a corporation, that glass mirrors gained popularity. These mirrors were made from highly polished glass with a very thin metal backing, usually made from an alloy of tin and mercury. In very expensive mirrors, silver metal was used as the reflective backing.

Today, mirrors have a silver or aluminium layer that is sealed by a painted or plastic outer layer to protect the metal.

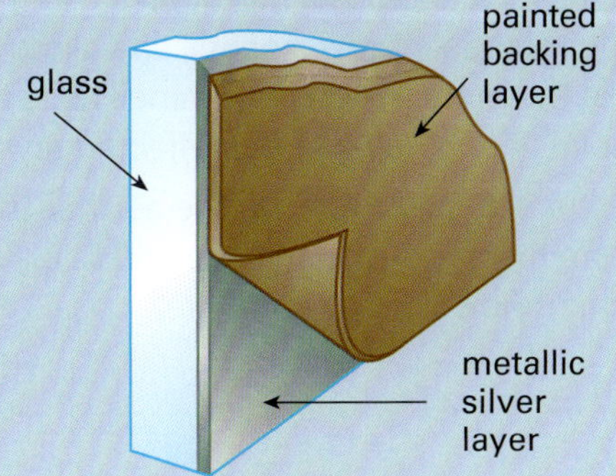

ISBN 978 1 4202 3735 1

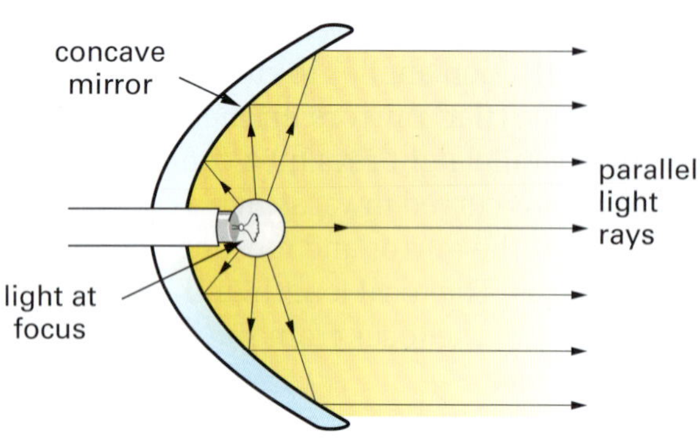

Figure 9.4 When a light source is placed at the focus of a concave mirror, a beam of parallel light rays is produced. This is why concave mirrors are used as reflectors in torches, car headlights and floodlights.

Figure 9.5 Convex mirrors are used at dangerous road intersections. Light rays from wide angles strike the mirror, giving a wider-angle image than would be seen using a plane mirror.

INVESTIGATION 9.1 Reflection

Aim

To investigate the reflection of light.

Materials

- ray box kit and power pack
- pencil, ruler and protractor

Risk assessment and planning

- Your teacher will tell you how to set up the power pack and ray box kit correctly.
- Make sure you ask your teacher to check your set-up before you turn on the power.

Method

1 Use the diagram on the right as a guide to set up the ray box and plane mirror on a clean page of your notebook. Draw a pencil line along the back of the mirror. Then draw pencil lines along the incident light ray and the reflected ray.
Use a protractor to draw the normal, then measure the angle of incidence and angle of reflection. How do they compare?

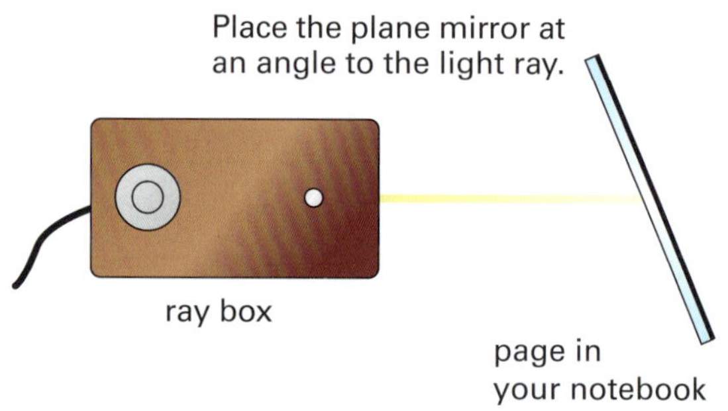

2 Change the angle of the mirror and repeat step 1. Measure at least three different angles.

3 Replace the plane mirror with a concave mirror and shine parallel rays of light directly at it.
Describe what happens. Draw a diagram of the set-up and mark on it the axis and focus (see Figure 9.3 on the previous page).

4 Use a ruler to find the focal length of the mirror.
Record your results.

5 Try using a convex mirror.
Draw a diagram of what happens when parallel light rays strike the mirror. Does this mirror have a focus?

ISBN 978 1 4202 3735 1

Refraction of light

Air and water are *transparent* substances. This means that light passes through them. Substances such as paper, wood and brick do not allow light to pass through them and are called *opaque*. However, when light passes from one transparent substance to another, for example from air to water, strange things happen.

ACTIVITY

1 Put a coin in a coffee mug (or a beaker) and place the mug on a bench.

2 Hold a ruler vertically on the bench. Place your eye level with the zero mark on the ruler and have your partner position the mug so that you can see only the far edge of the coin.

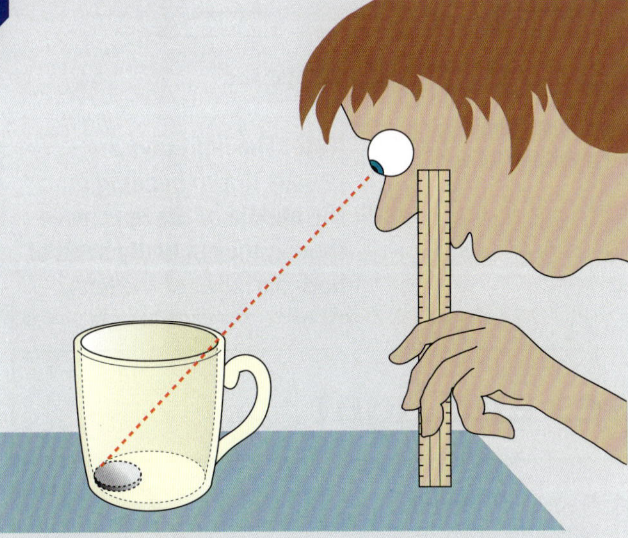

3 Have your partner slowly pour water into the mug until it is full.

What happens to the coin as the water is added?

4 Move your eye down the ruler until you see the edge of the coin again.

How far down the ruler did you move your eye?

This effect shows another property of light—refraction.

The activity showed that when you put a coin in a mug and add some water, the coin seems to change position. This is caused by light bending when it passes through different transparent substances at an angle. This bending of light is called **refraction**. Refraction is another property of light.

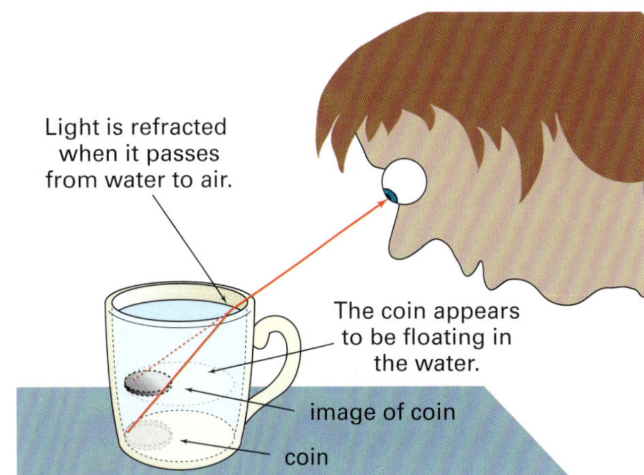

Figure 9.6 Refraction

You can see how light refracts as it passes through the glass block in the photo below. Notice that the angle of refraction is less than the angle of incidence. This is because the refracted light ray bends towards the normal.

In general, when light passes from air to another transparent substance such as water, glass, plastic, diamond or alcohol, it bends towards the normal—the angle of refraction is always less than the angle of incidence.

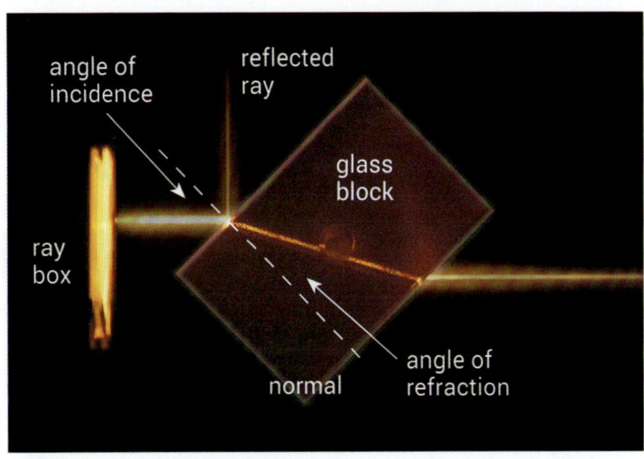

Figure 9.7 Light is refracted as it passes through a glass block. Notice that some light is also reflected by the block.

ISBN 978 1 4202 3735 1

The amount of refraction of a light ray depends on the type of substance. For example, light bends more when it passes from air to glass than it does when it passes from air to water.

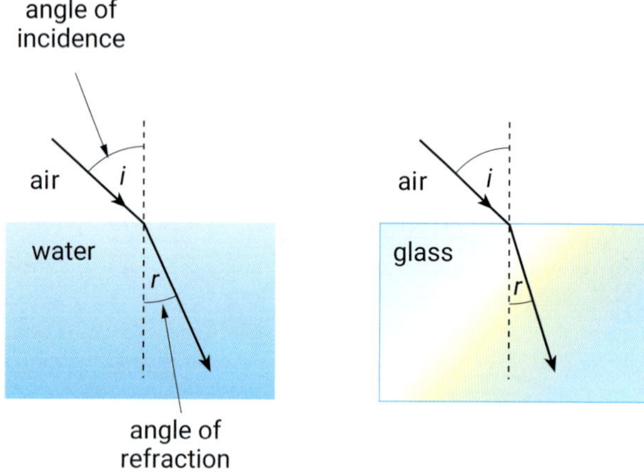

Figure 9.8 Refraction in water and glass

Lenses are pieces of glass or plastic curved on one or both surfaces. They refract light in certain patterns; for example, lenses that refract light inwards are called *converging lenses*. *Diverging lenses* refract light outwards.

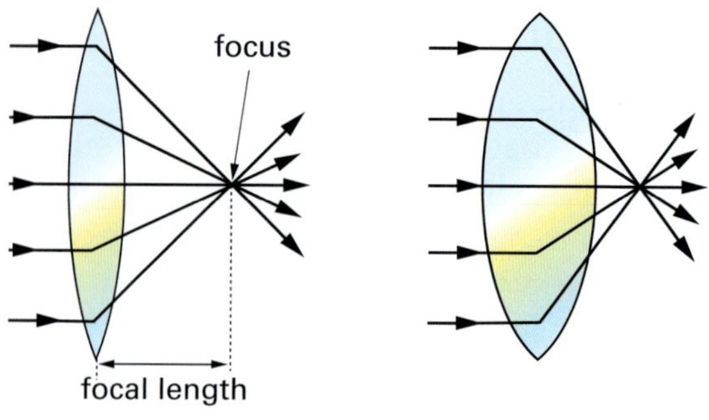

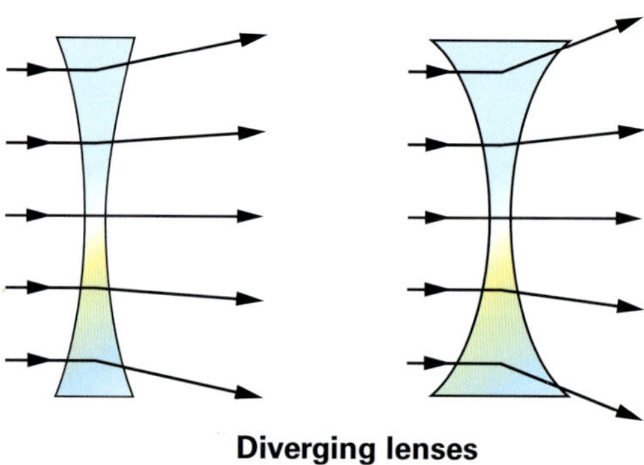

Figure 9.9 Converging and diverging lenses

Note: The light rays are usually drawn bending in the middle of the lens, even though they actually bend at both surfaces of the lens.

INVESTIGATION 9.2 Lenses and light

Aim

To observe how lenses refract light and form images on a screen.

PART A Ray box lenses

Materials

- ray box kit and power pack

Method

1 Set up the ray box and a converging lens as shown in the diagram on the right.

2 Find the focal length of the lens.

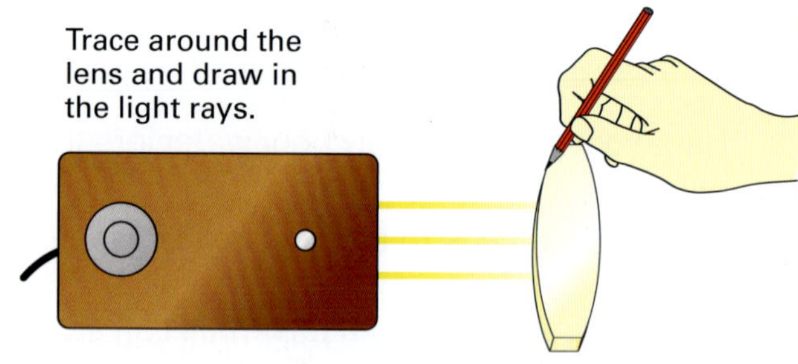

3 Replace the lens with a different converging lens and find its focal length.

What happens when you use a diverging lens instead of a converging lens? Can you find the focal length of a diverging lens? Try it.

ISBN 978 1 4202 3735 1

Discussion

1 Does a fatter converging lens refract light more or less than a thinner one? Suggest a reason for your answer.

2 Do diverging lenses have a focus? Explain your answer.

PART B Images with lenses

Materials

- metre ruler
- candle
- round glass converging lens and lens holder or plasticine
- screen (white hardboard or cardboard)

Method

1 Hold the converging lens near a window and focus the image of a distant object on the screen. Measure the distance between the lens and the screen. This is the *focal length*.

Record the focal length.

2 Light the candle. Then place the lens in a holder (or plasticine) and position it so that the distance from the lens to the candle is twice the focal length of the lens (2*f*).

3 Move the screen backwards and forwards until you get a clear image of the candle flame.

screen
lens
lens holder
2*f*
candle

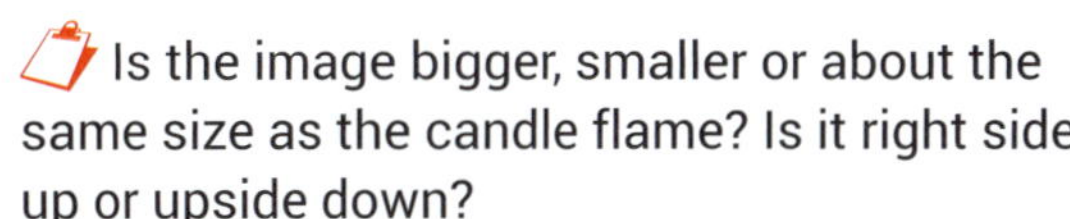

Is the image bigger, smaller or about the same size as the candle flame? Is it right side up or upside down?

4 Move the lens further away from the candle and describe what happens to the image. Do this for two or three distances.

Record your observations.

5 Now move the lens closer to the candle but not up to the focus.

What happens to the image?

6 Move the lens inside the focus. Remove the screen and look through the lens at the candle flame.

Describe the image.

Conclusion

Summarise your results in a table. Put the distance between the lens and the candle in one column and the image description in another. The distances are:

- at twice the focal length (2*f*)
- further than 2*f*
- between 2*f* and *f*
- closer than *f*.

PART C Predicting images

Ray diagrams are scale models used to predict the position and size of the image. The ray diagram below uses a lens of focal length 20 cm. To find the image, draw one light ray straight through the centre of the lens, and another through the focus to the lens and then parallel to the axis. The image is where these two lines meet.

Take some measurements using your set-up from Part B, and then check the image distances with those predicted using ray diagrams.

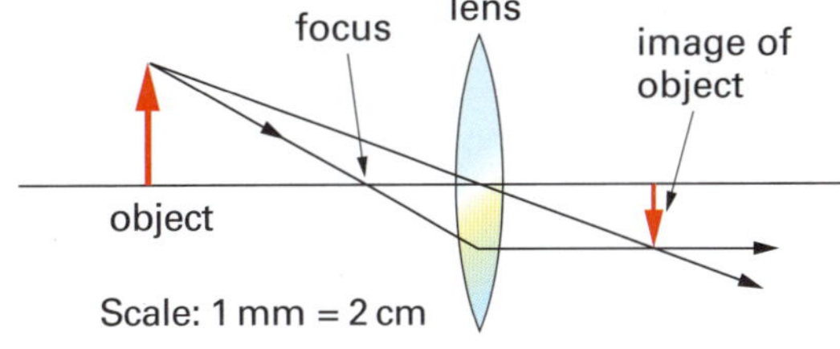

ISBN 978 1 4202 3735 1

SCIENCE AS A HUMAN ENDEAVOUR

Inventing the microscope

Over 400 years ago, a young Dutch child named Zacharias Janssen was playing with his father's lenses when he suddenly screamed. The church spire he saw was so large it seemed to be coming towards him. When he grew up, Zacharias became a spectacle-maker like his father Hans. He remembered the fright he received as a child and began combining lenses to make small objects bigger. In 1595, he invented the first compound microscope, probably with help from his father. It consisted of two tubes, with one fitted neatly inside the other. There was a lens at each end, and one lens magnified the already enlarged image from the other. The microscope was focused by sliding the tubes back and forth.

The English scientist Robert Hooke improved on Janssen's microscope. When he examined a thin slice of cork (the bark of an oak tree) through his microscope, he was astonished to see a block-like pattern inside the cork. Hooke called these blocks *cells* because they looked like the little cubicles where monks studied and prayed. Amazed by this, other scientists were soon examining parts of all kinds of plants and animals. They found that all the specimens they examined were built from rows and rows of cells. These cells were different shapes and sizes, and some could even move about.

A Dutchman named Antonie van Leeuwenhoek (LAY-wen-hook) read about Hooke's work and began making his own single-lens microscopes. When he placed a small live fish in front of the lens and held it up to his eye, he could see blood surging through the blood vessels in the fish's tail. He took a drop of water from the bottom of a pot plant and observed it with his microscope. To his amazement he saw tiny single-celled animals that he called 'cavorting beasties'. Leeuwenhoek was the first person to observe bacteria—in plaque he collected from his own teeth and from 'two old men who had never cleaned their teeth in their life'. He even noticed that the bacteria in the plaque were killed when he drank hot coffee.

After Hooke and Leeuwenhoek, microscopes were made with higher and higher magnifications, but it was not possible to go beyond a magnification of ×1500. In 1931, a German named Ernst Ruska made an *electron microscope*. Instead of light he used a stream of electrons that could be focused using electrical or magnetic fields. Electron microscopes can magnify objects up to one million times, and for the first time scientists could observe viruses and the details of cell structure. Scanning electron microscopes can take 3D pictures of the various structures in the human body; for example, blood cells and muscle fibres. The photo below shows a head louse with an egg (nit) clinging to a human hair.

Question

Use the information on this page to write one or two paragraphs that describe how developments or improvements in technology have transformed science.

ISBN 978 1 4202 3735 1

Optical fibres

The photo below shows a surgeon using an *endoscope* to examine a patient's stomach. The endoscope has a long flexible tube containing **optical fibres** that is inserted through the patient's mouth and oesophagus.

The endoscope contains bundles of optical fibres, each one about 10 micrometres in diameter. The optical fibres act like a flexible torch. Light shines through one end of the optical fibres in the endoscope and it comes out the other end, no matter how much the tube is twisted or bent.

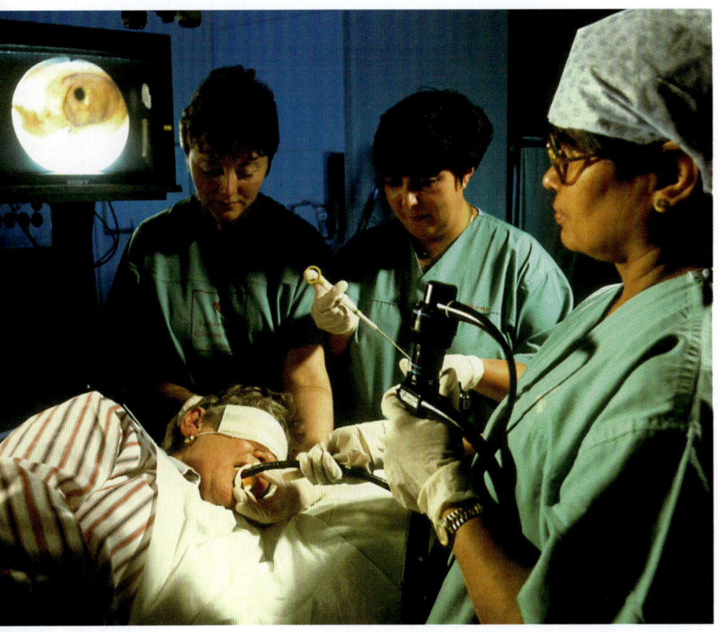

Figure 9.10 A doctor using an endoscope

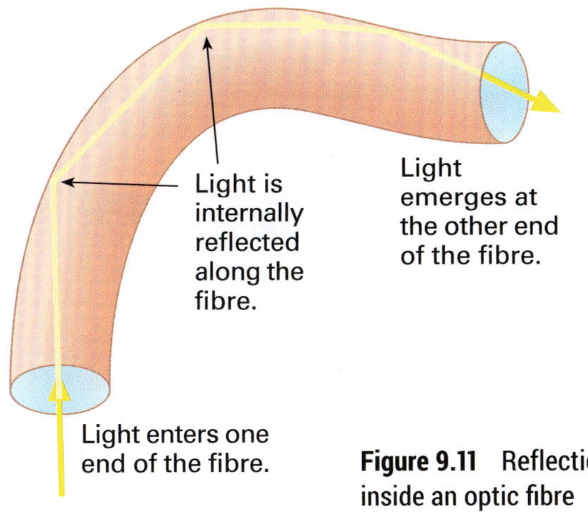

Figure 9.11 Reflection inside an optic fibre

ACTIVITY

You will need a ray box kit and power pack for this activity.

1 Use the diagram below as a guide to set up the ray box and a triangular prism.

2 Shine one beam of light onto the side of the prism, and slowly rotate the prism until the light beam is totally internally reflected.

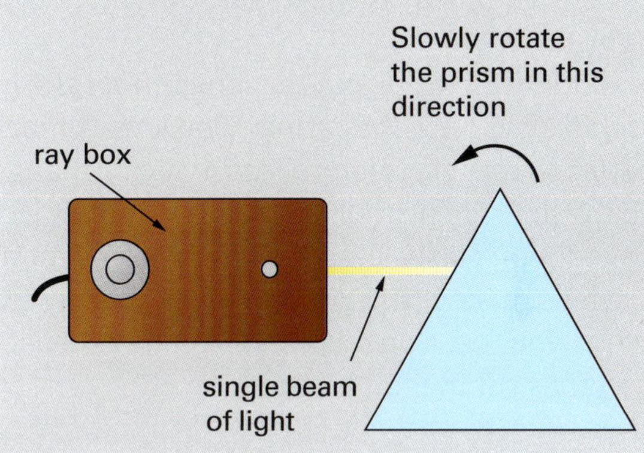

An optical fibre uses the principle of **total internal reflection** to transmit the light. Figure 9.11 shows how the light ray comes in from one end, hits the side of the fibre and is reflected back into it. None of the light escapes from the fibre. This is why it is called total internal reflection.

Endoscopes have two bundles of optical fibres inside the flexible tube, and each bundle consists of thousands of fibres. One bundle transmits the light from the surgeon's end to the patient so that the surgeon can see. The other bundle carries the image back to the surgeon via a microscope, video screen or computer.

Optical fibre communications

Optical fibres have replaced most metal cables used in communications. Electrical signals from computers, telephones and televisions are converted to pulses of light using a laser. These light pulses are then sent along optical fibres that can transmit the signal over large distances.

Optical fibres are much lighter, and the same thickness of fibres can transmit thousands more messages than copper wire cables can.

ISBN 978 1 4202 3735 1

How the eye focuses light

Light enters the eye through the transparent cornea that covers the front of the eye. The coloured part in the front of the eye is called the iris. This is a ring of muscle that changes in size and thus controls the amount of light that enters the eye. The light passes through the pupil, the lens and the jelly-like substance inside the eye and finally hits the retina. The retina contains structures called *vision receptors*, which detect light.

Both the cornea and the lens refract the light and focus it onto the retina. The lens is much better than a glass lens because it can change shape to focus near objects and distant objects. To focus on close objects, tiny muscles around the lens make it thicker and more sharply curved. When you focus on distant objects, the lens becomes thinner and flatter.

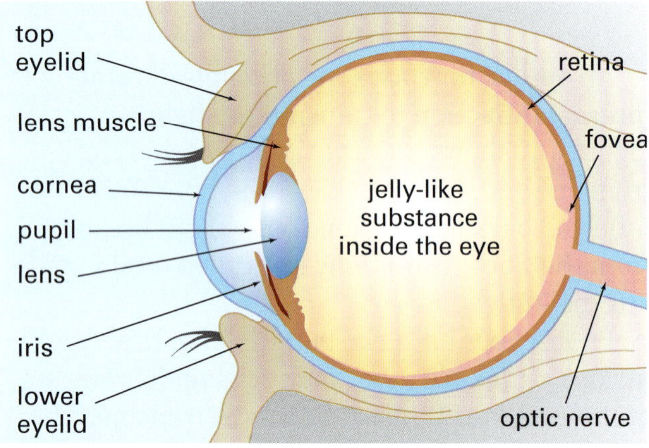

Figure 9.12 A cross-section of the human eye

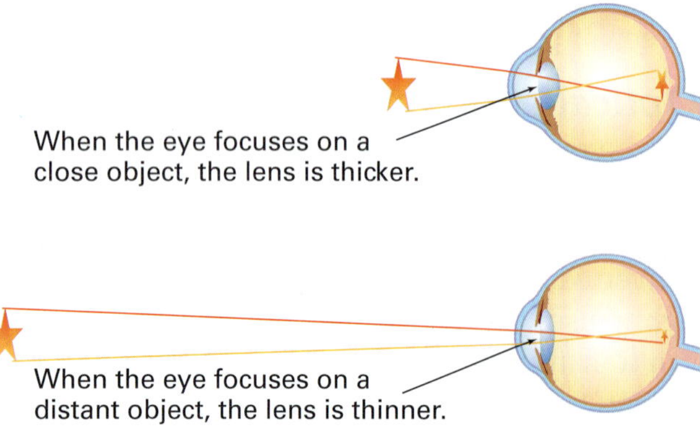

Figure 9.13 The lens in the eye focuses light onto the retina.

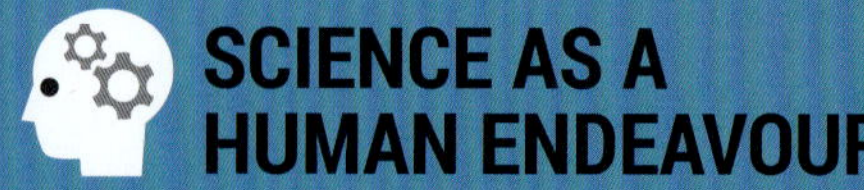

SCIENCE AS A HUMAN ENDEAVOUR

Corneal transplants

When a person's cornea becomes cloudy as a result of disease, injury or infection, their vision is reduced, often making them blind. In a surgical operation called a *corneal transplant*, a damaged cornea can be replaced by a donated healthy one.

In a corneal transplant, an eye surgeon, called an ophthalmologist (OFF-thal-MOL-o-gist), cuts a circular section of the damaged cornea using a tool that works in the same way as a round pastry cutter (see photo). The damaged section of cornea is removed and replaced with the same-sized section of a healthy cornea. The new cornea is held in place by hair-like stitches.

Corneal transplants are the most successful of the organ donation transplants. Over 90% of all patients have restored vision after the operation. The greatest risk with corneal transplants is tissue rejection. This is where the body of the patient rejects the donor's eye tissue. The eye swells and the two types of tissues never bind together.

EXPLORE ONLINE

Use the internet to find more information on corneal transplants and organ donation of eyes. Try entering the following words in the search engine: *corneal transplant* and *eye organ donor*.

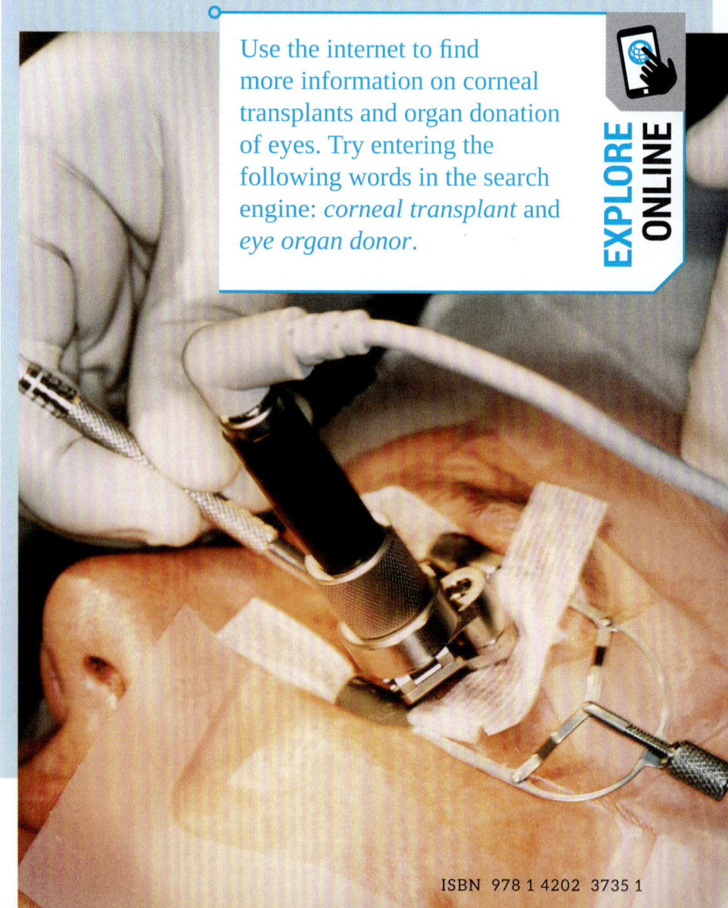

ISBN 978 1 4202 3735 1

Correcting eye problems with lenses

Some people have vision problems because their eyes do not focus on the retina and a blurry image results. There are two common eye problems: **long-sightedness** and **short-sightedness**.

Long-sightedness

The image focuses behind the retina and the person has trouble focusing on close objects. This is common in older people where the lens muscles have become weak or the lens has hardened. In young people, a short eyeball will cause this problem.

Long-sightedness can be corrected by wearing glasses with converging lenses. These bend the light rays inwards and make the image form on the retina.

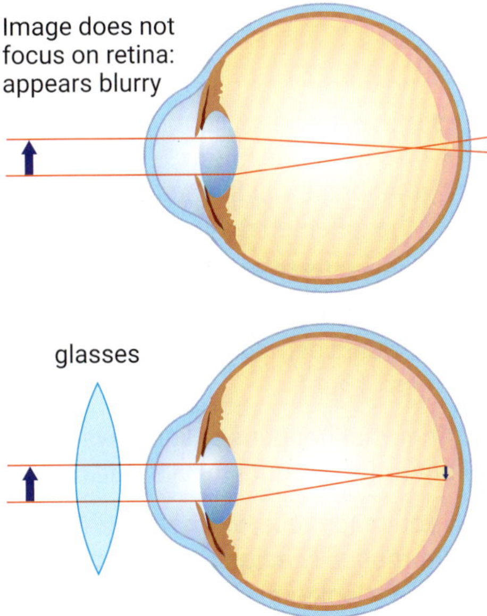

Figure 9.14 Long-sightedness causes light to focus behind the retina. It can be corrected by wearing glasses with a converging lens.

Short-sightedness

The image is focused in front of the retina and the person cannot focus on distant objects. An out-of-shape, long eyeball will cause this problem.

Short-sightedness can be corrected by wearing glasses with diverging lenses. These bend the light outwards, and make the image form on the retina. Many people prefer to wear contact lenses instead of spectacles to correct their eye problems. Contact lenses are very thin pieces of special plastic which are worn on the outer surface of the cornea. The shape and size of the contact lens is specified by an optician who also fits them and tests to see that the person's vision is corrected.

Your eyeball changes shape as you grow. Between the ages of 12 and 15, the body grows rapidly. During this time the eyeball also grows, usually becoming longer. This may cause eye problems for some teenagers, leading to headaches and dizziness, particularly when doing a lot of reading. These problems are quite common and are easily corrected by an optician.

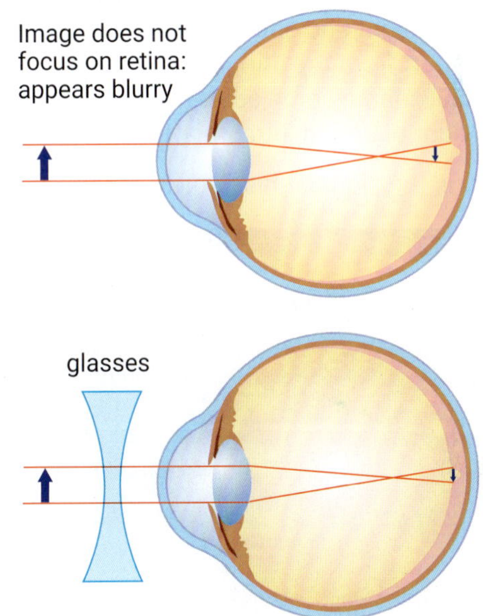

Figure 9.15 Short-sightedness causes light to focus in front of the retina. It can be corrected by wearing glasses with a diverging lens.

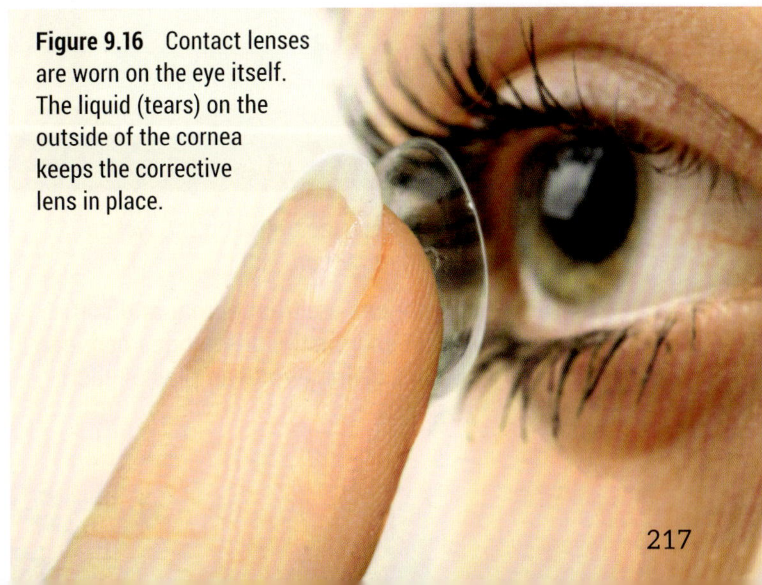

Figure 9.16 Contact lenses are worn on the eye itself. The liquid (tears) on the outside of the cornea keeps the corrective lens in place.

ISBN 978 1 4202 3735 1

1 Match these words with their descriptions:

image	reflection	incident ray
converging	plane mirror	refraction
focus	convex mirror	diverging

a a mirror that is flat and not curved
b when sound or light strikes a surface and bounces off
c the ingoing ray of light
d the point at which light rays meet after passing through a converging lens
e a picture of an object formed after light rays have been reflected or refracted
f light rays that come together

2 a How could you demonstrate to someone that light and sound are forms of energy?
b Apart from the things mentioned on page 208, name some other things that convert light or sound into other forms of energy.

3 Why is the sign on this van written like this?

4 Explain the difference between the words *reflection* and *refraction*.

5 Some substances are transparent and some are opaque. Describe the differences between the two terms, giving two examples of each.

6 Copy the drawings below and show what happens to the light rays when they pass through the lenses. In each case label the focus.

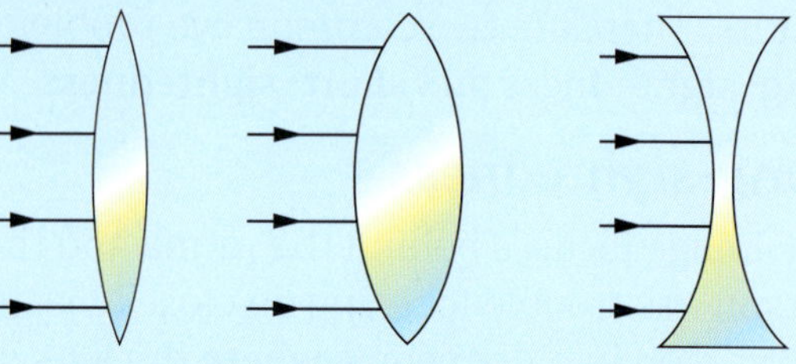

7 Two light beams strike a plane mirror as shown in the diagram below. Copy the diagram and show the normal and the reflected rays.

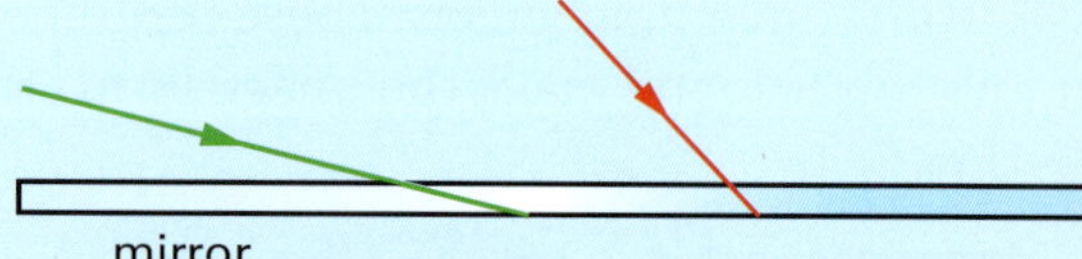

8 Two plane mirrors are placed at right angles to each other. A light ray is shone onto mirror 1 at 30° to the mirror.
a At what angle will the beam strike mirror 2?
b Will the reflected ray from mirror 2 be parallel to the original incident ray? Explain.

9 Reflection is a property of light. Briefly describe how you would demonstrate two other properties of light.

10 Describe what happens to light rays from when they enter the eye until they hit the retina.

11 When you read these words, the lens in each eye automatically adjusts to focus on the words. Now look out of a window. Immediately you focus on distant objects.
Describe what happens to the lens in your eye when you do this.

12 Some substances are transparent, some are opaque, while others are translucent. Use a dictionary to find out what the word *translucent* means.

13 a Describe what causes short-sightedness. Use a diagram to support your answer.
b Now draw a new diagram to show how this condition can be fixed.

ISBN 978 1 4202 3735 1

CHALLENGE

1 The diagram below shows a solar hot water heater using a semicircular shiny, silvered reflector.

a Explain how you think the heater works.

b Where would you place the water tube to get the maximum heating efficiency?

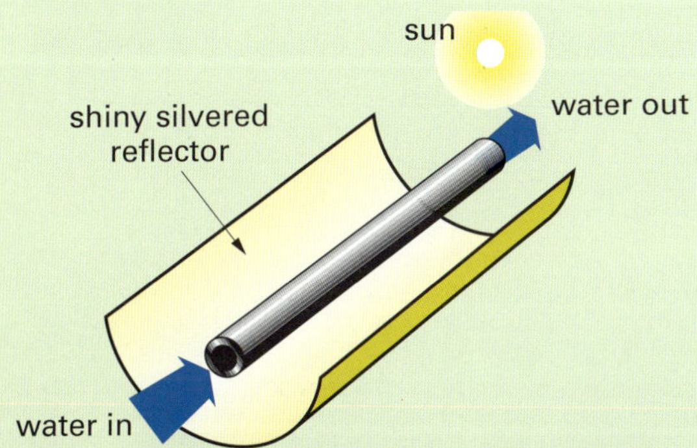

2 A pencil 15 cm long stands vertically on a bench 30 cm from a small spotlight. A screen is placed 1 m away from the spotlight and a shadow of the pencil forms on it.

a Which property of light is being shown here?

b How tall will the shadow be?

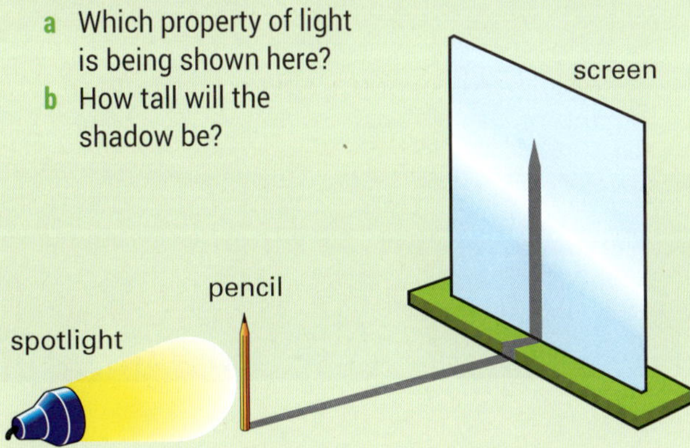

3 What type of mirror (plane, concave or convex) would be best to use as a rear-vision mirror with a wide field of view. Your teacher will give you the mirrors to help you with your decision.

4 A pencil 20 cm high was placed 60 cm in front of a concave mirror of focal length 20 cm. The ray diagram below shows that the image of the pencil is upside down and smaller than the object.

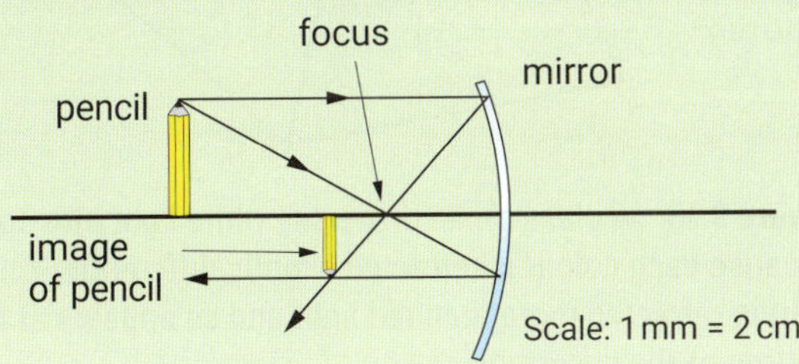

Do a similar drawing to find out what happens to the image if the pencil is placed 100 cm in front of the mirror.

5 A converging lens has a focal length of 20 cm. An object 15 cm high is placed 40 cm from the lens. Use a ray diagram to describe what the image will be like.

6 Design the following items using mirrors.

a Make a periscope that can be used to see things around corners.

b Use a concave mirror to make a solar cooker.

7 This diagram shows a light ray hitting the surface of a transparent substance. The light ray refracts at the surface and bends towards the normal.

How much the refracted ray bends depends on the type of substance. The table below shows different substances and their refractive index. The higher the refractive index, the greater the refracted ray is bent.

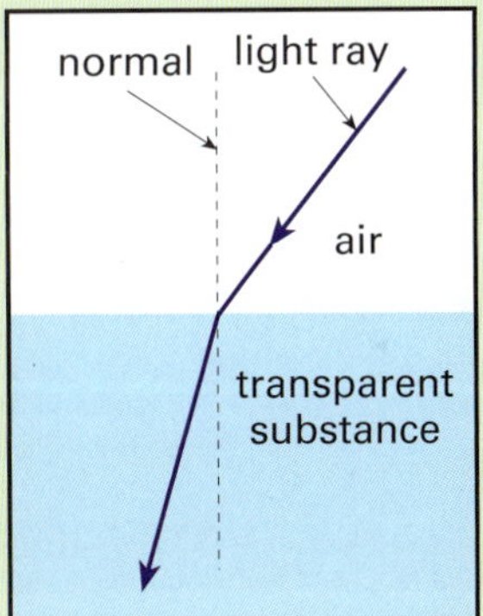

Substance	Refractive index
air	1.00
water (at 25 °C)	1.33
ethanol	1.36
glass	1.52
diamond	2.40

a Does light bend more when it goes from air to water, or when it goes from air to glass? Justify your answer.

b Glycerine has a refractive index of 1.47. Does glycerine bend light more or less than glass does?

c Predict what happens when a light ray passes from air to glass to water and out to air again. Draw a diagram of your prediction.

ISBN 978 1 4202 3735 1

9.2 Light and colour

Figure 9.17 A rainbow

Forming a rainbow

Rainbows make a ribbon of colours in the sky after rain. They form when sunlight passes through the raindrops. The drops split the white light from the sun into a **spectrum** of colours.

The colours that make up the spectrum are continuous and blend into each other, but for convenience we say there are seven colours—red, orange, yellow, green, blue, indigo and violet. The splitting up of white light into this spectrum of colours is called **dispersion**. This occurs because each colour is refracted slightly differently when it passes through a raindrop.

White light is also dispersed into separate colours when it passes through a glass prism.

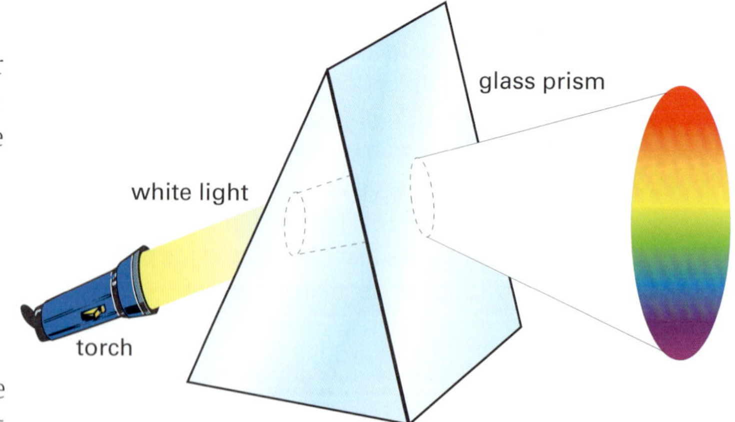

Figure 9.18 A glass prism disperses white light into colours because each colour is refracted slightly differently. Violet light is refracted more than red light and so appears at the bottom of the spectrum.

ISBN 978 1 4202 3735 1

Why are things coloured?

Why is a leaf green, milk white and a tomato red?

When white light hits a leaf, most of the colours in the white light are absorbed. Only the green light is reflected, and it is this colour that reaches your eye. So you see the leaf as a green colour.

Figure 9.19 A leaf reflects the green colour in white light and absorbs the others, so it appears green.

Figure 9.20 Milk reflects all the colours, so it appears white.

Figure 9.21 A red tomato reflects only red light and absorbs the others.

Filters made of coloured glass or plastic can also change the colour of light. When white light hits a red glass filter, the glass allows the red light to pass through and absorbs all the other colours. The colour of the filter tells you what colours it transmits (allows to pass through).

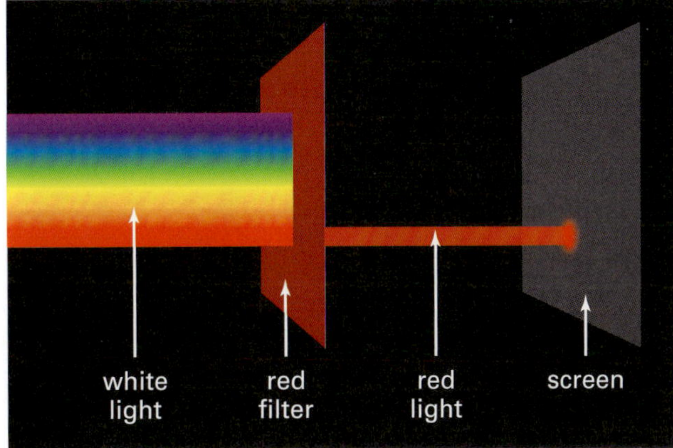

Figure 9.22 Coloured filters transmit their own colour and absorb the other colours.

What happens when you view a red tomato in green light? Since the tomato reflects only red light and absorbs all the others, green light is absorbed by the tomato. This means no light reaches your eyes, and the tomato therefore looks black.

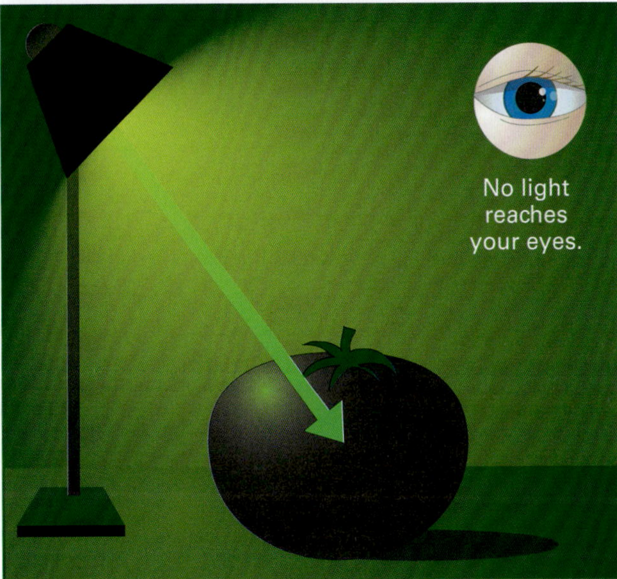

Figure 9.23 A red tomato will look black in green light because it absorbs all colours except red.

ISBN 978 1 4202 3735 1

INVESTIGATION 9.3 Colours

Aim

To observe the effects of filters and coloured cards on white light.

Materials

- ray box kit and power pack, with coloured filters
- piece of white paper
- pieces of coloured card (red, green, yellow, blue)

Risk assessment and planning

This investigation is best done in a darkened room.
Make sure you ask your teacher to check your set-up before you turn on the power.

PART A Coloured filters

1 Place the ray box on a sheet of white paper. Shine a full beam of light onto a triangular prism and turn the prism until the spectrum of colours is formed.

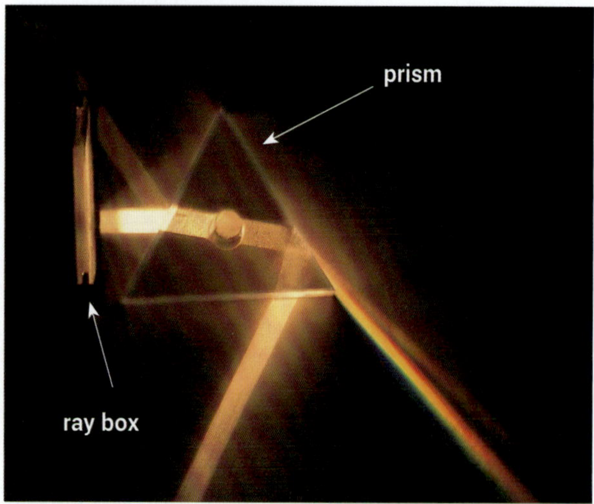

2 Predict what will happen when you put a red filter between the light and the prism.
Try it and record your observations.

3 Try other filters and record your observations.

PART B Coloured objects

1 Shine a full beam of light from the ray box onto a piece of red card. Then place different coloured filters in the ray box and record the colour of the card in each light.
Draw up a table and record your results in the table.

2 Repeat step 1 with the other coloured cards.

Discussion

1 How do the results in Part A step 1 help you decide which colour of the spectrum is refracted the most? Which one is refracted the least?

2 Which colours are transmitted and which are absorbed when white light is shone through a yellow filter?

3 Which colours of the spectrum would you see if white light was shone through a red filter and then a green filter? Explain your answer.

4 Explain, using the words *absorb* and *reflect*, why a blue card is blue in white light.

5 What would you see if you placed the blue card in red light? Explain your answer.

ISBN 978 1 4202 3735 1

Making colours

There are two main ways of making colours. The first way is to shine different coloured lights together. The other way is to mix different coloured paints or pigments together.

Your teacher will set up three slide projectors (or light boxes) in a darkened room. Each projector will have a different coloured filter, red, green and blue, and the spots of colour will overlap on a white screen.

- What colour do you see when red light and green light overlap? As a challenge, try to suggest why this happens.
- What colours would you see if the screen was red instead of white?

Addition

Making colours by adding different coloured lights together is called **addition**. White light can be made by shining red, green and blue lights together as shown below.

Figure 9.24 Blue, red and green light add together to make white light.

Subtraction

The method of making colours by mixing various paints together is called **subtraction**, because each paint colour subtracts or absorbs colours from white light. For example, blue paint reflects blue light and absorbs the rest.

Suppose 5-year-old Emily has seven pots of paints, and each one is a different colour of the spectrum. When she mixes them all together, she ends up with a black mess. What happens is that each of the seven paints absorbs its colour from white light. When they are all mixed together, all the white light is absorbed and none is reflected. So the mixture looks black.

Figure 9.25 below shows how green is the only colour reflected when blue and yellow paints are mixed. All the other colours are absorbed.

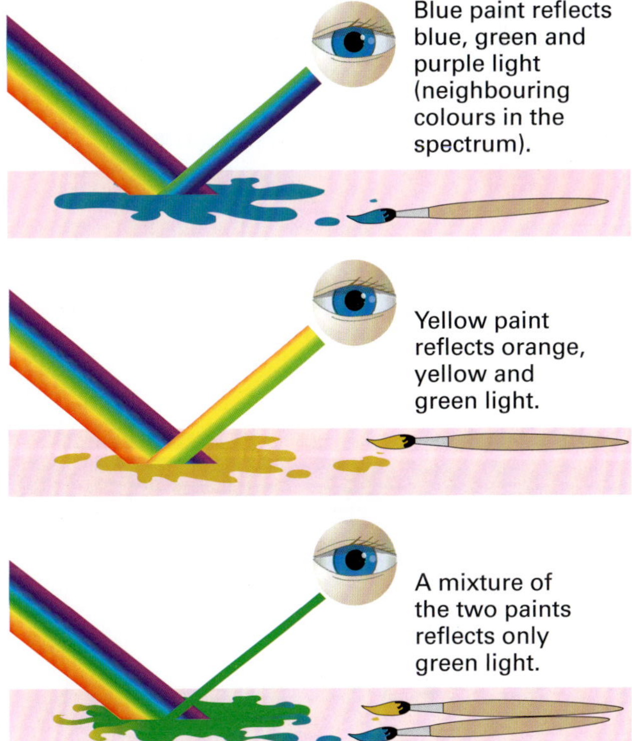

Figure 9.25 What happens when paints are mixed?

ISBN 978 1 4202 3735 1

Seeing colours

The retina that lines the inside of the human eye contains receptors that are sensitive to colours and give you colour vision. There are other receptors that are sensitive to shades of light and give you black-and-white vision.

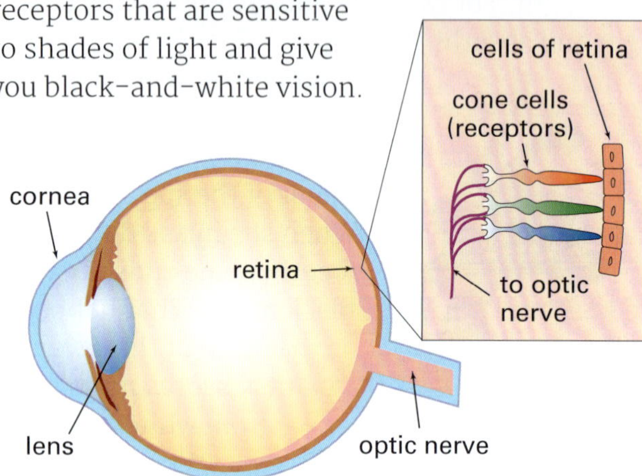

Figure 9.26 Receptors in the retina can detect red, green and blue light.

The colour vision receptors are called *cone cells*. There are three types of cone cells—one type is sensitive to blue light, another to green light and the third to red light. The diagrams below show how you see colours.

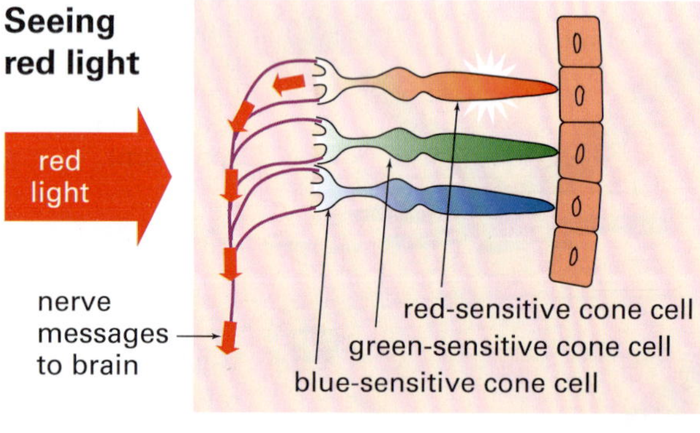

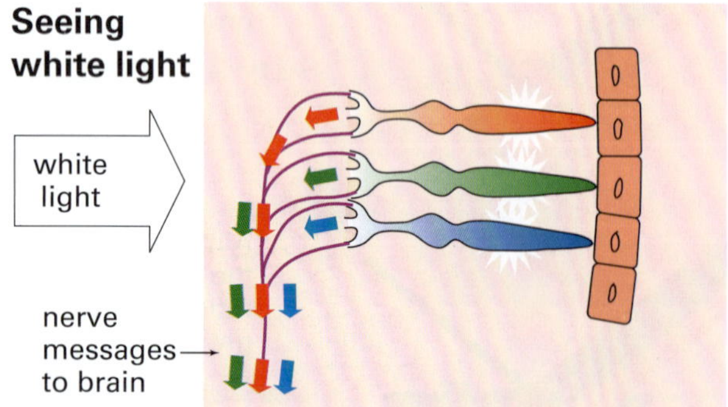

Figure 9.27 How the eye sees colours

Colour blindness

Colour blindness is a condition that causes people to have trouble distinguishing between certain colours. The most common form of colour blindness is red–green colour blindness. People with this condition cannot see, or they confuse, shades of red, green and brown.

The condition is usually inherited, which means it is passed on from parents to children. In Australia, about 9% of males and about 0.4% of females have some form of colour blindness.

Each of the cone cells in the retina contains a type of light-sensitive pigment. One type of pigment is sensitive to blue light, another to green light and the third to red light.

In people with defective colour vision, one or more of the light-sensitive pigments functions poorly, or, in severe colour blindness, is absent altogether. Generally it is the red-sensitive and green-sensitive pigments in the cones that function poorly, giving rise to red–green colour blindness.

Check out the following websites.

Colour blindness
This website describes colour blindness and lists some of the everyday problems colourblind people put up with.

Ishihara tests for colour blindness
This website contains the Ishihara colour charts to test for red–green colour blindness.

Use the websites above to answer these questions.

1 Does a colourblind person see only in black and white and shades of grey?
2 Is there a cure for colour blindness?
3 Describe three everyday frustrations for colourblind people.
4 Use the Ishihara colour charts to test your colour vision.

ISBN 978 1 4202 3735 1

EXTRA FOR EXPERTS

Lights, colour, action!

Colour is a vital part of any stage play or film, and it can be used to produce a response from an audience. For example, the blue light on the characters in Figure 9.28 gives an impression of coldness and sadness, while the red light in Figure 9.29 gives warmth and excitement to the scene.

The coloured lights for theatre or film-sets are made by using coloured filters or gels that are attached to spotlights. The lights can be used to flood the stage with a particular colour, as in the two scenes in the photos, or they can be used to focus on particular characters.

The three characters in Figure 9.30 are dressed in different colours. Notice how the character dressed in yellow seems to have 'disappeared' in blue light, and you tend to focus your attention on the other two characters.

The character in yellow seems to disappear because her yellow clothes absorb the blue light and do not reflect any light. The character in blue, on the other hand, reflects all the blue light and does not absorb any light.

Using different coloured lights and characters dressed in different coloured clothes, the lighting director can create special effects on stage where characters come in and out of attention.

Questions

1 Suggest why red light flooding the stage gives a feeling of excitement and action. Do any other colours give the same effect?

2 Look at the singer in the middle in Figure 9.30. Explain why the colour of her dress and hair change in blue light.

3 Suppose you were designing the lighting for a scene that is set around a haunted house. Which coloured filters would you use? Why?

4 Why would it be best for the characters in the scene in Figure 9.28 to wear neutral-coloured clothes (such as white or grey) rather than reds, oranges and yellows? Use the words *reflect* and *absorb* in your answer.

Figure 9.30 The colours of objects on stage change with different coloured spotlights.

Figure 9.28 The blue light used in this scene gives a feeling of serenity and sadness.

Figure 9.29 The red light used in this scene gives a feeling of warmth and excitement.

ISBN 978 1 4202 3735 1

EXTRA FOR EXPERTS

Why is the sky blue?

When a beam of light passes through smoke or dust, some of it bounces off the tiny particles and is reflected towards your eyes. This is why you can see the beam. This bouncing of light from particles such as smoke or dust is called **scattering**. You cannot see a beam of light in clean air because the particles of air are too small to scatter the light.

The air around the Earth contains tiny bits of dust. These are too small to see, but they are big enough to scatter light. Blue light is scattered more by the dust than red light is. As the light from the sun comes through the atmosphere, the blue light is scattered. This scattered blue light bounces from dust particle to dust particle, spreading blue light through the whole sky. This is why the sky normally appears blue.

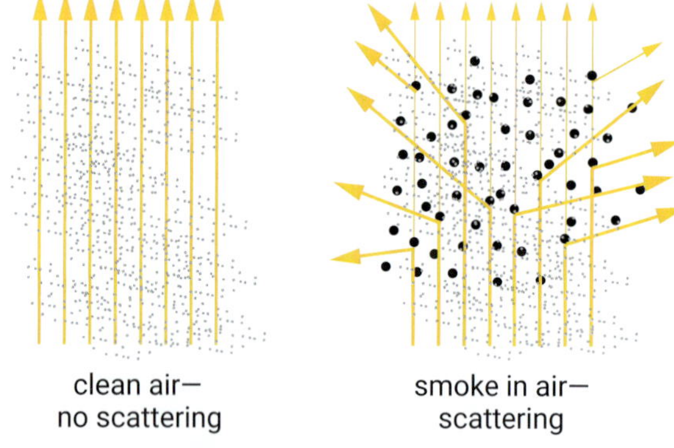

Figure 9.31 Light is scattered by particles in the air.

Why are sunsets red?

When the sun is low on the horizon, the light has more air to pass through as it travels through the atmosphere. Also the lower part of the atmosphere close to the horizon contains much more dust, so the blue light is scattered and the red light reaches your eyes. This is why sunsets are red. The dustier or smokier the atmosphere, the redder the sunset.

Questions

1 Why do suspensions scatter light but solutions do not?

2 a Why can you see the beam of light from a car's headlights when driving at night in fog?
  b Suggest why yellow lights are more effective than white lights when driving on foggy nights.

3 Suggest why sunsets are redder on cloudy, polluted or dusty days than on fine, clear days.

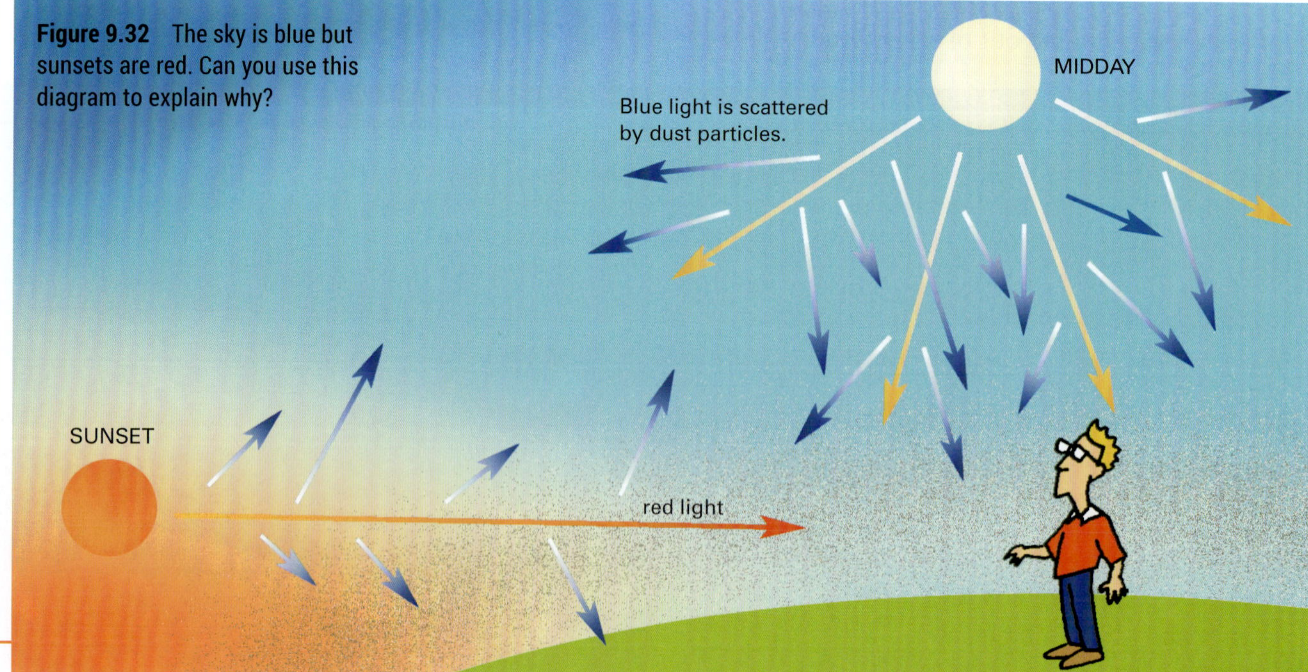

Figure 9.32 The sky is blue but sunsets are red. Can you use this diagram to explain why?

ISBN 978 1 4202 3735 1

CHECK

1 Go back to your answer for the second question in Getting started on page 207. Use the words *dispersion* and *spectrum* to explain why you see a rainbow of colours.

2 Use the words *absorbed* and *reflected* to explain why a banana looks yellow in white light.

3 What colour would a bunch of green grapes be in red light? Why?

4 A beam of white light shines through a blue filter. Use the words *transmitted* and *absorbed* to explain what happens to the colours in the white light.

5 A beam of white light passes through a filter and then through a prism. The prism disperses the light, and the different colours shine on a white screen. Use the information in the diagram below to work out the colour of the filter.

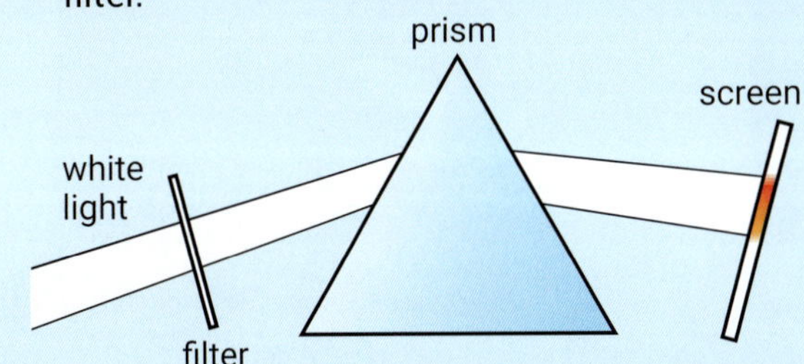

6 A combined beam of red and blue light hits a glass prism. On the other side of the prism, two separate beams of light are observed.

a Why did this happen?

b Which beam, A or B, on the diagram below is red? Give a reason for your answer.

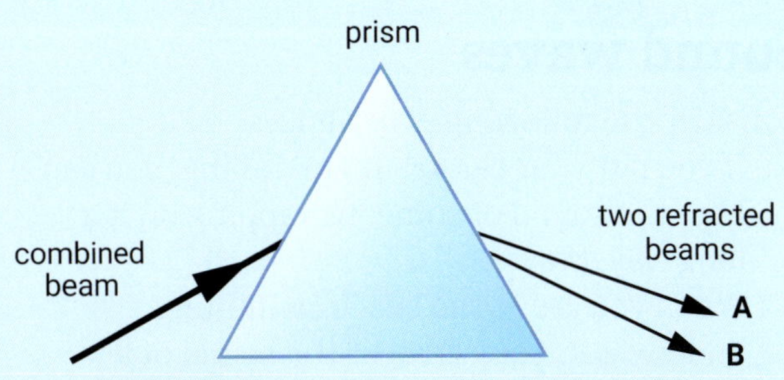

7 Why is the method of creating colours by mixing various coloured paints called subtraction?

8 Suppose green light is shone into your eyes.

a How do you see green light?

b Suppose a red light and a green light are shone together into your eyes. Predict what colour you would see. Which type of cone cells would not be detecting a colour?

CHALLENGE

1 A beam of light shines through a green filter. A red filter is then placed in front of the green filter. What will happen?

2 Some people believe that there are only six colours in the spectrum. Use library resources to find out which colour is in dispute.

3 When white light was detected by a light probe connected to a datalogger, graph 1 was obtained. It shows that white light contains equal intensities of all the spectral colours.

a When a coloured filter was placed in the path of the light, graph 2 was obtained. Infer the colour of the filter.

b Predict and draw the shape of the graph you would get if a violet filter was placed in the beam of white light.

Graph 1

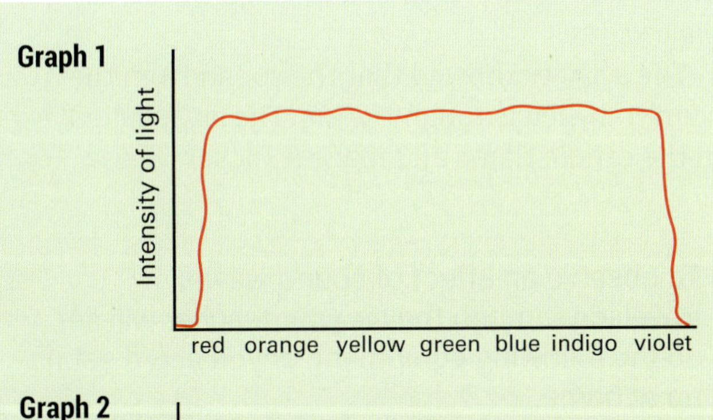

Graph 2

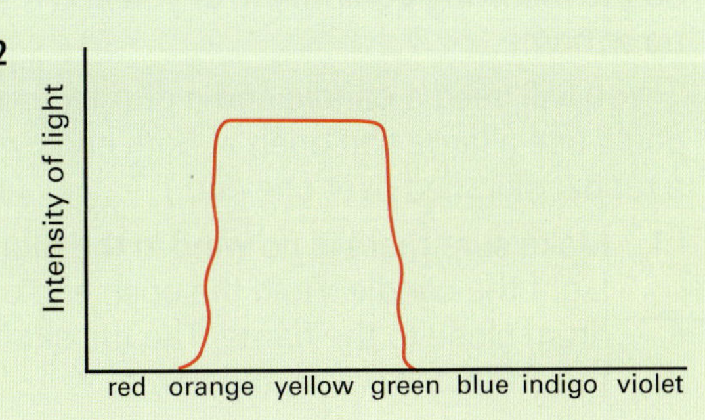

ISBN 978 1 4202 3735 1

9.3 Light and sound as waves

In the first section you learnt that sound and light are both forms of energy. How do these forms of energy travel from place to place?

Sound waves

Consider the following experiences.

- If you put your ear to a metal railing, you can hear the sound of someone tapping on it a long way away.
- When you are at the beach swimming underwater, you can hear the sound of a motorboat more than a kilometre away.
- The photo below shows an electric bell inside a large jar. The bell is heard when the switch is pressed. However, if all the air is pumped out of the jar, you cannot hear the bell. (Your teacher may set this up for you.)

Sounds travel in solids, liquids and gases. This is why you can hear sounds in a metal railing, in water and in air. But when the air is pumped out of the jar, no sounds are heard.

Sounds are made by vibrating objects. The vibrating strings on a guitar make sounds, as does the vibrating skin on a drum when it is struck. These vibrations are carried through the air as *sound waves*.

The activity showed that the sound from the drum travelled through the air. The air 'pushed' on the candle flame and made it flicker. These 'pushes', or sound waves, are the way sounds travel in air.

You may also have noticed that a soft tap on the drum made a soft sound and produced a small flicker in the flame. A harder hit on the drum made a louder sound and produced a larger flicker in the flame.

Figure 9.33 A vacuum jar: When the air is pumped out can you hear the bell?

ACTIVITY

To observe an effect of sound waves travelling through the air, your teacher will set up the following equipment, or you could set it up at home.

You will need a candle and a drum. (You could use a large can open at both ends with a rubber skin tied over one end.)

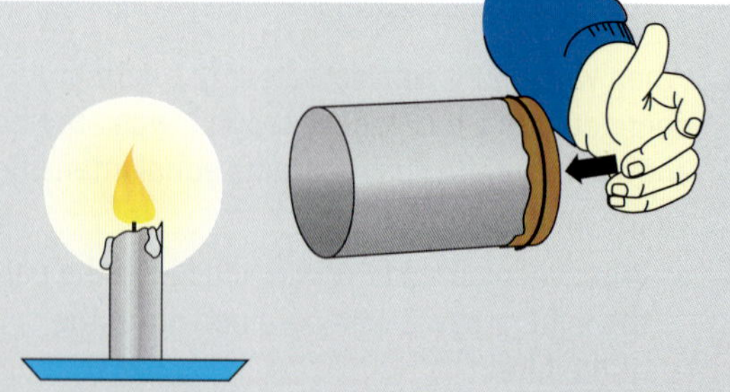

1 Make sure there is no wind in the room. Light the candle. Hold the open end of the drum close to the flame. Tap the skin on the drum and watch the flame.

2 For a more dramatic effect, hit the drumskin with a drumstick.

How do you think sound waves are responsible for the movement of the candle flame?

ISBN 978 1 4202 3735 1

Sound waves are made up of bands of high and low air pressure. The energy from the vibrating source is transferred from one air particle to another as the sound waves travel.

This is called a **longitudinal wave**, meaning the particles travel back and forth in the same direction as the sound energy that is being transferred.

Sound waves spread out in all directions through the air from the source of the sound. As they do this, the energy in the waves gradually decreases and the sounds become fainter.

The air is made up of particles of various gases.

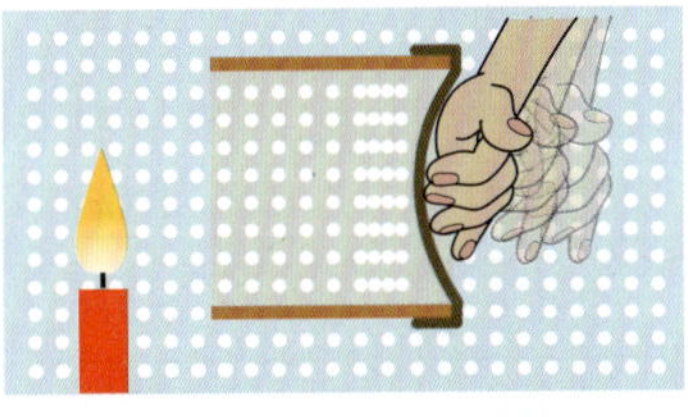

When the drum is struck, the particles are pushed together. This increases the air pressure.

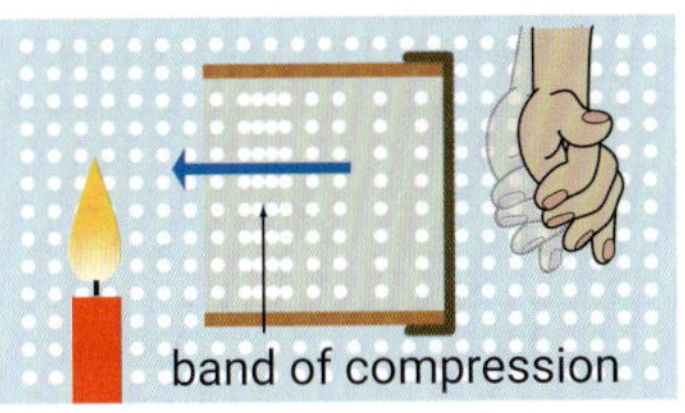

Each particle pushes on the next one as the wave moves through the drum, but the air as a whole does not move.

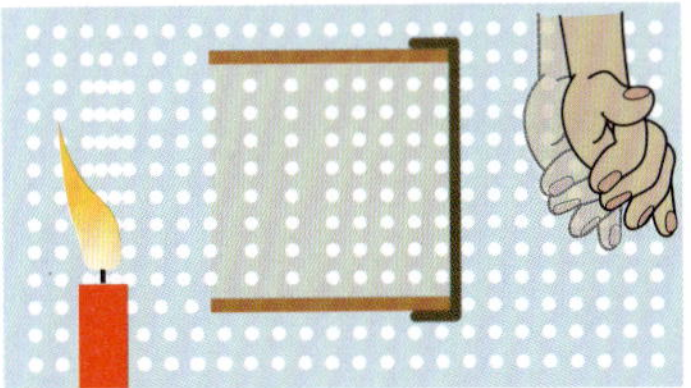

The candle flame flickers when the wave of increased air pressure hits it.

Figure 9.34 How sound travels as a longitudinal wave.

ACTIVITY

Your teacher may show you how you can model sound waves in a slinky spring. The waves travel through the spring as compressions.

Measuring sound

When the vibrations of sound are fast, they are heard as a high-pitch sound. This means that the pressure waves are close together. When the vibrations are slow and the pressure waves are further apart, the sound is heard as low-pitch sound. The pitch of sound is also known as its **frequency**.

We can therefore say that a high-pitch sound has a high frequency, and a low-pitch sound has a low frequency. Frequency can be represented by a wave as in Figure 9.35.

The loudness of sound can also be measured. The loudness depends upon the amount of energy carried by the wave, and is shown in Figure 9.35 as the **amplitude**.

The closer together the waves get the higher the pitch or frequency. And the larger the height of the wave, or amplitude, the louder the sound is. The loudness of sound is measured in **decibels**, dB (DESS-ee-bels).

Compression
Rarefaction
Direction of wave
Longitudinal wave
Amplitude (loudness)
Wavelength (pitch)

Figure 9.35 The pitch and loudness of sound can be shown on a wave diagram.

ISBN 978 1 4202 3735 1

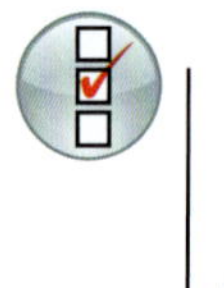

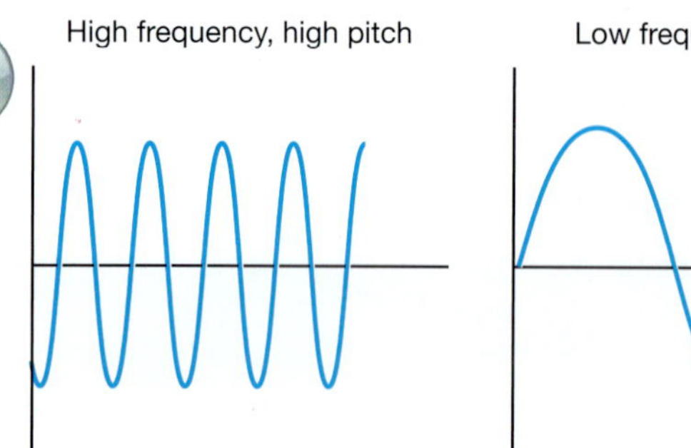

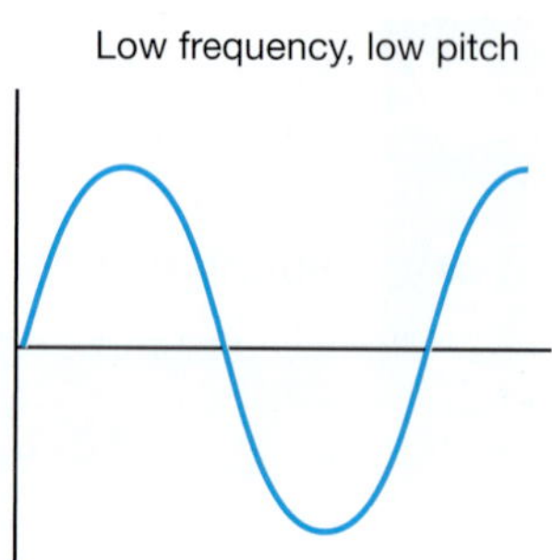

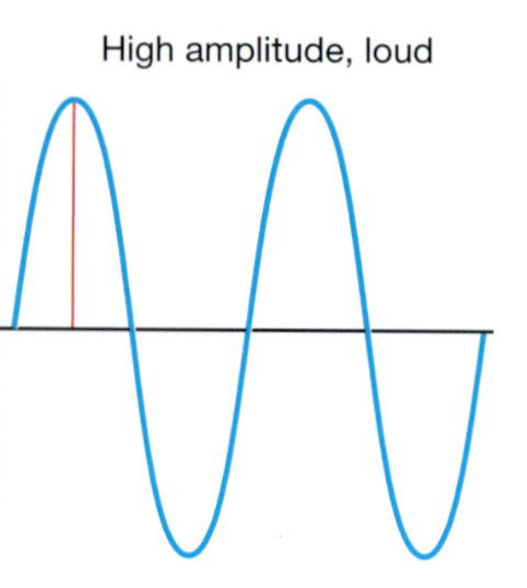

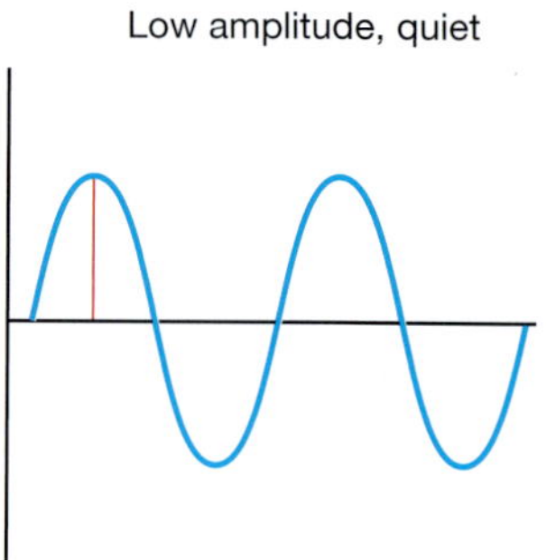

Figure 9.36 Frequency and amplitude change with the shape of the wave.

1. Place your fingers over the voice box in your throat. Now say 'AHH' in a deep voice. Can you feel your throat vibrating? Now say 'AHH' in a higher voice.
   Did you notice any difference in the way your throat vibrated?
2. Your teacher will set up a number of stations around the room. Each station will have an object that is capable of making a sound, or carrying sounds from one point to another. (Objects may include rulers, tuning forks, paper to roll up, reeds, stethoscopes and bottles.)
   Work in a small group and try to answer the following questions for each object.
   How do you get the object to make a sound? (Hint: You can strike, flick or blow!)
   For each object make a generalisation about the way it vibrates and the sounds it produces.
3. Your teacher may set up a sound generator connected to a computer. On the screen you will be able to see wave patterns that represent sounds.
   Describe the wave patterns made by:
   a. high-pitched sounds compared to low-pitched sounds
   b. loud sound compared to soft sounds.

   Include drawings in your descriptions.

The speed of sound

Sound travels at 330 metres per second in air at 0 °C. However, it travels even faster in other substances.

Substance	Speed of sound (m/s)	Speed of sound (km/h)
air (at 0 °C)	330	1 188
air (at 15 °C)	342	1 231
oxygen (at 0 °C)	317	1 141
water (at 0 °C)	1 410	5 076
water (at 15 °C)	1 450	5 220
lead (at 20 °C)	1 200	4 320
copper (at 20 °C)	3 500	12 600
iron (at 20 °C)	5 100	18 360
granite (at 20 °C)	6 000	21 600
wood (at 20 °C)	about 5 000	about 18 000

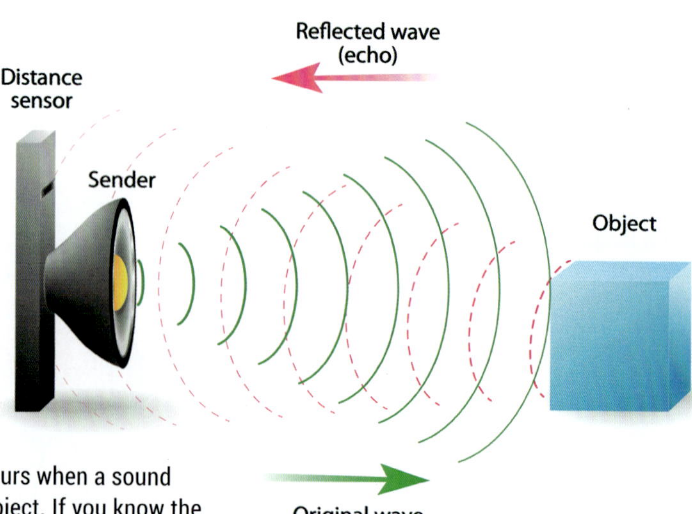

Figure 9.37 An echo occurs when a sound wave is reflected off an object. If you know the speed of sound and the time the echo takes to return, you can calculate the distance.

ISBN 978 1 4202 3735 1

Light waves

Light from the sun is a type of radiation that comes to us in the form of waves called **electromagnetic waves**. Unlike sound waves, light waves do not transfer their energy through the particles of gases, liquids or solids. Light waves can travel through the vacuum of space.

The various types of electromagnetic waves are different because they have different wavelengths.

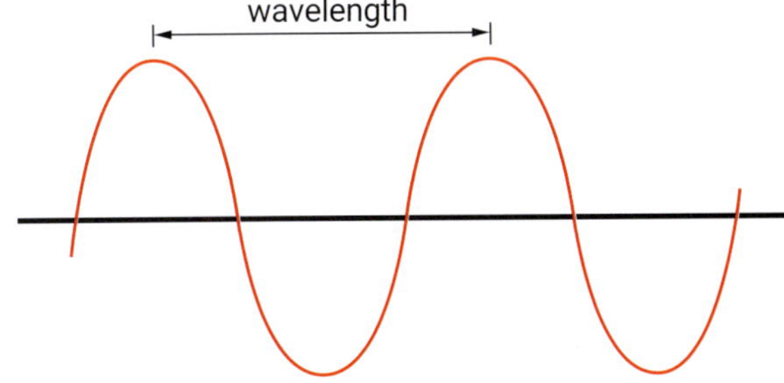

Figure 9.38 Light is represented as a wave.

Light is just a small part of the **electromagnetic spectrum**. Microwaves, infrared radiation and X-rays are other parts.

Visible light has a wavelength of about 0.000 000 5 m, while radio waves have very long wavelengths of about 10 m. The radiation with the shortest wavelength is gamma radiation, a very high-energy radiation that causes injury to the cells of living things. Generally, the shorter the wavelength of the radiation, the higher the energy of the waves.

ACTIVITY

Models for light and sound

On page 229 you used two models to explain how sound behaves—a theoretical particle model and an actual spring model.

Suggest ways of making an actual particle model to show how sound travels. You might like to use marbles or styrofoam balls attached to pieces of string.

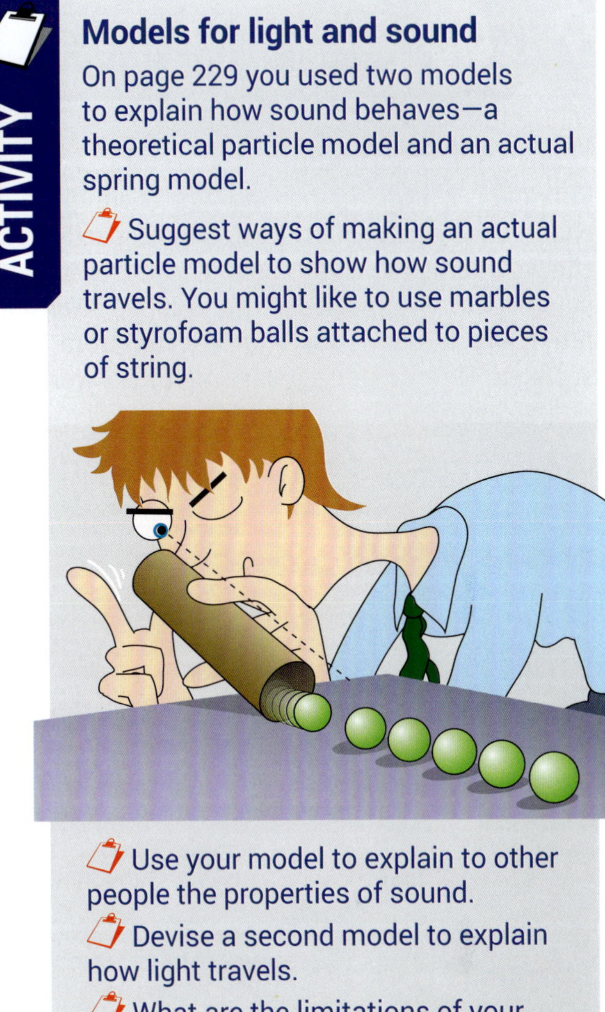

Use your model to explain to other people the properties of sound.

Devise a second model to explain how light travels.

What are the limitations of your models?

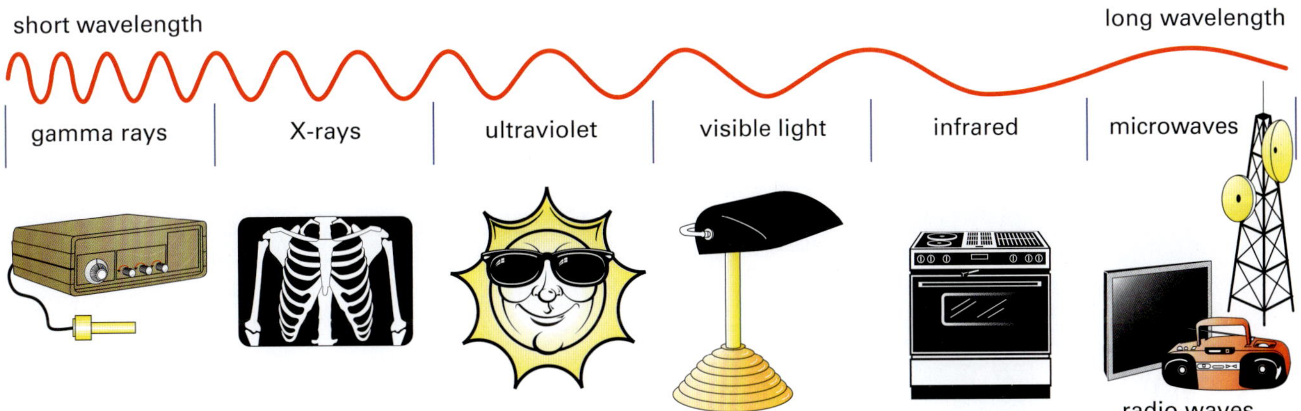

Figure 9.39 The electromagnetic spectrum. The short wavelength waves carry more energy than long wavelength ones.

ISBN 978 1 4202 3735 1

SCIENCE AS A HUMAN ENDEAVOUR

Why wear polaroids?

Polaroid is a brand name for a type of sunglasses that reduce glare from reflected light, particularly the light reflected from water. The lenses in these sunglasses contain special filters called *polarising filters*, which reduce the light reflected from surfaces. How do these polarising filters work?

Figure 9.40 A scene taken without a polarising filter on the camera lens (top) and the same scene taken with a polarising filter (bottom).

Polarised light

When you flick a rope up and down, you can make regular vertical waves in the rope. If you flick it sideways, you can make horizontal waves.

Light rays behave like the waves in a rope. A ray of light contains waves that vibrate vertically, horizontally and in all planes in between. So if you switch on a light, billions of light rays are emitted, all vibrating in different planes.

A polarising filter is a transparent substance that allows light waves that vibrate in only one particular plane to pass through.

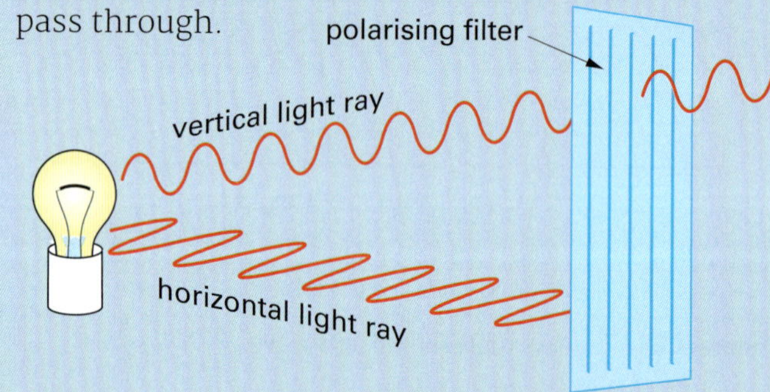

Figure 9.41 Only the light rays that vibrate vertically are able to pass through this polarising filter.

When light is reflected from water or wet roads, it is often polarised; that is, the reflected light waves are in one plane only. Most of these reflected polarised waves vibrate horizontally. So the lenses in polarising sunglasses contain filters that allow only vertically polarised light through and block the horizontal waves. This is how your polarising sunglasses reduce glare when you are at the beach or driving along wet roads.

Inquiry

1 Go outdoors and look at the reflections from still water, glass, a wet surface or grass through a polarising filter. Observe what happens when you slowly rotate the filter.
2 Reflected light from water or glass is polarised, but the reflected light from shiny metals is not. Use a polarising filter to test this.
3 Hold two pairs of polarising sunglasses up to the light, one in front of the other. Slowly rotate one and you will find that at a certain position no light can be seen. Why?

ISBN 978 1 4202 3735 1

Light waves and refraction

In the first section of this chapter, you learnt that light refracts when it passes from one transparent substance to another. This is because light slows down as it passes from air to glass.

Light, like all types of electromagnetic radiation, travels at incredible speed—about 300 000 000 m/s or 3×10^8 m/s. This is about a million times faster than the speed of sound. No wonder you see the lightning before you hear the thunder of a distant thunderstorm!

The speed of light in glass is 1.98×10^8 m/s —about two-thirds of what it is in air. When light passes from air to glass at an angle, it slows down and bends towards the normal.

How far away is that thunderstorm?

You can use the fact that light travels nearly one million times faster than sound in air to calculate how far away a thunderstorm is.

It takes sound 3 seconds to travel 1 km in air. So when you see the lightning flash, count the seconds by saying 'one thousand, two thousand ...' then calculate how far away the storm is.

How a rainbow forms

A drop of water has the same effect on light as a prism does—it is dispersed into the spectrum of colours. But why is violet light refracted more than red light?

It has been found that different colours of light have slightly different speeds in the same substance. For example, the speed of red light in water is 2.280×10^8 m/s, while that of violet light is slower, at 2.255×10^8 m/s. This slight difference in speed means that violet light bends more than red light when it passes through a drop of water.

When sunlight hits a raindrop at a particular angle, the white light is dispersed into the spectral colours. These colours come out of the raindrop at different angles. Because of this, your eye only sees one colour from each drop (see the top diagram). The red light in the rainbow comes from the droplets highest in the sky and the violet light from the droplets lowest in the sky. So red should be on top of the rainbow and violet underneath. Check this in the rainbow photo on page 220.

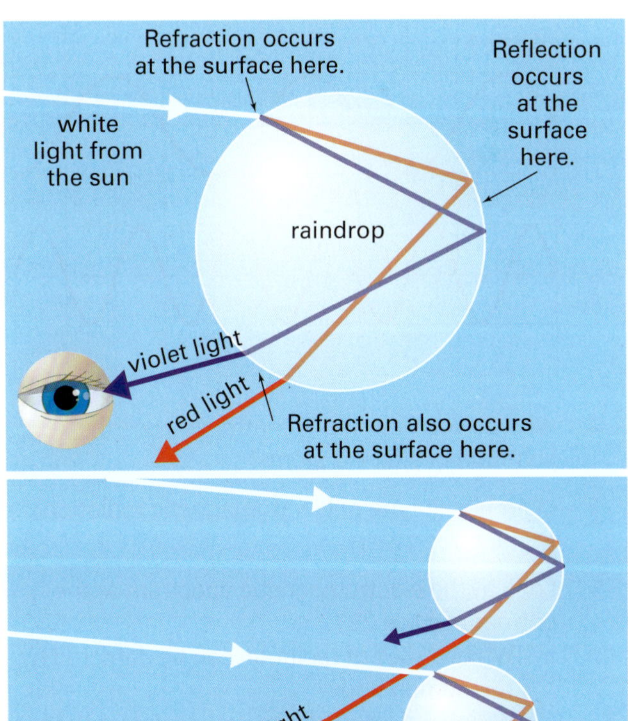

Figure 9.42 How a rainbow forms.

ISBN 978 1 4202 3735 1

1 Sound cannot travel through:
A wood.
B fresh water.
C outer space.
D the ocean.
E the Earth's crust.
Justify your answer.

2 Decide whether each of the statements below is true or false by referring to the table of speeds of sound on page 230. For each case, give reasons for your decision.
a Sound travels faster through gases than through liquids.
b Sound travels faster through warm air than through cold air.
c Sound travels at the same speed through all gases.
d Sound travels faster through metals than through non-metals.

3 Two waves were drawn on centimetre square graph paper.
a Which wave has the longer wavelength?
b What is the wavelength of wave A?

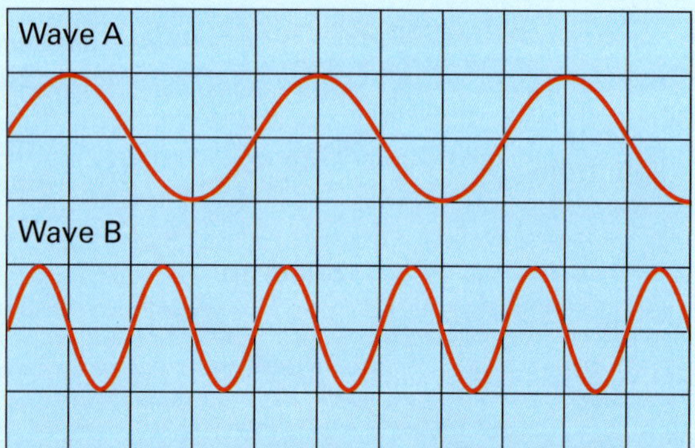

4 Why do you hear thunder after you see the lightning in a far-off storm?

5 Light is one type of electromagnetic radiation. Name three others.

6 Look at the electromagnetic spectrum at the bottom of page 231.
a Which types of radiation can be detected by the human body?
b Which receptors do you use to detect them?
c Which types of radiation can be used for communicating with other people?
d Which type of radiation is commonly referred to as heat?

7 Using your knowledge of sound and light, write a paragraph outlining the similarities and differences between them.

8 A combined beam of yellow and blue light was shone onto a prism. Two separate beams emerged from the other side. Use your knowledge of light waves to explain why beam B is blue.

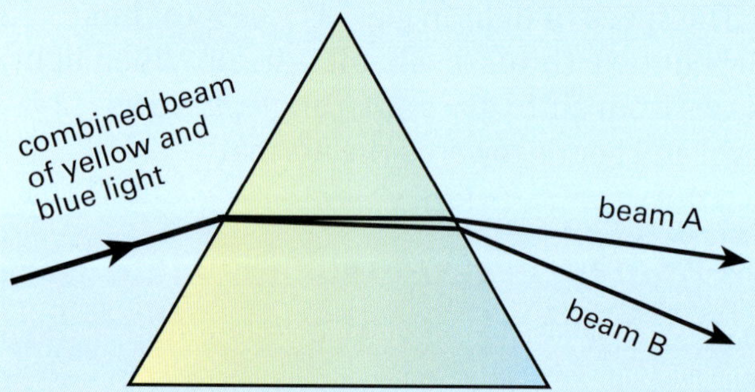

9 How does the candle and drum demonstration on page 229 show that sound is a form of energy?
Use the particle model to explain why there is more energy in a loud sound than in a soft one.

10 Design an experiment to show that light, unlike sound, does not need a substance such as air in which to travel.

11 A string telephone can be made from two metal cans and some string. Suggest why:
a the telephone works only when the string is stretched tight
b the telephone does not work when a third person touches the string.

12 a Use a diagram to describe what the amplitude and frequency of a sound wave are.
b Draw a wave diagram to show the difference between a loud sound and a quiet sound.
c What does the term 'longitudinal wave' mean?
d What is the unit for sound?

ISBN 978 1 4202 3735 1

1 Sam is a long way away from you and he is trying to tell you something. He rolls up a piece of cardboard in the shape of a cone and speaks through it. You can now hear him. Explain in terms of sound waves why this happens.

2 Suggest why you can hear sounds better when the wind is blowing towards you than when it is blowing away from you.

3 In a science-fiction movie, the goodies destroy an enemy spacecraft in deep space with laser guns, and they hear it explode as they fly past. What is wrong with this scene?

4 The diagrams below show two different sounds. Different sounds have different wavelengths. The wavelength of sound is the distance between the bands of compression of the particles.

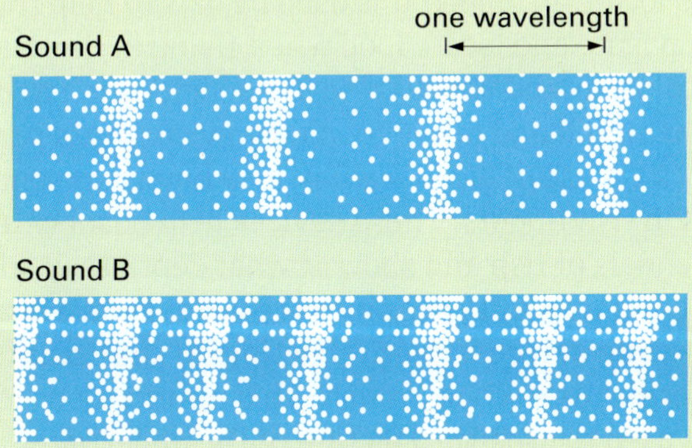

a Which sound has the shorter wavelength?
b High-pitched sounds have shorter wavelengths than low-pitched sounds.
Which sound has the higher pitch: A or B?
c Hold a ruler over a bench and flick it. It vibrates and makes a sound. Notice how the ruler vibrates. Increase the length of the ruler over the bench and flick it again. Look at the way it vibrates and listen to the pitch.
How do you think the wavelength, the speed of vibration and the pitch of the sound are related?

5 The diagram below shows a ray of light passing from air through three different transparent substances.

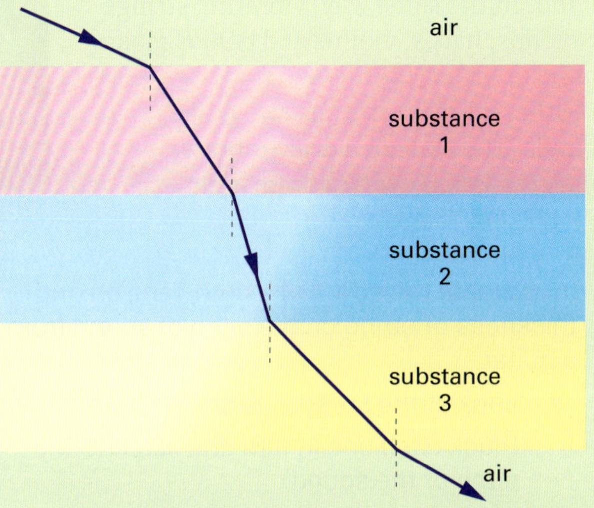

a Does light travel faster or slower in substance 1 than in air? Give a reason for your answer.
b In which substance is the speed of light closest to that in air?

6 On page 229 the particle model was used to explain how sound waves travel in air. Use the model to explain why sound travels faster in liquids than in gases, and even faster in solids.

7 A person fires a gun and hears an echo from a cliff after 5 seconds. If the temperature is 15 °C, use the speed of sound on page 230 to calculate how far away the cliff is.

8 Suppose someone is talking about you in the next room. When you put your ear to the wall, you can hear what the person is saying.
a Try to explain in terms of waves why you can hear sounds through the wall but cannot see light through it.
b Which types of radiation can pass through walls? (Hint: Refer to the electromagnetic spectrum on page 229.)

ISBN 978 1 4202 3735 1

9.4 Applications of sound

Making music

You have probably strummed a guitar, even though you may not be able to play one. If you look closely at the strings, you will see that they vibrate when they make a sound. When they are still, you hear no sound. In the same way, vibrating strings in pianos, harps, violins and banjos make sounds.

Other vibrating things also make sounds. The skin on a drum vibrates when it is struck with a drumstick.

Figure 9.43 All instruments make sound by causing a vibration. How does each of the different instruments in this orchestra cause a vibration?

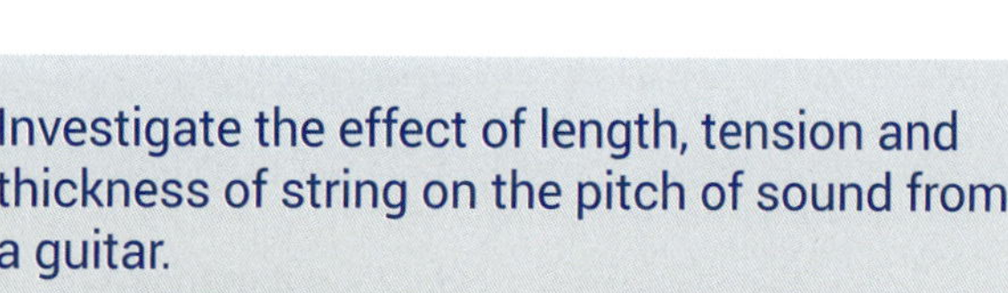

Investigate the effect of length, tension and thickness of string on the pitch of sound from a guitar.

- **a** Look at the six strings on the guitar.

  Pluck each one in turn and listen to the pitch of the sound.

  Make a generalisation about the thickness of the string and pitch.
- **b** Turn one of the tuning keys. This alters the tension in the string. What happens to the pitch of the sound when you alter the tension in the string? Does this happen for all the strings?
- **c** What is the purpose of the frets on the guitar? Investigate this question and write a brief report of your findings.

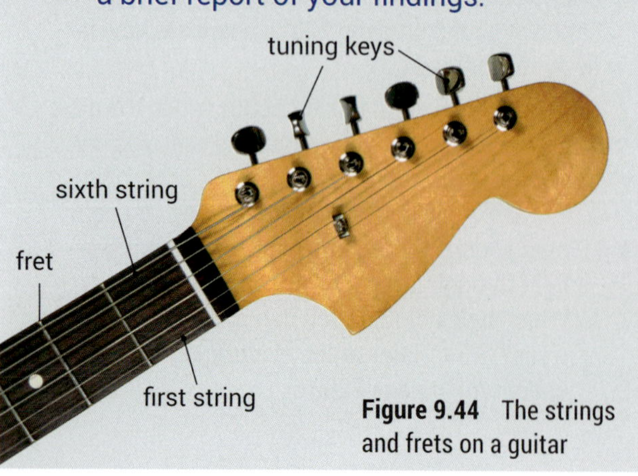

Figure 9.44 The strings and frets on a guitar

Tissue paper vibrates when you place it over a comb and blow through it. An empty bottle makes a sound when you blow across its mouth because the air inside the bottle vibrates. This is how musical instruments such as the trumpet, tuba, saxophone, flute and clarinet work.

Echolocation

Echolocation is the use of sound waves and echoes to locate an object. It is used by bats to help them navigate and locate food when flying in the dark. Dolphins and whales also use echolocation to help them find and target food when visibility is poor.

One use of echolocation, also known as sonar, is in fish finders. These devices help boats locate schools of fish and show at what depth they are. Sound waves are sent into the water from a boat, and are detected after reflecting off the ocean bottom or a school of fish. This information is used to produce a picture of the underwater landscape. The same technique is used to map the ocean floor. The distance or depth is calculated using the speed of sound in water and the time taken for the sound to return. The longer the sound takes to return, the greater the depth.

ISBN 978 1 4202 3735 1

Figure 9.45 Sound waves travel through water and will reflect off a school of fish back to a detector on the boat.

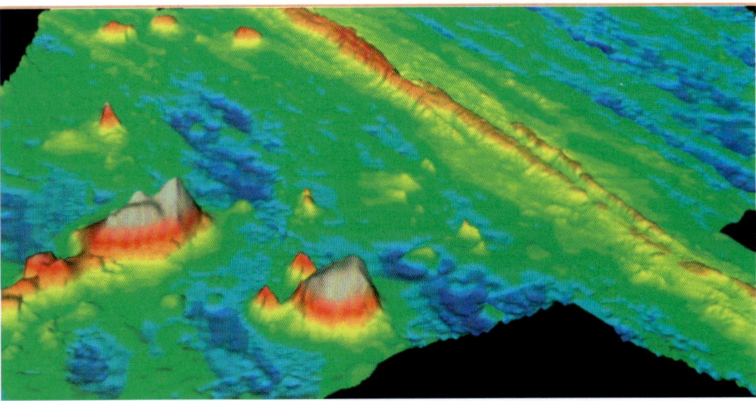

Figure 9.47 The colours in this map of the Pacific Ocean floor—produced using sonar—show the depth, from dark blue (deepest) through light blue, green, yellow and red to white (shallowest). Some underwater volcanoes can be seen.

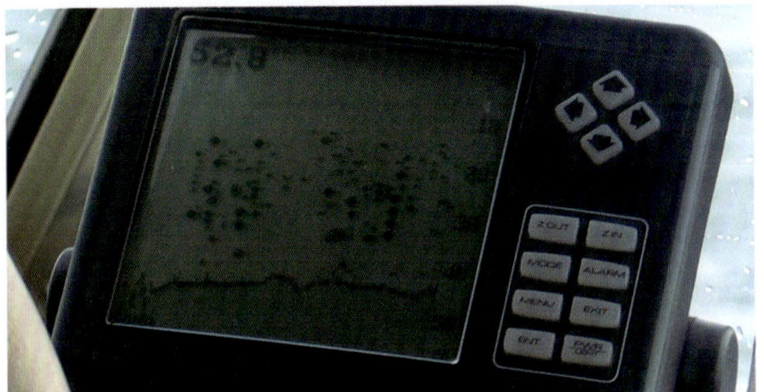

Figure 9.46 This recreational fish finder shows the ocean floor and schools of fish.

The human ear

The ear is the sense organ that detects sounds. Hearing is a very important sense in humans. Without it, communicating with others is very difficult. Young children learn to speak by listening to other people speak. As well as detecting sounds, the ear contains structures that control your sense of balance. With these you can tell what is up and down without using your eyes. They also maintain your balance when walking and running. When sound waves enter your ear, you hear sounds. The diagram below explains how this happens.

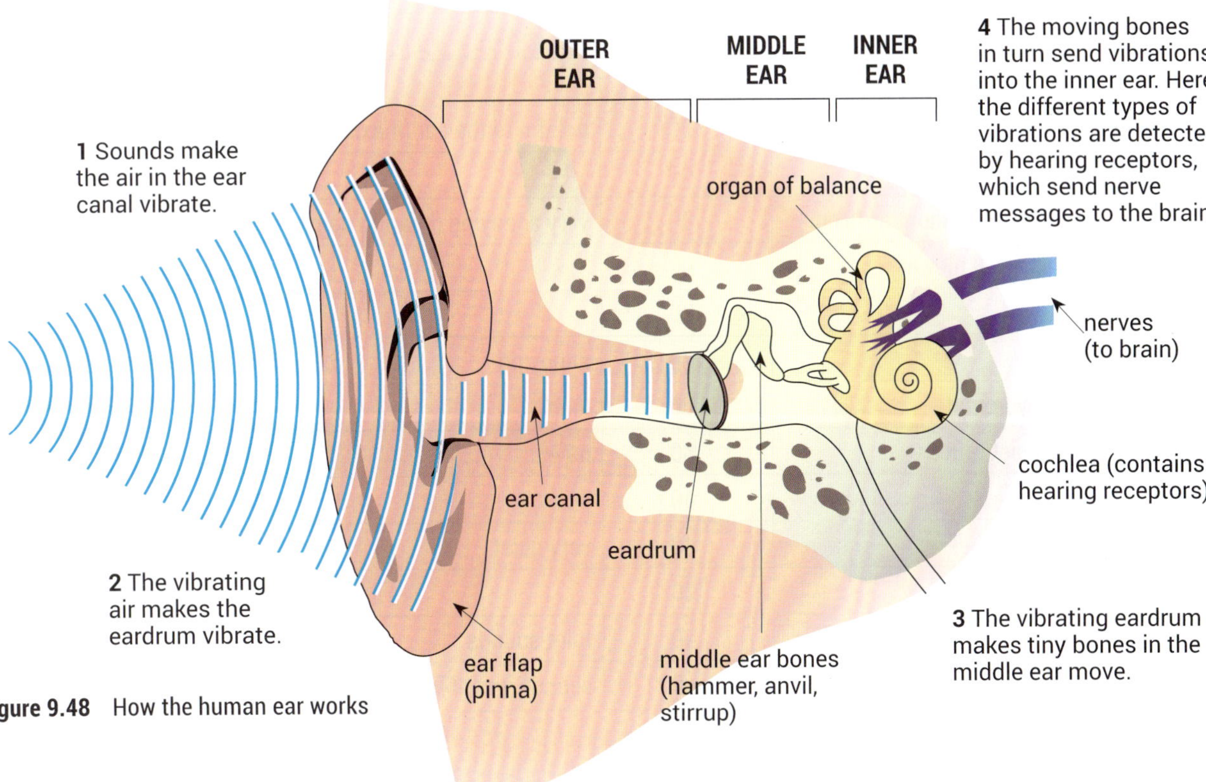

Figure 9.48 How the human ear works

ISBN 978 1 4202 3735 1

Problems with hearing

The volume of sound is measured in decibels. Zero decibels is the quietest sound that can be heard by the most sensitive ears. When sounds exceed 150 decibels they are so intense that they can cause your eardrums to break.

Cars and trucks are a source of noise pollution. The table below shows the noise levels produced by cars and trucks as heard by a person on the footpath.

Source of noise	Decibel level
car travelling at 60 km/h	65
car travelling at 100 km/h	75
truck travelling at 60 km/h	100

Very few people can tolerate working in a noise level of 70 to 100 decibels. Loud noise levels like these have been blamed for health problems such as high blood pressure and nervous tension.

Loud noises (85 decibels or more) for any length of time can cause permanent damage to the inner ear and premature hearing loss can result. The graph below shows how the average person suffers a hearing loss of the high-pitched sounds. For people who work in noisy situations, the graph could be quite different and show a larger hearing loss.

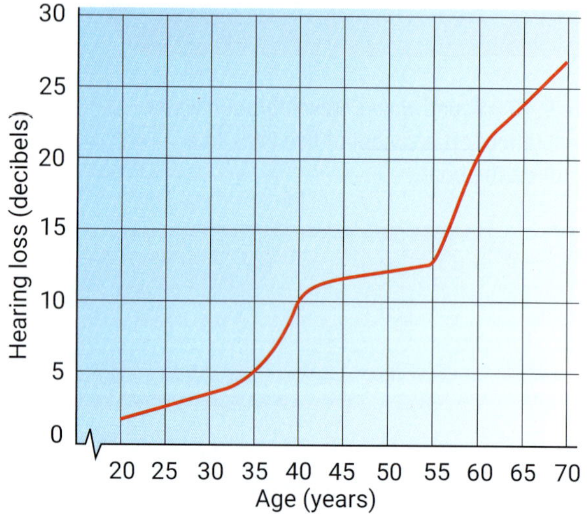

Figure 9.49 Graph of hearing loss with age

Sound source	Decibel range	Effect on person
	150	ear drum breaks
explosions, gun firing	140	
siren 1 m away, jet take-off at 10 m	130	
riveting machine	120	painful to the ear (maximum loudness of human voice)
car horn at 1 m	110	
Rock band, noisy motor cycle, shout at 15 cm	100	very annoying (hearing damage after 8 hours)
pneumatic drill at 10 m, lawn mower	90	
train at 10 m	80	
city traffic, noisy party	70	annoying
noise office, school playground	60	loud
normal conversation, quiet office	50	quiet
library	40	
soft whisper 3 m away	30	very quiet
	20	
	10	faint sounds
	0	sounds just able to be heard

ISBN 978 1 4202 3735 1

Deafness occurs in humans when they cannot hear certain sounds. In hearing tests, sounds are played into your ears and your hearing is compared with what the 'average' ear can hear. If you cannot hear the sound, you are said to have a hearing loss. If the sound has to be increased 30 decibels before you can hear it, then you have a 30 decibel hearing loss. Listening to loud music, especially through headphones can cause such hearing loss.

Loud noises or ear diseases can damage the hearing receptors in the inner ear. Ear infections in young children can cause damage to the eardrum and to the tiny bones that transfer the sounds to the inner ear. Because of this, serious ear infections should be checked by a doctor.

Improved hearing

For thousands of years, people with hearing loss used horns from cattle to improve their hearing. These devices channel more sound waves into your ear so that you hear more clearly.

Electric hearing aids were invented around 1900 and have become more efficient with the use of electronic parts. Figure 9.50 shows how a modern digital hearing aid works.

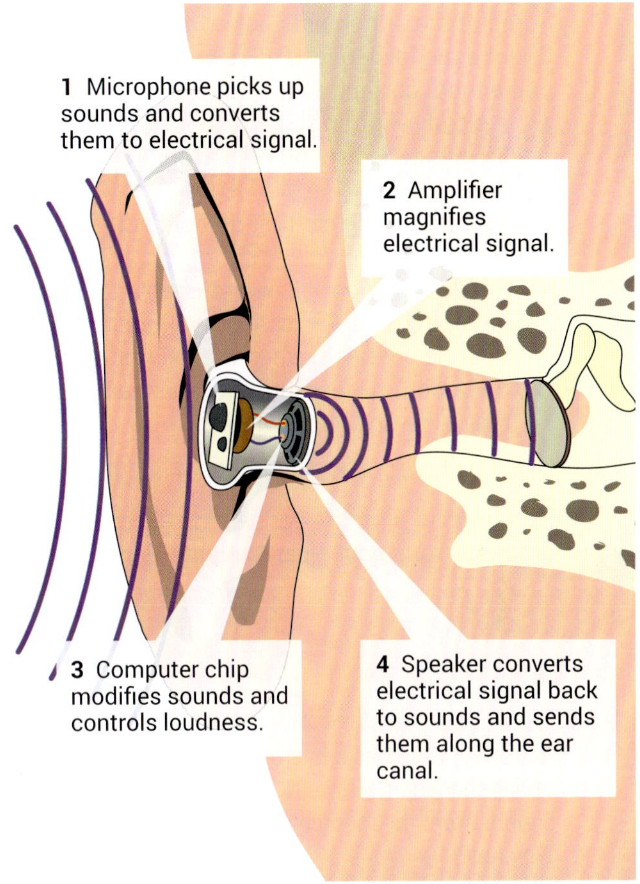

Figure 9.50 How an in-the-ear hearing aid works.

SCIENCE AS A HUMAN ENDEAVOUR

The bionic ear

Rosie is 16 years old and is completely deaf as a result of a tumour in both ears. The tumour destroyed some of the hearing receptors in her inner ear. Because of this, hearing aids do not help her hear sounds. Doctors suggested that she should consider having an artificial listening device called a cochlear implant, or bionic ear, inserted in her inner ear.

The cochlear implant was developed by Professor Graeme Clark at the University of Melbourne to help deaf people hear. The diagram shows how it works. Today more than 80% of the world's cochlear implants use the device made by the Australian company Cochlear Limited.

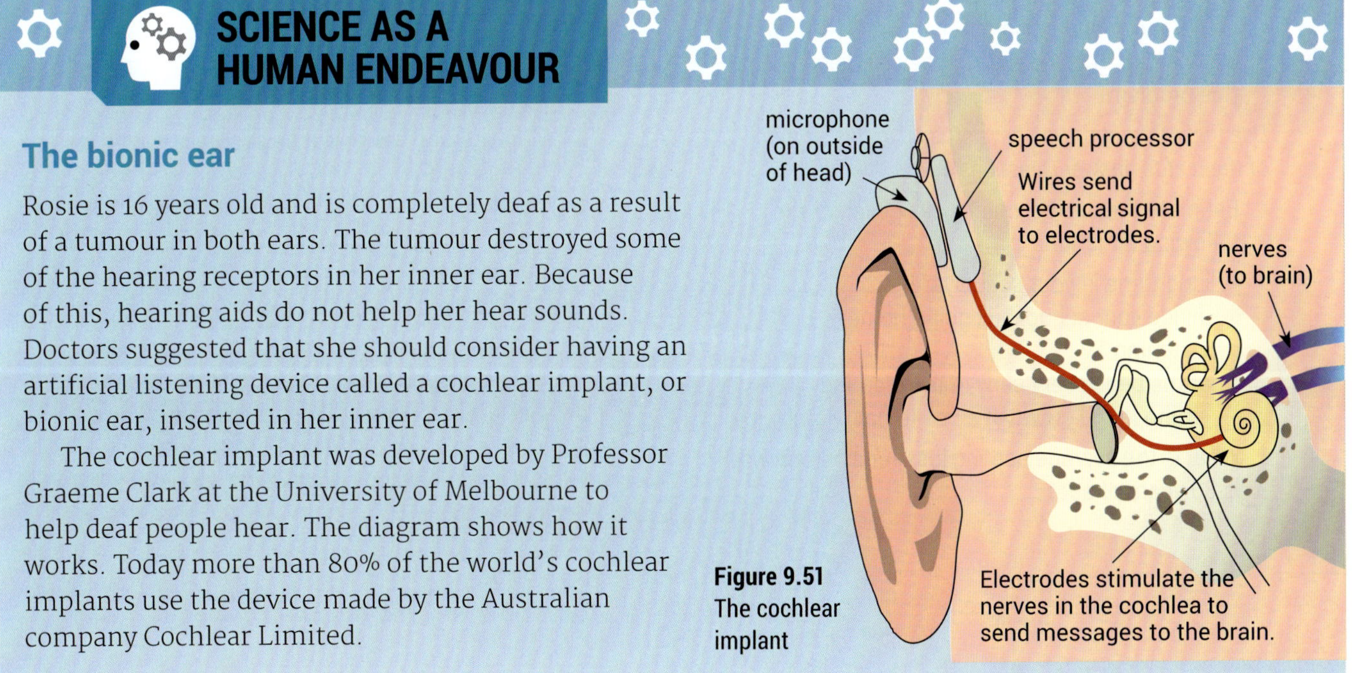

Figure 9.51 The cochlear implant

ISBN 978 1 4202 3735 1

CHECK

1 Some of the statements below are false. Choose the false ones and rewrite them to make them correct.
   a A slowly vibrating violin string makes a lower-pitched sound than a rapidly vibrating one.
   b The part of the ear that detects sound is found in the middle ear.
   c An older person will be better able to hear high-pitched sounds than a younger person.
   d Sounds are heard when their vibrations hit the eardrum, pass into the inner ear and make the tiny bones vibrate.
2 Describe how you hear sounds.
3 Explain how an in-the-ear hearing aid is different from a cochlear implant.
4 a What is the unit for sound measurement?
   b What are the minimum and maximum values that can be heard by humans?
5 How can people who work in noisy factories avoid damaging their hearing?
6 Between which hours are noisy machines not allowed to be operated? Do you think these times are reasonable? Give your ideas.
7 How does the speed of a car affect the loudness of the sound it produces?
8 Explain why listening to music through headphones at high volume for hours a day could be dangerous?
9 a Explain the term *echolocation* in your own words.
   b Outline one use of echolocation by people.

CHALLENGE

1 Astronaut Kate has a microphone on the outside of her spacesuit. However, she cannot hear any sounds from the moon buggy she is riding in. Why not?
2 You pluck one string on a guitar and it makes a sound. How could you cause that same string to make a higher-pitched sound?
3 When you drive into the mountains or fly in an aeroplane your ears sometimes 'pop'. Find out why they do this. How does chewing lollies help this problem?

Use the information you have studied in this section of the text to answer Questions 4–8.

4 Estimate the decibel readings for each of the following situations:
   a lunchtime in your schoolyard
   b inside your school bus going home
   c watching TV
   d at the beach very early in the morning
   e sitting in the crowd at a basketball match
5 What is the average hearing loss for a person who is 35 years of age?
6 In which 5-year age span is the average hearing loss greatest?
7 What hearing loss occurs from age 40 to 55?
8 Look at the table on the right. It shows the hearing losses of five people.

Person and age	Hearing loss (decibels)
Emily–30	3
Max–65	18
Sunil–25	15
Jessica–50	18
Daniel–42	15

   a Which person has a hearing loss 50% greater than average?
   b Which person has about an average hearing loss for their age?
   c Which person may have had a serious middle ear disease during childhood? Give reasons for your answer.
9 Imagine you are a member of parliament and that you are preparing noise pollution laws. What factors would you take into account when setting decibel levels for city and country areas? Would some area in cities be different from others? Explain your decision and give the maximum decibel levels.

ISBN 978 1 4202 3735 1

MAIN IDEAS

Copy and complete these statements to make a summary of this chapter. The missing words are on the right.

1 Reflection is a ______ of light and sound. Another property of light is that it travels in ______.

2 The ______ states that the angle of ______ is equal to the angle of reflection.

3 ______ of light occurs when a beam of light passes from one ______ substance into another, e.g. from air to water. The amount of refraction depends on the substances.

4 White light can be ______ by a prism into the colours of the ______.

5 A coloured object reflects some colours and ______ the rest. The colour you see depends upon the colours that are reflected.

6 Different colours can be made by mixing different coloured lights (addition) or by mixing paints (______).

7 Sound waves are produced by ______ objects and travel through gases, liquids and solids.

8 Light is a form of ______ radiation that can travel as waves through a ______. The speed of light is much greater than the speed of sound.

9 The unit for the loudness of sound is the ______.

10 Sound energy travels in a ______ wave.

11 On a sound wave diagram, the loudness is represented by the ______ of the wave.

spectrum
dispersed
amplitude
incidence
property
refraction
absorbs
subtraction
longitudinal
transparent
decibel
vacuum
straight lines
law of reflection
vibrating
electromagnetic

CH•9 REVIEW

1 A ray of light hits a mirror. The path of the light ray after it is reflected is shown by light ray:

A AW.
B AX.
C AY.
D AZ.

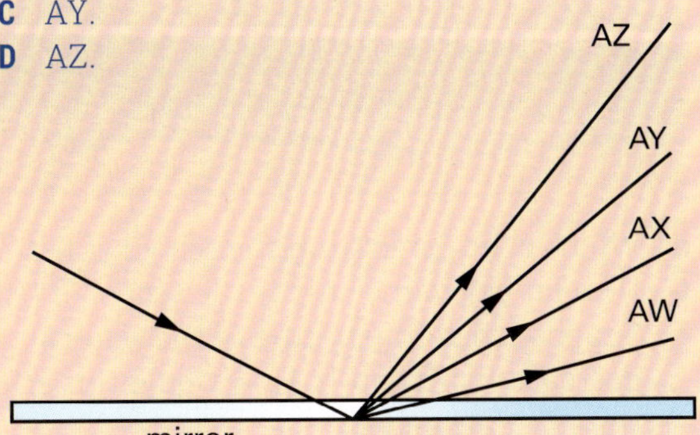

2 Ian is using a fine spray to water his seedlings. When he sprays the water into the air, he sees the colours of a rainbow. The rainbow is caused by the:

A reflection of light.
B transmission of light.
C absorption of light.
D dispersion of light.

3 Each of the drops of water in Question 2 is acting as a:

A glass prism.
B lens.
C plane mirror.
D concave mirror.

ISBN 978 1 4202 3735 1

4 Three parallel light rays shine through a transparent object and are refracted as shown below. Which shaped object will cause this refraction?

A a rectangular glass block
B a converging lens
C a diverging lens
D a triangular glass prism

5 Three parallel light rays shine onto a converging lens. The scale drawing below shows the results. Each square of the grid is 5 cm × 5 cm.

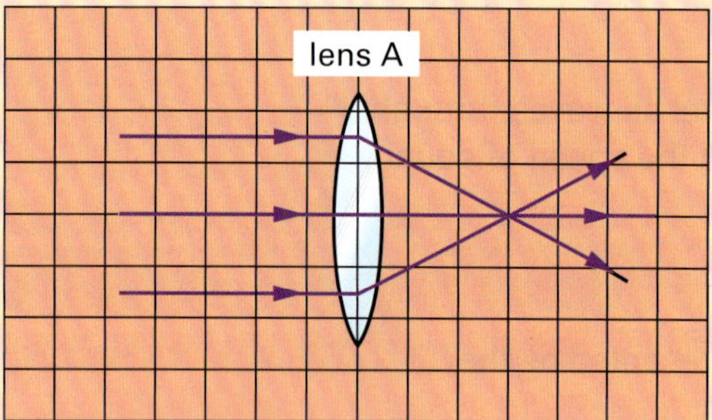

a What is the focal length of the lens?
b Suppose lens A is replaced with another converging lens, B. This lens has a focal length of 25 cm. What is the shape of lens B compared with that of lens A?

6 A beam of light consisting of red, green and violet light shines on a white screen. A coloured filter is placed over the beam. Green light is seen on the screen.

a Which colours are being transmitted?
b Which colours are being absorbed?
c Infer the colour of the filter.

7 Which colour light is shining in your eye when all three types of receptors in your retina are sending messages to your brain? Explain your answer.

8 A combined beam of three different coloured lights is shone through a prism. Two of the coloured lights are green and red. The other coloured light is either yellow or blue. The diagram below shows the results. If beam C is green, work out the colour of beams A and B. Explain your answer.

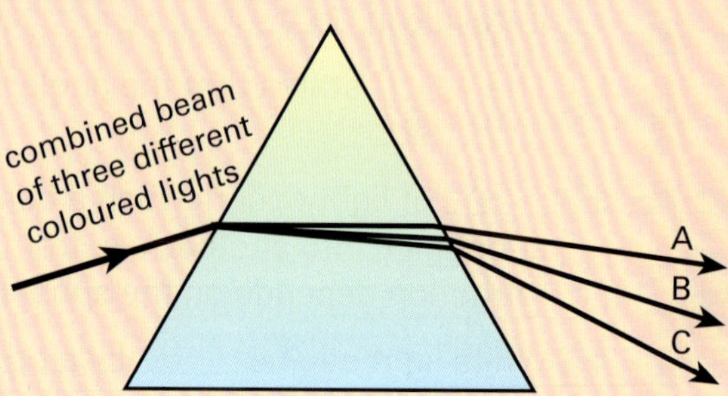

9 Echo sounders send sound waves through water to determine its depth. They can also be used to find the depth of schools of fish. Suppose a reflected sound wave returns after 0.1 seconds, and a second one returns after 0.2 seconds. The fisherman believes that one echo came from a school of fish. (The speed of sound in water at 15 °C is 1450 m/s.)

a Which echo came from the school of fish?
b How far below the ship is the school of fish?
c Suppose the temperature of the water decreases with depth. How would this affect the calculations?

ISBN 978 1 4202 3735 1

10 Label parts 1 to 8 of the ear on shown on this diagram.

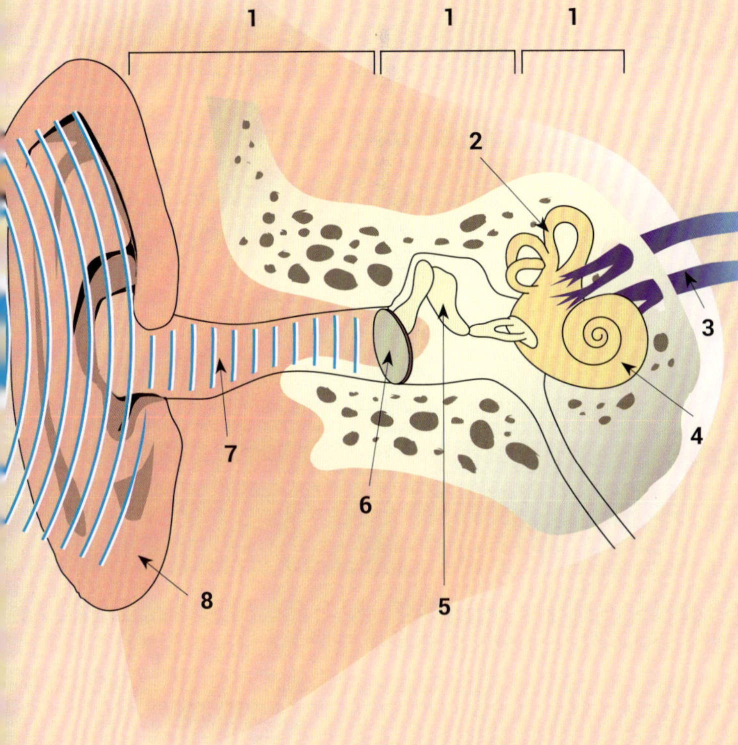

11 Draw a diagram or diagrams of sound waves and label the following:

a amplitude
b frequency
c compression and rarefaction
d longitudinal wave
e wavelength

12 Correctly identify the following wave diagrams as high pitch or low pitch.

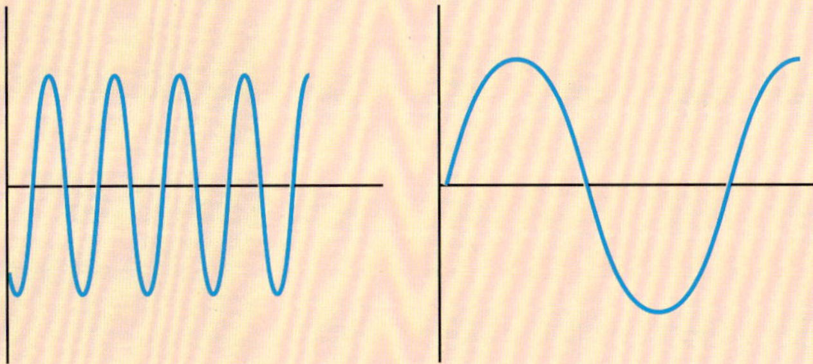

13 Explain how a violin can make a low-pitch or a high-pitch sound.

14 **a** Loud sound can be dangerous to your hearing. At which loudness level does sound become dangerous?
b Explain how the amount of time spent listening to loud sounds can also affect how dangerous a sound is.
c Describe three ways in which you could protect yourself from hearing damage if you attend a Grand Prix race.
d Imagine that you have permanent hearing loss and are finding it hard to hear. Describe one technology that could help improve your hearing.

15 Draw diagrams to compare short- and long-sightedness. Explain how you would fix each condition.

Check your answers on page 300.

ISBN 978 1 4202 3735 1

IN THIS CHAPTER

Science Understanding

- recognise that cells reproduce by cell division
- learn about sexual reproduction, pregnancy and childbirth in humans
- distinguish between asexual and sexual reproduction
- compare reproductive systems of various organisms

Science Inquiry Skills

- construct data tables and graphs to display data
- use correct scientific terms to describe parts of animals and plants and how these parts function

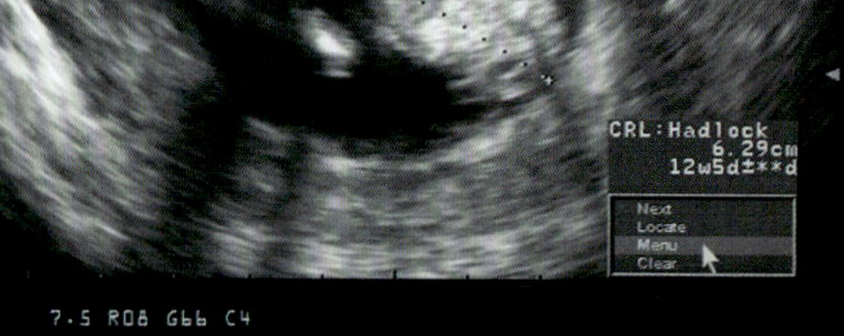

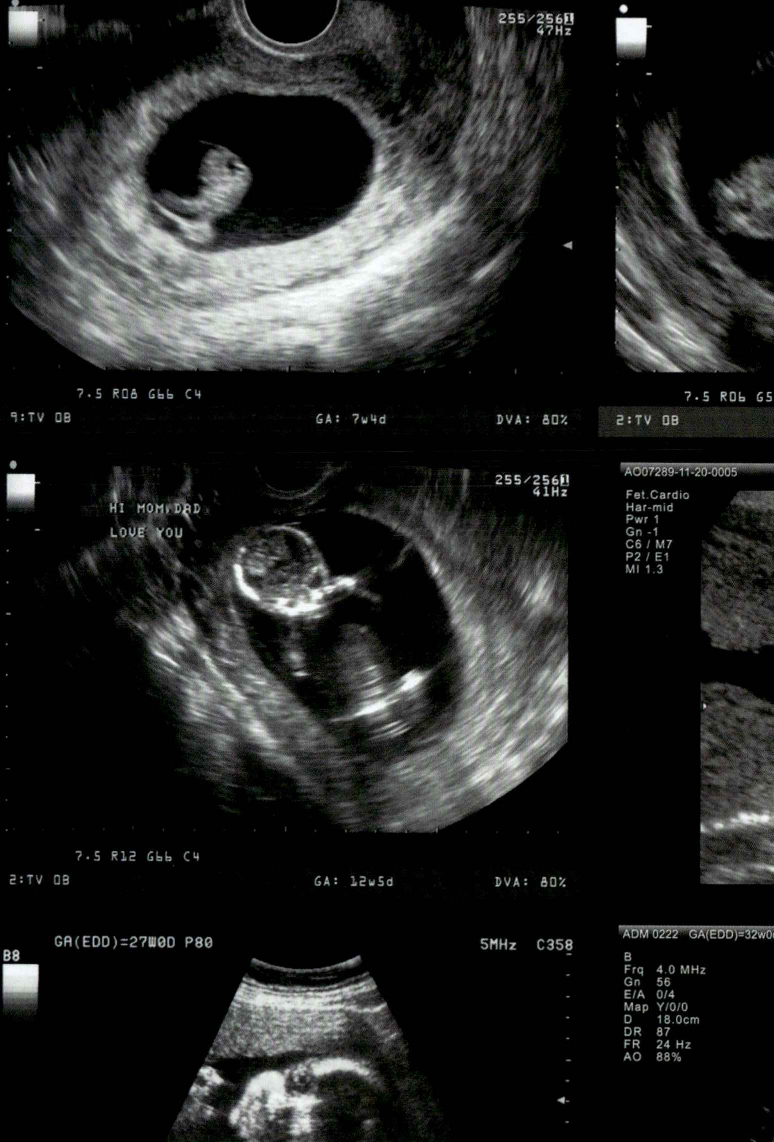

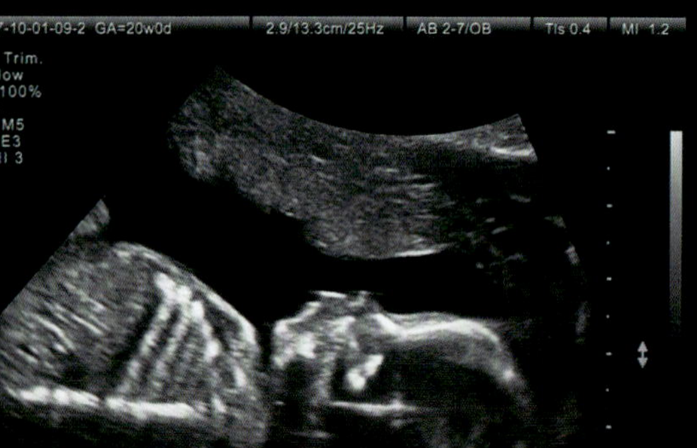

CH•10 Growth and reproduction

ISBN 978 1 4202 3735 1

GET STARTED: *QUESTION*

- A human baby is about 50 cm long at birth. At death, the person may be 1.7 metres tall. Do humans grow steadily throughout their life? Make a profile of a person's life, showing when growth occurs.
- The photo below shows newly hatched turtles racing down the beach towards the water. Adult turtles lay between 250 and 1000 eggs at a time. Penguins lay between one and three eggs at a time. What is the role of the adult animal in each case. How does this affect the survival of the young?
- Some seeds of plants have spikes or burrs on them, some seeds are sticky and others have fleshy, edible fruit around them. How do these features help in the survival of the plants?

ISBN 978 1 4202 3735 1

10.1 Growth

Figure 10.1 A house is made of individual bricks.

This house is made of small units called bricks. The bricklayer starts at the base and adds more and more bricks until the house is complete.

As you learnt in Chapter 4, living things are also made of small units. These units are called cells. Most cells are very small and can be seen only with a microscope. The cells in Figure 10.2 are cheek cells that line the inside of the mouth.

The house contains about 10 000 bricks, mainly of one type. Your body has different types of cells, and has a total of about 50 million million million cells!

The inside of cells is filled with a jelly-like substance called cytoplasm (SIGH-toe-plaz-um). The dark rounded object in each of the cheek cells is called a nucleus.

Living things grow by making new cells. When a bean seed is planted in moist soil, the roots begin to grow and become larger and longer. This growth occurs because certain cells in the roots multiply and make more cells by a process called **cell division**.

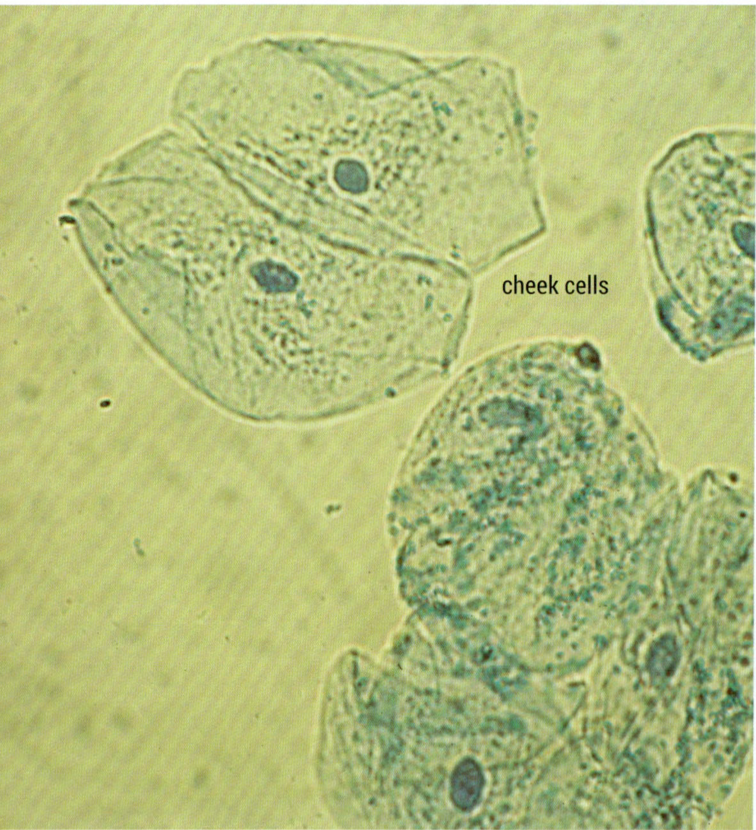

Figure 10.2 Living things are made of cells.

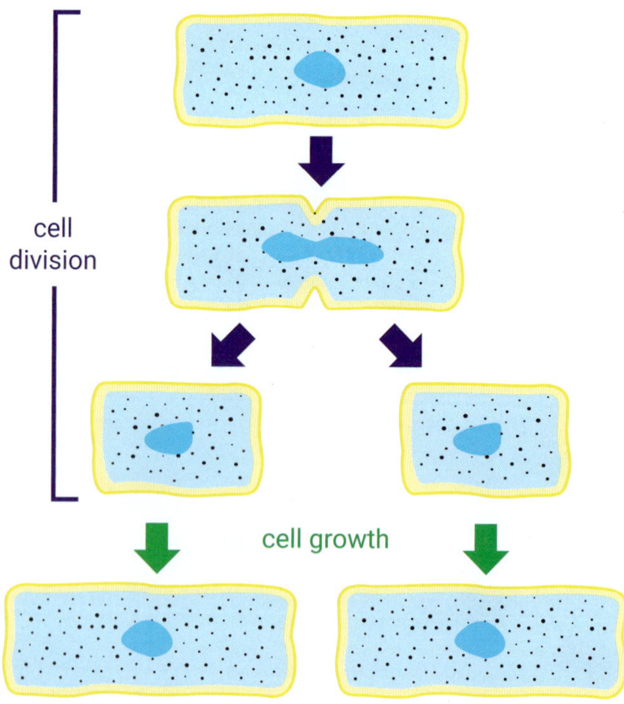

Figure 10.3 All living things grow when cells divide to make new cells. Each of these new cells then grows in size and becomes a mature cell.

ISBN 978 1 4202 3735 1

Growth and cell division

All living things grow by making new cells. Your body grows rapidly in stages up to the age of about 15. During this time your bones grow thicker and longer. For example, your thigh bone or *femur*, the largest bone in your body, grows to about three times the length it was when you were born. Bone cells in the enlarged rounded ends of the femur divide to make new cells.

The growth of your skin also occurs by cell division. This happens in cells below the surface of the skin. So, as the bones and other parts of your body grow larger, your skin also grows. However, unlike bones, which stop growing at adulthood, cell division continues in your skin until death. The skin continually loses cells from its surface and replaces them with new ones.

How bones grow

The ends of bones are covered in a soft but tough substance called cartilage. This cartilage prevents bones from rubbing against each other, but it is also one of the places of bone growth.

Bones grow in length by replacing the cartilage at each end with bone. Figure 10.4 shows an X-ray of the hand of a 1-year-old child. The spaces between the bones are filled with cartilage. The amount of cartilage gradually decreases as the amount of bone increases. This process continues throughout childhood. Figure 10.5 shows the bones of a 12-year-old. Notice that the amount of cartilage between the joints has decreased.

Bones also grow wider during childhood as more bone is added to both the outside and inside of the bone shaft.

The substances that make bones hard and strong contain calcium and phosphorus. This is why your diet should contain enough of the foods containing these substances. Green leafy vegetables, beans and peas, milk, cheese and other milk products are good sources of calcium, while phosphorus is found in breads, cereals, nuts, meat, fish and eggs.

You body uses bones as a store of calcium. Throughout your life, your bones gain and lose calcium. In adults the gain of calcium is usually equal to the loss, whereas in growing children bones gain calcium.

In people over 50, the body tends to lose calcium. Women, in particular, sometimes suffer from osteoporosis, or porous bones, where the bones have lost so much calcium that they become weak and can easily break.

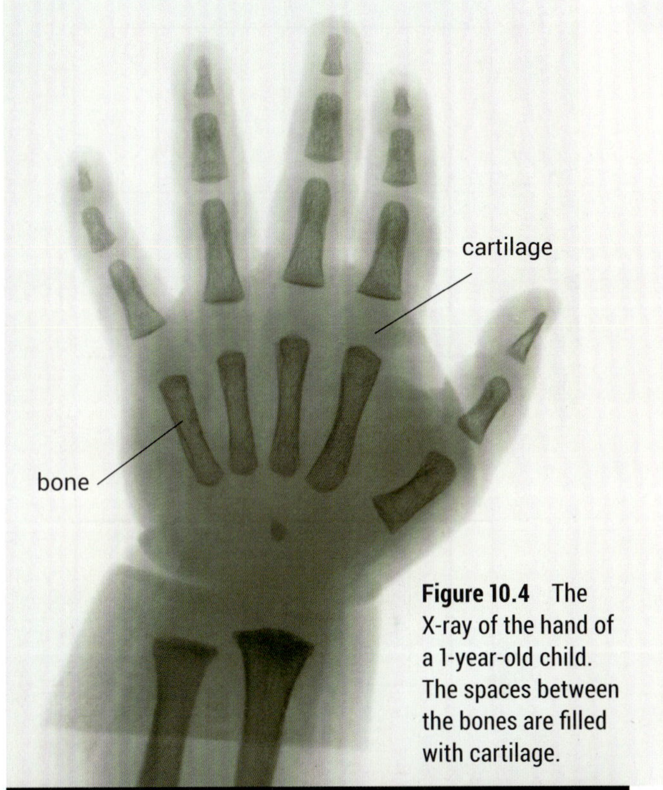

Figure 10.4 The X-ray of the hand of a 1-year-old child. The spaces between the bones are filled with cartilage.

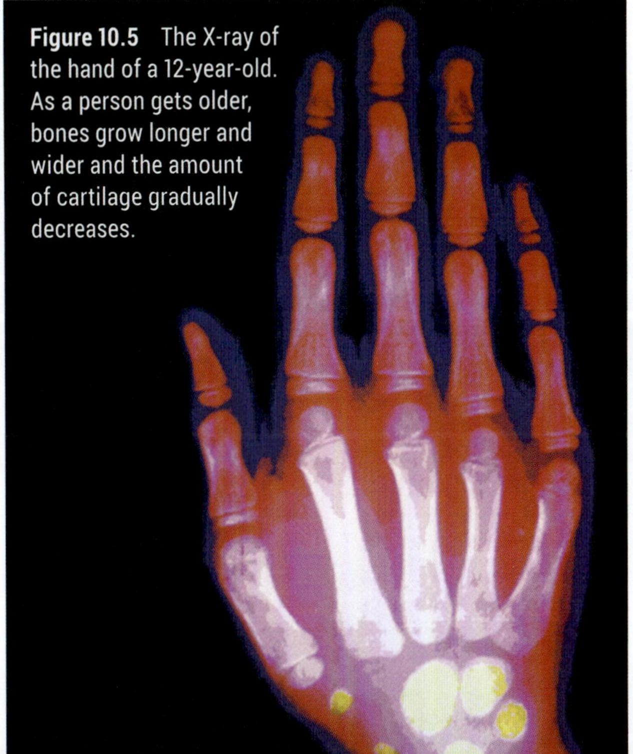

Figure 10.5 The X-ray of the hand of a 12-year-old. As a person gets older, bones grow longer and wider and the amount of cartilage gradually decreases.

ISBN 978 1 4202 3735 1

ACTIVITY

Human growth

In 1759 French count Philibert de Monteillard measured the length of his newborn son. Then every year until his son's 18th birthday in 1777, de Monteillard recorded his height. This was one of the first recorded studies on growth patterns in humans.

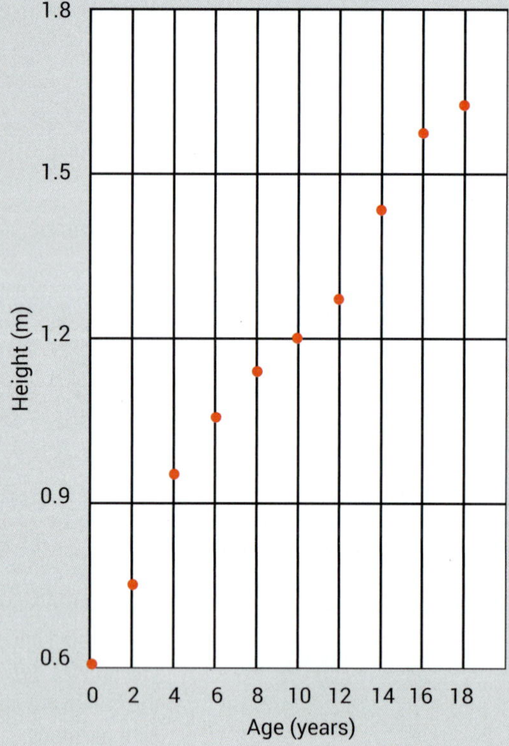

- Compare your height to the height of de Monteillard's son at your age.
- What was the length of de Monteillard's son when he was born?
- What is the range of heights in your class? Does anyone match de Monteillard's son's height?
- Is his growth rate constant up until his 18th birthday? What is his approximate growth rate in metres per year between the ages of 6 and 10?
- Make some generalisations about the information in the graph.

Plant growth

A Year 7 class was investigating the growth of corn plants.

Five corn seeds were planted in pots containing different growing substances—sawdust, sand, soil and soil with fertiliser.

Each pot was given the same amount of water every 2 days, and the pots were kept in the same location in a sunny place in the classroom. After the seeds sprouted, the heights of the seedlings were measured every 2 days and the average height of the seedlings in each pot calculated. Here are the results.

Day	Average height of seedlings (cm)			
	Pot 1 sand	Pot 2 sawdust	Pot 3 soil	Pot 4 soil + fertiliser
1	0.5	0.4	0.6	0.5
2	0.9	0.8	1.0	1.0
4	3.2	3.0	3.5	4.0
5	5.5	5.5	7.2	7.8
8	9.0	8.2	11.0	12.0
10	11.0	10.2	15.5	16.6
12	13.2	12.0	18.5	20.0

- Plot a bar graph of the results at day 12. Hint: On your graph, put the height on the vertical axis and the type of pot on the horizontal axis.
- Suggest reasons for the different heights at day 12.
- Why were the seedlings in all pots given the same amount of water and kept in the same place?
- Make some generalisations about the information in the table.

ISBN 978 1 4202 3735 1

Growth stages

In humans, there is steady growth in height of the body from birth to about 18 years. Then the body stops growing, and in old age some people experience a decrease in height.

There are five major stages in the growth of the human body.

Stage	Approx. age (years)	Growth in height	Other body growth
childhood	0–12	steady increase in height	
adolescence	13–17	rapid spurts of growth in height	muscles grow larger and stronger
adulthood	18–50	very little change in height	body usually increases in weight
middle age	50–74	no change in height	some loss of muscle strength
old age	75–	some decrease in height	body usually loses weight

In adolescence, the body experiences a rapid growth in height, weight and shape. During this time the sex organs mature. This change is called **puberty** (PEW-ber-tee). You will learn more about this in the next section of this chapter.

Throughout the growth stages, the human body doesn't change shape as dramatically as it does in other animals. Many animals have a definite life cycle in which there is an immature form and an adult form.

Life cycles

Frogs

Frogs have three distinct stages in their life cycle —the egg, tadpole and adult stages.

An adult female frog lays hundreds or even thousands of eggs in a jelly-like mass in water. The eggs hatch and become tadpoles. During this stage, tadpoles have gills and breathe under water. They generally eat very small plants and algae in the water.

At the end of the tadpole stage, the tadpole undergoes a very big change. This change to its internal and external structure is called a **metamorphosis** (meta-MOR-fo-sis). The tadpole grows legs, loses its tail and gills and develops lungs so it can breathe air on land. Unlike tadpoles, frogs are carnivores and eat other animals such as insects.

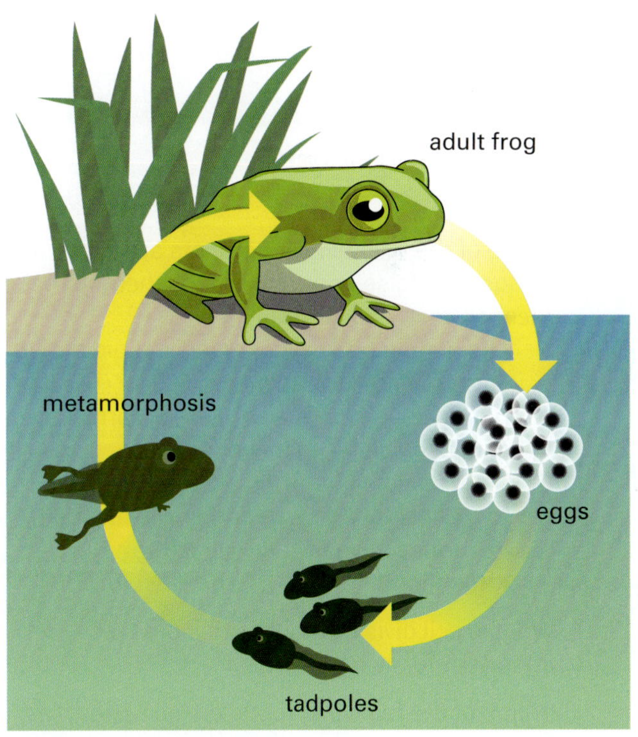

Figure 10.6 The life cycle of a frog

Insects

Insects, like frogs, have definite stages in their life cycles. Most insects have four stages—egg, larva, pupa and adult stages. The larva (grub or caterpillar) undergoes a metamorphosis through the pupal stage to form an adult insect. For example, in a butterfly's life cycle, an egg hatches into a caterpillar. The caterpillar eats plants and has no wings. It grows larger over time and changes into a pupa. The pupa is usually inactive and undergoes a dramatic change to form an adult butterfly. Butterflies have wings and have a long tubular mouthpart, which they use to suck nectar from flowers or plant sap.

ISBN 978 1 4202 3735 1

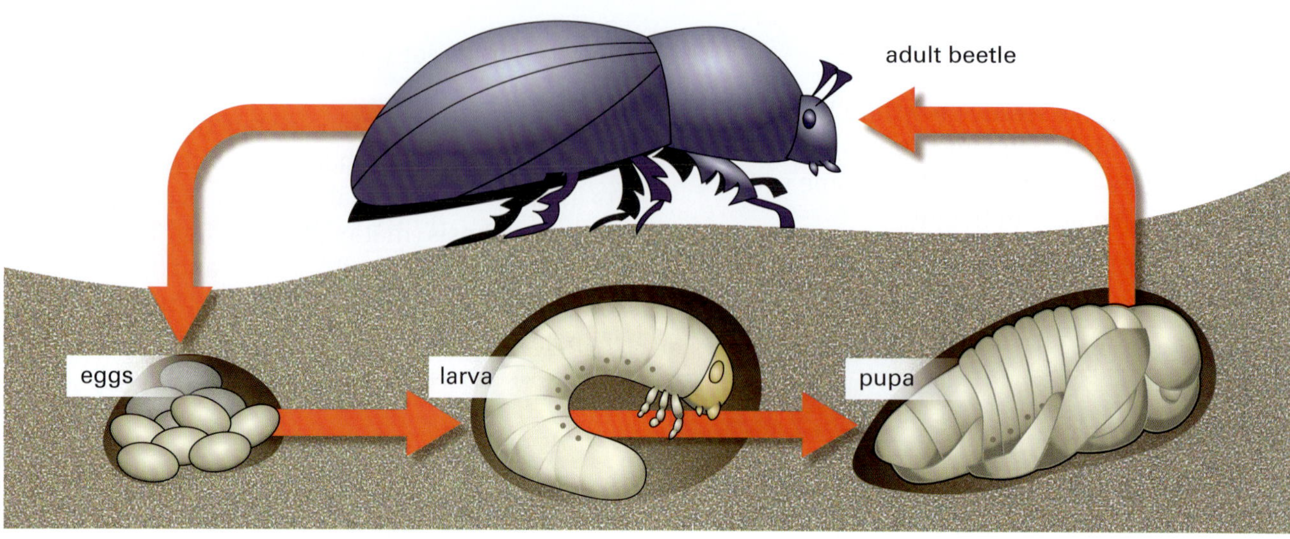

Figure 10.7 The life cycle of a beetle

CHECK

1 Some of the following statements are false. Choose the false ones and rewrite them to make them correct.
   a The human body experiences a rapid growth spurt during adolescence.
   b During childhood the amount of cartilage increases between the joints.
   c Most insects start from an egg, change to a pupa, then a larva and finally an adult.
   d In humans, bones continue to grow throughout life.

2 What is cell division? Why is this process important for living things?

3 Which of the following foods are good sources of calcium? Give reasons for your choice.

bacon	yoghurt	cottage cheese
soft-drink	ice-cream	sausages

4 Describe the metamorphosis in the life cycle of a frog. Do humans undergo metamorphosis in their life stages?

5 Name two places in the body of a 1-year-old child where cell division occurs. Would this be the same for a 50-year-old person. Why?

6 The larva of a butterfly has strong biting and chewing mouthparts, whereas the adult has a sucking mouthpart.
   a Describe the food each animal eats.
   b How is eating different foods an advantage to the survival of the insect?

CHALLENGE

1 Cell division in microscopic organisms can occur rapidly if the conditions are suitable. For example, bacteria can divide into two every 20 minutes. Assuming one bacterium divides every 20 minutes, and none die, how many bacteria would there be after 6 hours?

2 A boy broke his forearm in an accident. The X-ray shows the break. His arm was set in a cast.
   a Suggest reasons for the cast.
   b Why does the cast have to remain on the arm for at least 8 weeks?

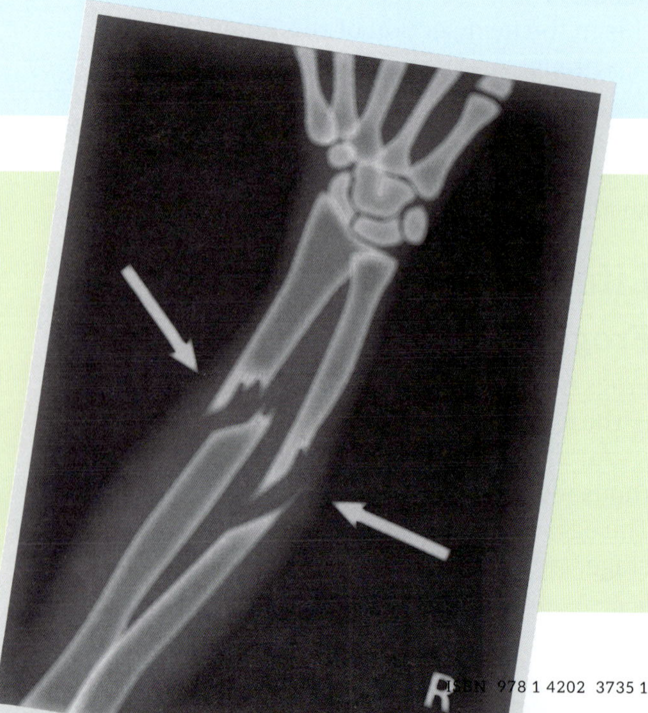

ISBN 978 1 4202 3735 1

10.2 Reproduction

You may have heard of the question 'Which came first, the chicken or the egg?' Whatever the answer, we do know that each chicken, and every other living thing, begins life as a single cell.

Figure 10.8 The chicken or the egg?

Every living thing reproduces to make more of its own kind. Very small, simple organisms reproduce by splitting into two. They produce offspring that are identical to the original organism. This is called *asexual reproduction*. This type of reproduction can produce large numbers of organisms very quickly.

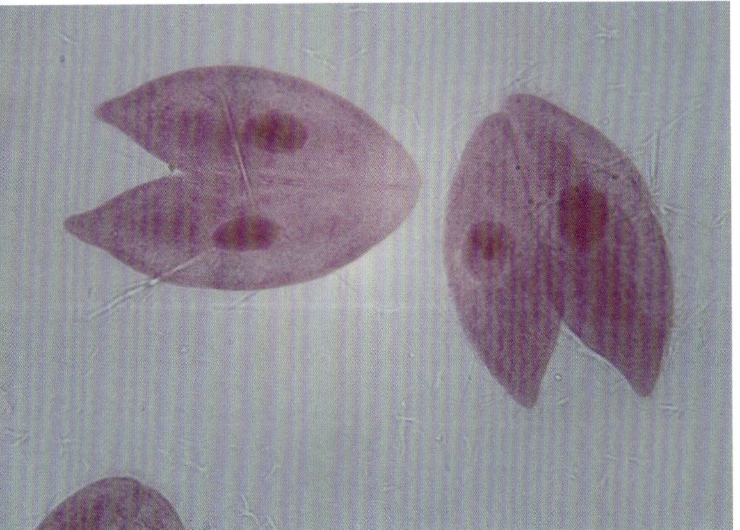

Figure 10.9 Microscopic organisms, like these paramecia, reproduce by splitting in two.

Sex cells

Larger organisms reproduce sexually. To do this, the two parents (one male and the other female) produce **sex cells**. These cells are different from other cells in the body. They can combine to make a cell that eventually becomes a new and independent organism.

The female sex cell is called an **ovum** or *egg cell*. Ova (plural of ovum) are made in organs called **ovaries** (OH-var-ees). The male sex cell is called a **sperm cell** and is much smaller than an ovum. Sperm are made in organs called **testes** (TES-teez).

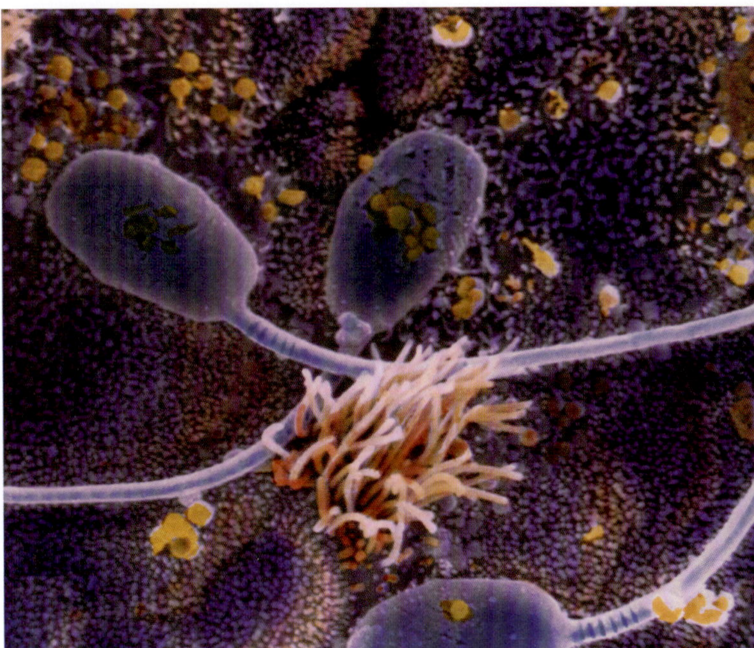

Figure 10.10 Human sperm cells. They use their tails to swim through liquid. The head of the sperm contains the nucleus.

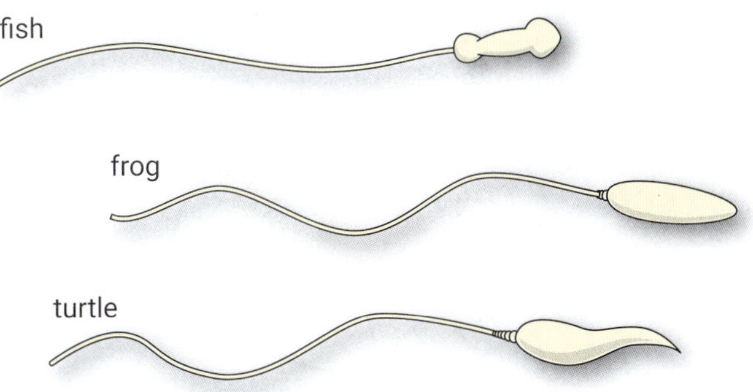

Figure 10.11 The shape and size of sperm cells is different in different animals.

ISBN 978 1 4202 3735 1

Fertilisation

When a sperm and ovum meet, the nuclei of the two cells join together, and a new living thing is formed. This process is called **fertilisation** (FUR-til-eyes-AY-shun).

Fertilisation can occur externally or internally. For example, in humans internal fertilisation occurs. The male deposits the sperm inside the female's body. The sperm then swim towards the ovum, where fertilisation occurs. Frog eggs are fertilised externally. Female frogs release their eggs in the water, and sperm released by the male swim to the eggs.

The sex cells of organisms are not all the same size and shape. Figure 10.13 shows the actual sizes of ova from five different animals.

The eggs of mammals are small compared with the eggs of birds, reptiles, amphibians and fish. They have only a small quantity of cytoplasm, because the eggs develop internally and receive nourishment from the mother within a few days of fertilisation. In birds and many other animals, the fertilised egg develops outside the mother's body and must contain enough food for the whole period of growth.

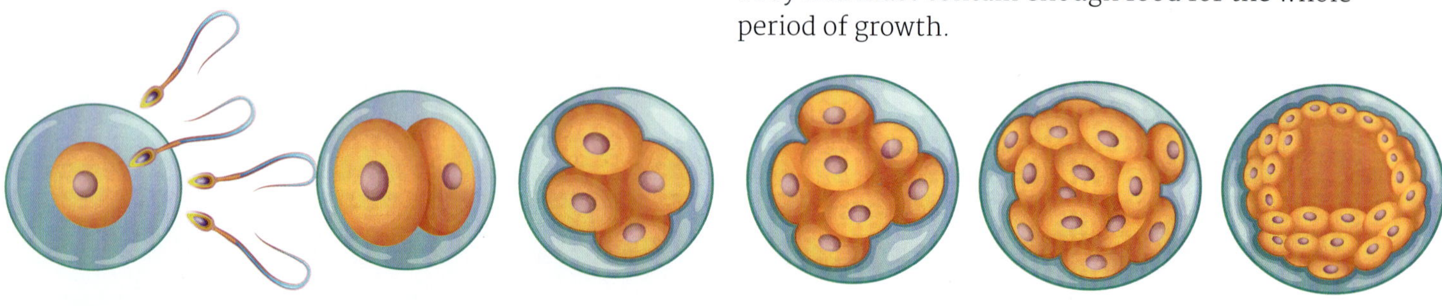

Figure 10.12 A human ovum is fertilised by one sperm cell. The ovum is much larger than a sperm cell, because it has much more cytoplasm. The cytoplasm contains food for the fertilised egg during the first few days of cell division and growth.

ACTIVITY

Break open a hen's egg in a flat glass dish or petri dish. Notice the yellow yolk and clear albumen (the 'white'). These make up the cytoplasm of the cell.

Look at the yolk carefully and you should see a tiny white patch. This is the nucleus of the egg. It is this part of the egg that develops into a young chicken.

Observe the 'ropes' in the albumen. These keep the young chicken in place in the egg and stop it from rolling over.

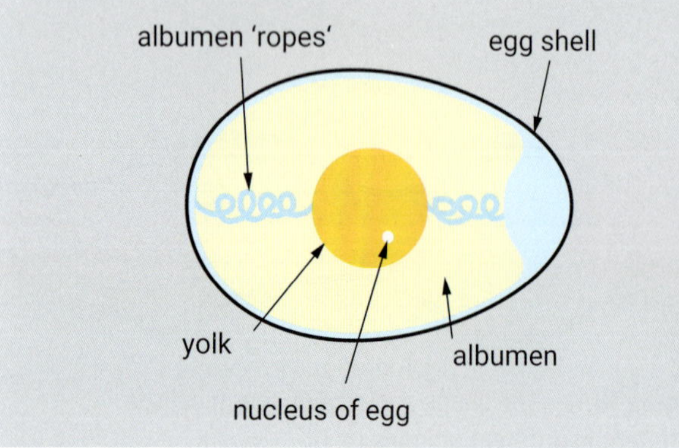

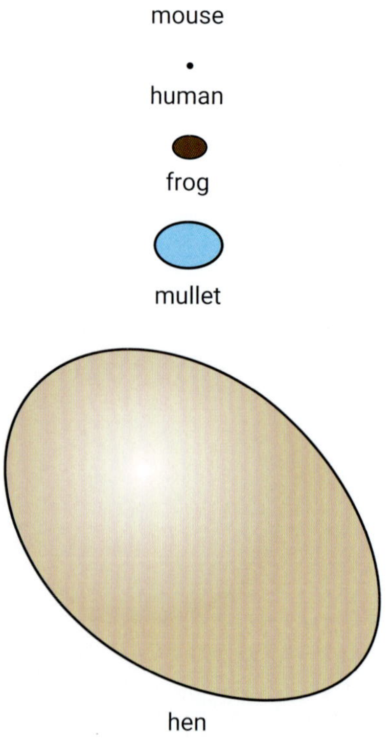

Figure 10.13 The sizes and shapes of the ova of five different organisms

ISBN 978 1 4202 3735 1

Female puberty

When a human baby girl is born, she has about 200 000 eggs inside her ovaries. However, the eggs will not develop until certain changes take place in her body. These changes begin at the start of the adolescence stage.

The time during which these changes take place is puberty. In girls, puberty occurs sometime between the ages of 10 and 13, but it can also vary with the girl's type of diet, her lifestyle and the amount of exercise she does.

At puberty, changes happen to the inside and outside of the body. The most noticeable changes are the growth of breasts and pubic hair, and the start of **menstruation** (MEN-strew-AY-shun) or periods. These changes occur over a number of years.

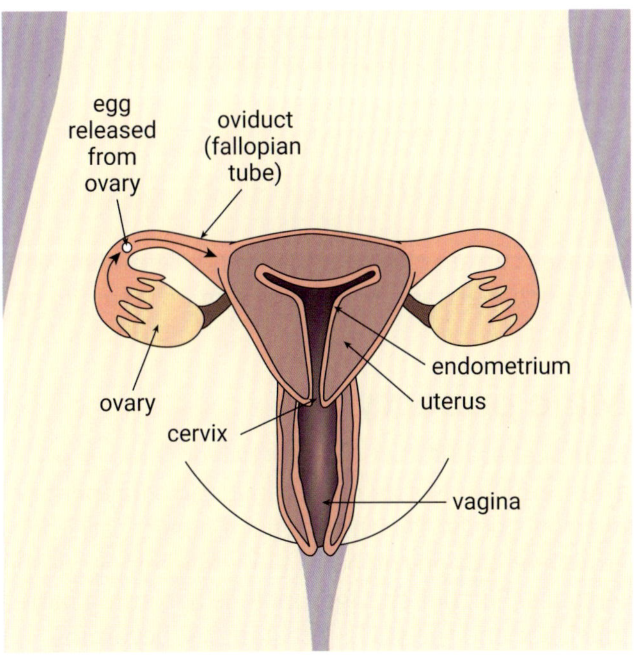

Figure 10.14 The female reproductive organs

Menstruation is the start of a cycle of events called the **menstrual cycle**. This cycle lasts about one month or 28 days. About halfway through the cycle, the ovaries release eggs. Usually only one egg is released each month from one of the ovaries. Unless it is fertilised, this egg eventually passes out of the **vagina** (va-JI-na).

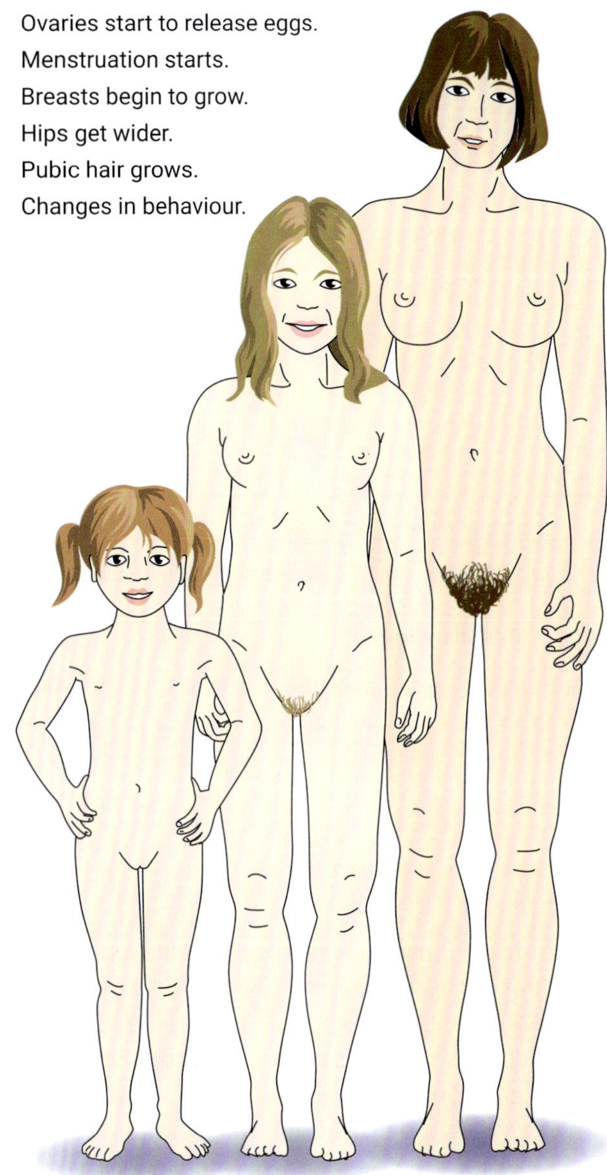

Figure 10.15 Changes in the female during puberty

The menstrual cycle

At puberty, a girl's body begins the menstrual cycle: a series of events that occurs about every 28 days or one month, and continues until a woman is about 50 years old.

The most noticeable event in the menstrual cycle is bleeding or 'having a period'. The bleeding is due to the loss of blood and cells from the lining of the **uterus**. This is not a bad thing, since the cells will be replaced with new cells ready for the arrival of a fertilised egg, if one is present

ISBN 978 1 4202 3735 1

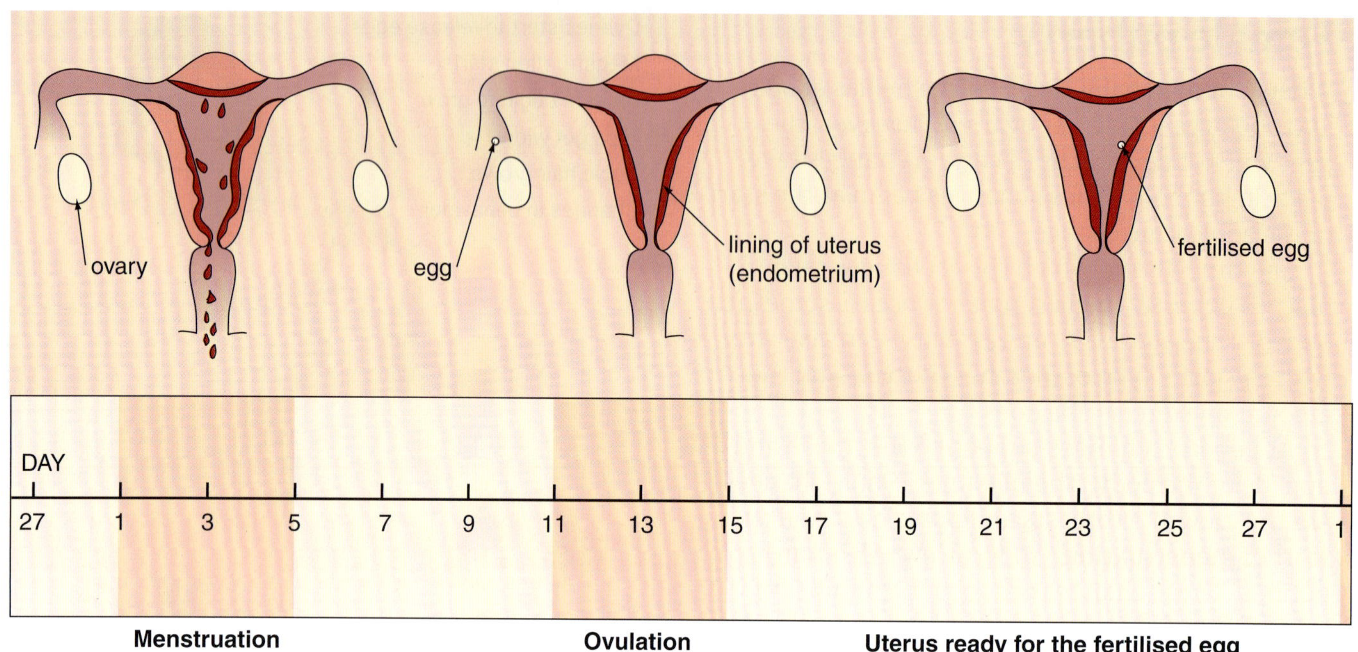

Figure 10.16 The menstrual cycle

next month. If an egg is fertilised, it attaches to the lining of the uterus, which steadily grows in thickness. Once this happens, the menstrual cycle is stopped by chemicals released by the ovaries and the embryo.

Twenty-eight days is only the average time for the menstrual cycle. Some women have a longer cycle, while for others it is shorter. It is not unusual for a period to be late or to be missed entirely. Worry, tension or stress can also cause periods to be irregular.

An important event in the menstrual cycle is the release of an egg from one of the ovaries. This is called **ovulation** (OV-you-LAY-shun). It is not noticed in a woman's day-to-day life.

Only the females of humans and some apes have a menstrual cycle. Other mammals, such as dogs, mice and kangaroos, have a breeding cycle in which no bleeding occurs. The most noticeable event in the breeding cycle of these animals is during ovulation. During this time, the female releases certain chemicals from her reproductive organs that attract males. The female is said to be 'on heat'.

Male puberty

Males generally start puberty two or three years after females. Noticeable changes also happen to a boy's body. The first signs of puberty are the growth of hair on the body, and enlargement of the testicles. Further changes include becoming more muscular and the voice 'breaking' or becoming deeper. Over the next few years, the fine hair on his face becomes coarser, thicker and darker.

At puberty, the testes start making sperm cells. They make hundreds of millions of sperm each day. Sperm is released out of the **penis** (PEE-nis) and this is called *ejaculation*.

When sperm are released from the testes, they travel up narrow tubes to two small glands (the seminal vesicles and bulbourethral glands) where they may be stored for a short while. These glands

ISBN 978 1 4202 3735 1

also make liquids that nourish the sperm cells and make up the **semen** (SEE-men). The semen is the liquid containing the sperm. If the semen is not released from the body, the sperm are broken down and the materials in the cells are reused by the body.

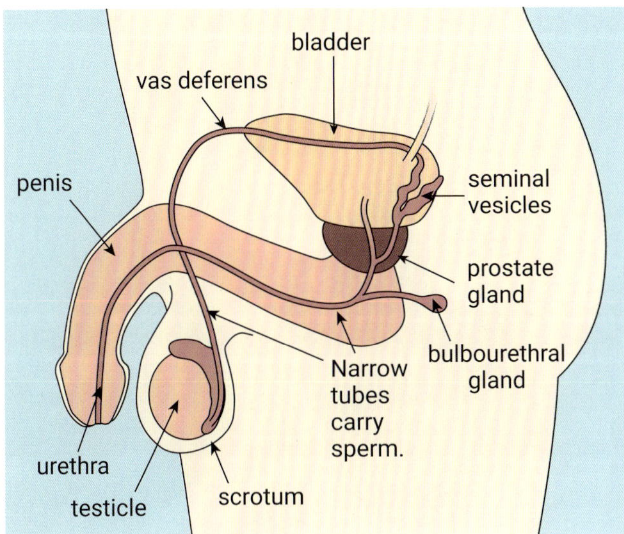

Figure 10.17 The male reproductive organs

Penis and testes grow larger.
Testes start making sperm.
Body becomes muscular.
Hair grows on face and body.
Changes in behaviour.
Pubic hair grows.
Voice deepens.

Figure 10.18 Changes in the male during puberty

Fertilisation, pregnancy and childbirth

For a man and a woman to have a baby, an egg inside the woman has to be fertilised by a sperm. To do this, the man's penis is placed in a woman's vagina. This is called *sexual intercourse*. The penis rubs against the walls of the vagina and after a while ejaculation occurs and semen swim up through the uterus (YOU-ter-us) to the **oviducts** (OH-vee-ducts). Of the millions of sperm released only a fraction will survive the journey. If an egg is present, only one sperm passes through the egg's outer membrane. Fertilisation then occurs. If an egg is fertilised, it attaches to the lining of the uterus (sometimes called the womb). This is a thick-walled organ that will hold and nourish the developing baby. The fertilised egg divides over and over and grows in size. At this stage, the developing baby is called an **embryo** (EM-bree-oh).

The part of the egg that is attached to the wall of the uterus grows to form the *umbilical cord*. At the end of the umbilical cord, a large spongy pad full of blood vessels grows to form the **placenta** (pla-SEN-ta).

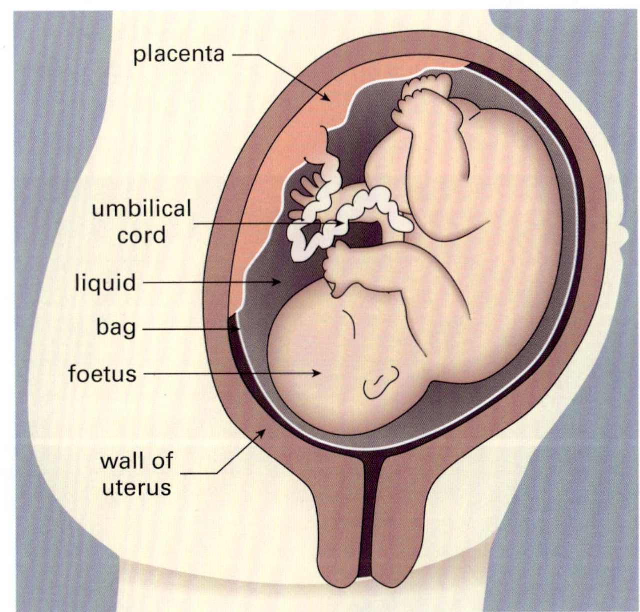

Figure 10.19 The placenta is attached to the wall of the uterus and passes food and oxygen from the mother to the baby.

ISBN 978 1 4202 3735 1

The embryo gets all its food and oxygen from the mother's blood. But the mother's blood and the baby's blood never mix. The food and oxygen from the mother pass through the wall of the uterus and into the blood in the placenta, then along the umbilical cord to the baby. Wastes from the baby travel back along the umbilical cord to the mother.

After about 14 weeks the embryo looks like a human baby and is now called a **foetus** (FEE-tus). It is kept warm and well protected inside the mother's uterus. The foetus is surrounded by a bag of liquid that protects it from bumps or jolts.

A human baby takes about 40 weeks to develop. When the baby is born, the placenta separates from the uterus. The baby's lungs then start working and take in oxygen from the air. The placenta is sometimes called the 'afterbirth' and is about the size of a dinner plate.

Protecting the foetus

Most germs in the mother's blood don't pass through the placenta into the baby's blood. One exception is the germ that causes German measles or rubella.

Harmful chemicals can pass into the baby's blood and for this reason the mother has to be careful with her diet and the medicines she takes. Alcohol, some drugs and the chemicals in cigarette smoke can pass into the baby's blood and may be very harmful.

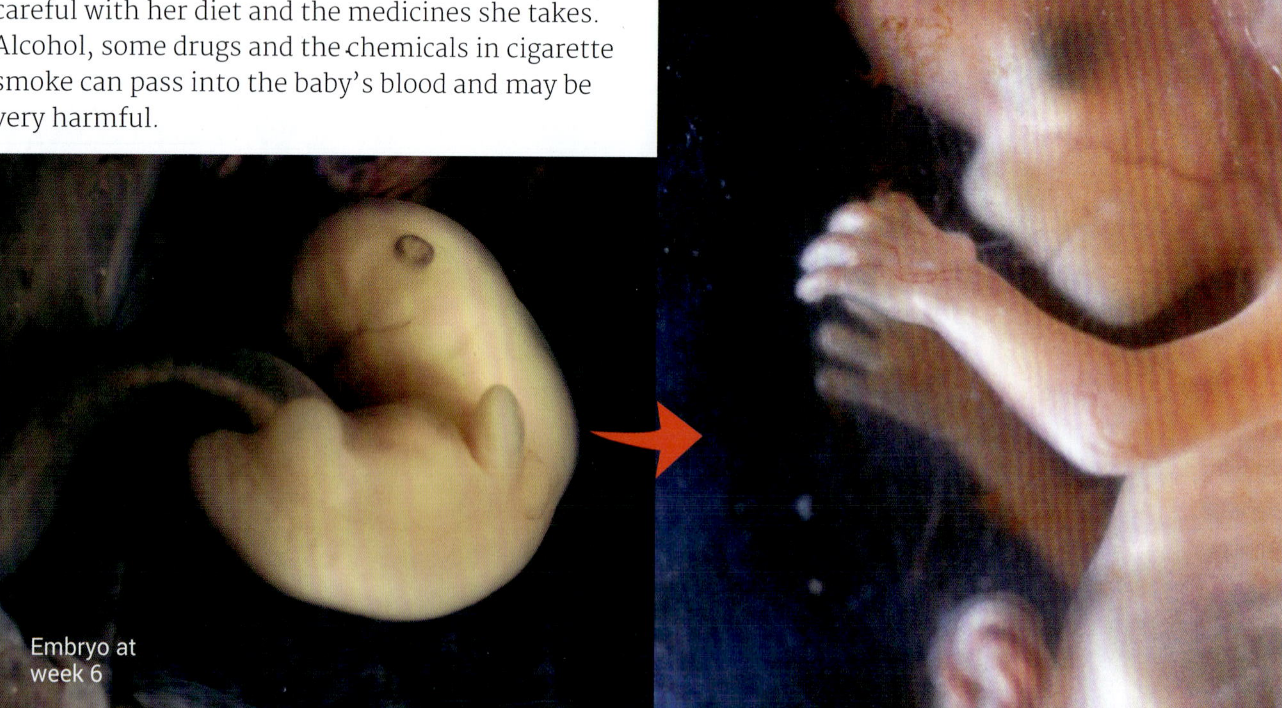

Figure 10.20 Stages in the development of a baby

ISBN 978 1 4202 3735 1

1 Copy and complete the following sentences.
   a ______ ______ is a process in which cells split into two.
   b In organisms that reproduce sexually, the male produces ______ and the female produces ______.
   c Fertilisation occurs when the ______ of sex cells combine.
   d In males the ______ produce sperm, and in females the ______ produce eggs.

2 Make a list of the changes that occur in a male and those that occur in a female during puberty.

3 Look at the Term and Meaning lists top right. Match the terms in the left-hand column with the meanings in the right-hand column. To do this, draw up a table like the one below.

Term	Meaning

Term	Meaning
ovary	where sperm are made
semen	when the nuclei of a sperm and ovum join
fertilisation	a liquid containing sperm
ova	the organ that produces eggs
testes	female sex cells

4 Describe the role of the placenta and umbilical cord in the developing baby.

5 Why do sperm cells have tails but eggs do not?

6 External fertilisation occurs in frogs and mullet. Suggest why these animals produce many more eggs than humans or mice do.

7 A male usually releases millions of sperm when it is mating with a female.
   a How many of these sperm do you think usually fertilise the female's ovum? Suggest a reason for your answer.
   b Suggest why the male makes and releases so many sperm.
   c Explain ejaculation and the journey sperm take to exit the penis.

1 Once a single sperm's nucleus has joined with the egg's nucleus, no more sperm can penetrate the outer membrane or wall surrounding the egg. Make an inference for this.

2 Calculate how many eggs are released by a woman in her lifetime. Assume that women reach puberty at 12 years of age and stop releasing eggs at 50.

3 Women usually give birth to one child at a time, but multiple births do occur. Identical twins occur when the egg splits into two just after fertilisation and each develops separately. Fraternal twins occur when two eggs are released from the ovary and each is fertilised by a different sperm.
   Use the information above to answer the following questions.
   a Identical twins are always the same sex and look almost exactly alike. Why?
   b Why is it possible to have fraternal twins with quite different features?
   c Suggest why twin births are much less common than single births.

ISBN 978 1 4202 3735 1

10.3 Reproduction and survival

All organisms reproduce, but not all organisms do this in the same way. In animals, sperm can fertilise eggs inside the female's body (*internal fertilisation*) or outside the female's body after she has laid her eggs (*external fertilisation*). In mammals, birds and reptiles fertilisation takes place internally; in most other animals the eggs are fertilised externally.

Figure 10.21 Fertilisation occurs externally in frogs. However, in some types of frogs, to make fertilisation more effective, the male clasps the female's back and produces sperm while she lays her eggs.

Caring for offspring

The young animals that hatch from eggs that are laid and fertilised externally are completely independent of each other and of their parents. For example, when a frog's eggs hatch, the tadpoles swim away from the leftover egg mass and have to find their own food and protect themselves from enemies.

The eggs of reptiles are fertilised internally, but most reptiles do not care for their young after the eggs hatch. For example, sea turtles lay their eggs in the sand on the beach. The eggs are covered up and left to incubate. When the young turtles hatch, they dig their way to the surface and then scramble down the beach to the water. On their journey to the water, many of the young turtles are eaten by birds and other animals.

Figure 10.22 Newly hatched turtles scramble towards the water. Many of the hatchlings die because there is no parental care and therefore no protection from enemies.

Birds and mammals produce considerably fewer eggs than reptiles, frogs and other animals. Young birds and mammals are generally dependent on their parents for food, warmth and protection from enemies. This increases the chances of survival of the young. For example, newly hatched birds cannot fly and cannot feed themselves and would certainly die without the protection of one or both parents.

The table at the top of the next page compares the method of reproduction and the parental care of four different types of animals.

Figure 10.23 These newborn puppies are completely dependent on their mother for food, warmth and protection.

ISBN 978 1 4202 3735 1

	Bream (fish)	Green tree frog	Magpie (bird)	Common wombat (mammal)
Number of eggs produced each year	about 5 million	up to 2000	three or four	one
How eggs are fertilised	externally in water (sea)	externally in water (ponds and creeks)	internally	internally
Parental care	None—the eggs are left in the water, the young hatch and have to find food and protection.	None—the eggs are protected by a mass of jelly, but after hatching, the tadpoles have to find food and protection.	The female sits on the eggs until they hatch, then feeds and protects the young until they can fly.	A bean-sized baby is born, which develops inside the mother's pouch for up to 10 months. It is then protected by the mother for another 10 months.

Parental care in seahorses

Many animals have peculiar reproductive behaviours—the seahorse is one such animal.

The seahorse is a bony fish (as distinct from non-bony fish such as sharks and rays), and in most bony fish fertilisation occurs externally.

The female seahorse has a long, hollow appendage called an ovipositor. In some types of seahorses, she uses this to place her eggs in the male's front belly pouch. Here he fertilises the eggs and protects them until they hatch. (Note the belly pouches on the two male seahorses in the photo.)

Reproduction and survival in flowering plants

Trees, shrubs, bushes, palms and grasses are examples of flowering plants. All of these plants reproduce sexually. Flowers contain the reproductive organs that make the sex cells.

Pollen contains sperm cells and is made in the anthers. The ova, or eggs, are made in the ovaries. To fertilise the ova, pollen has to travel to the stigma of the flower (part of the female reproductive organs). The process in which pollen travels to the stigma is called **pollination** (pol-in-AY-shun). Pollination occurs when pollen is carried by wind, or by insects and birds that are attracted to flowers by the bright colours or sweet nectar.

The pollen tubes carrying the sperm then grow down the style and the sperm eventually fertilise the ova in the ovary.

Asexual reproduction

Some flowering plants are able to reproduce asexually as well as sexually. For example, a strawberry plant has flowers and produces fruit (strawberries) containing seeds. The plant can also send out runners from which new strawberry plants grow. This form of asexual reproduction produces new plants with features identical to those of the original plant.

ISBN 978 1 4202 3735 1

INVESTIGATION 10.1 Observing flowers

Aim

To dissect a flower and identify its parts.

Materials

- a few different types of flowers, e.g. hibiscus
- forceps
- single-edged razor blade
- stereomicroscope or hand lens

Risk assessment and planning

Carefully read through Parts A and B, and select the materials you will need for each part.

Make a list of the safety precautions you will need to take in this experiment.

PART A Observing flowers

Method

1. Use the diagram of a flower below to identify the following parts of one of your flowers—petal, sepal, stigma, anther, filament and ovary.
2. Repeat for other flowers.

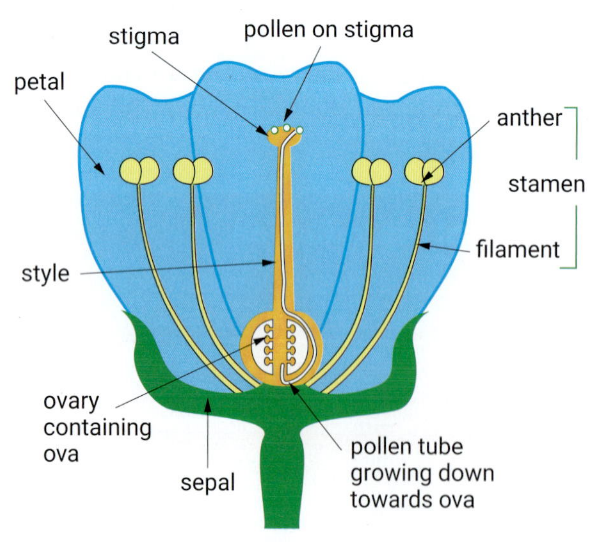

PART B Dissecting a flower

Method

1. Touch the end of the stigma with your finger or a pencil. Notice that it is sticky.
2. Use forceps to gently hold a flower while you cut it in half by cutting down the stem.

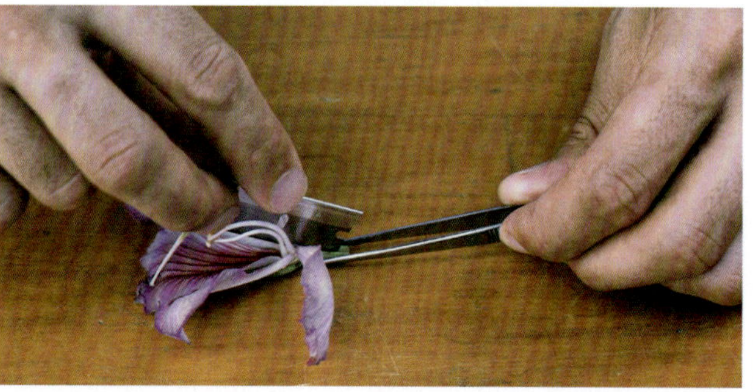

3. Look at the ovary. It contains a number of rounded objects called *ovules*. Each ovule contains an egg (*ovum*).
4. Use a stereomicroscope or hand lens to observe the ovary and ovules.

   Record your observations. Draw the arrangement of the ovules in the ovary.
5. Cut an anther in half and observe the pollen grains with the stereomicroscope. Repeat this for other flowers.

   Record your observations.

Discussion

1. Why is the stigma sticky?
2. Different types of flowers have different shapes and sizes of pollen. Suggest a reason for this.
3. Infer the functions of the sepals.
4. The petals on most flowers are brightly coloured. Suggest a reason for this.
5. What is meant by the word *pollination*? How is it different from fertilisation?

ISBN 978 1 4202 3735 1

Seeds and dispersal

After the ova have been fertilised and the seeds develop, the petals, sepals and stamens of the flower wither and fall off. The ovary becomes a fruit with the seeds inside it (or sometimes on the outside of it, e.g. a strawberry). In some fruits, such as apples, the wall of the ovary thickens to form an edible fruit. In others, such as eucalypts, it is hard and woody.

Seeds must be spread away from the adult plant to give the plants that grow a better chance of survival. This is called dispersal. There are four main methods by which fruit disperse their seeds.

1 The seeds fall out of the fruit and are carried away by the wind.
2 Animals eat the fruit, and the indigestible seeds pass out of the animal in its droppings. In this way, the seeds can be spread many kilometres away from the adult plant.
3 Some seeds are sticky or have hooks or spikes that get caught in the fur or hair of animals. These seeds may be carried a long way before they fall off or are rubbed off.
4 Some fruit explode, throwing out the seeds.

ACTIVITY

1 Collect about 10 different types of fruit or the seeds from the fruit.
2 Draw up a data table and classify the seeds into groups, depending on the way you infer they are dispersed. Include a brief description of the way each group of seeds is dispersed, in your data table.
3 Find more fruits or seeds, classify them and add them to your table.
4 Take digital photos of the seeds or fruit and present your report in a PowerPoint presentation. Or design a poster to record and display your results and talk about your findings to the class.

SCIENCE UNDERSTANDING

Nick Hansa operates a large native plant nursery. For many years he has studied plants and their methods of reproduction and seed dispersal.

He often goes looking for the seeds of rare or endangered native plants. To do this, he needs to know the type of seeds the plants produce.

For plants whose seeds are very small and are normally dispersed by wind, he covers the seed pods with special bags before the seeds mature. When the seed pods open, the seeds fall into the bag and are collected.

Larger seeds are collected on the ground after they have fallen from the plants.

EXPLORE ONLINE

To learn more about seeds check out these websites.

Fruit and seed dispersal
Has great photos and interesting descriptions

Seed dispersal
Contains video clips showing types of seed dispersal

ISBN 978 1 4202 3735 1

Growing plants from cuttings

Many plants, including flowering plants, are able to reproduce from parts of the adult plant. This is a form of asexual reproduction called *vegetative reproduction*.

Strawberry plants send out runners that produce new strawberry plants with leaves and roots. Potatoes are actually underground stems called *tubers*. The buds ('*eyes*') that develop on a potato can grow into new potato plants.

The advantages of vegetative reproduction are that a plant can multiply quickly in a place which suits it, and that it stops other plants from growing near it.

You can try growing plants from cuttings by following the instructions opposite and the hints below.

Helpful hints

1. Plants that are suitable for leaf cuttings are the ones that have soft, furry or velvety leaves: for example, African violet and coleus. You could also try begonia and snowflake (*Euphorbia leucocephala*).
2. Many types of shrub or small tree are ideal for growing plants from stem cuttings.
3. Daisies, fuchsias and native correas propagate easily from cuttings. For best results, use a good quality propagating mix.
4. When growing plants from stem cuttings, dip the stem into some plant cutting powder (root growth powder). This will promote root growth on the cutting.
5. Do not over-water the propagating mix. It is best to add a little water often.
6. A plastic bag stops the plants from drying out and dying from water loss. You can also buy mini-hothouse trays at plant nurseries to grow your plant cuttings in.

Leaf cutting

1. Place the cut end of the leaf stalk in a pot of moist propagating mix. Then tie a large clear plastic bag over the pot. Make sure the bag does not touch the cutting.

sealed plastic bag—not touching the leaf

leaf stalk

moist propagating mix

2. If the leaf has large veins, use a sharp knife to cut three or four of them as shown. Lay the leaf flat on a pot of moist propagating mix.

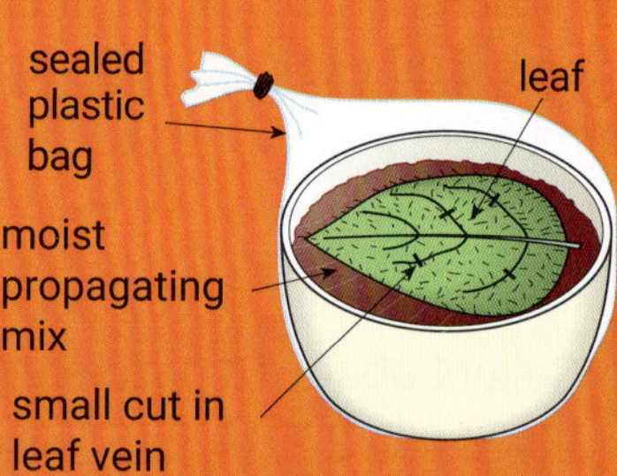

Stem cutting

Cut a stem about 10 cm long and remove all but two or three of the leaves at the top of the stem. Dip the cut end of the stem in plant cutting powder. Place the stem in a pot of moist propagation mix, then tie a large clear plastic bag over the pot. Make sure the bag does not touch the leaves.

ISBN 978 1 4202 3735 1

CHECK

1 Which of the following statements are true and which are false? Rewrite the false ones to make them correct.
   a Pollen contains the male sex cells and is produced in the ovary.
   b Fertilisation in most reptiles occurs externally.
   c Young reptiles are dependent upon their parents for food and protection.
   d All flowering plants reproduce sexually.

2 Describe the degree of care given to their young by most:
   a fish
   b frogs
   c birds
   d mammals

3 Suggest why the number of eggs produced per year by different types of animal decreases as the degree of parental care increases.

4 About three in every 100 000 eggs laid by a bream grow to be adult fish.
   a Suggest why the survival rate of the eggs is so low.
   b Use the table on page 259 to work out how many adult bream would be produced from the eggs laid by a bream in a year.

5 Use your own words to describe what the word 'disperse' means on page 261.

6 Suppose a particular type of plant can reproduce sexually (by seeds) as well as asexually (by sending out runners). List the advantages and disadvantages of each type of reproduction for the plant.

7 The coconut is a fruit with a very hard covering. It is hollow and does not sink in water. Suggest how coconut seeds are dispersed.

8 The photo shows a close-up of the seeds of the plant called cobblers pegs. Suggest how these seeds are dispersed.

CHALLENGE

1 Suggest why plants with bright flowers are mainly insect-pollinated, and grass flowers are usually wind-pollinated.

2 The seeds below are drawn at their actual size.
   a Which one(s) do you think would be dispersed by the wind? Give a reason for your answer.
   b Which one(s) might be caught on the fur of animals. Give a reason.

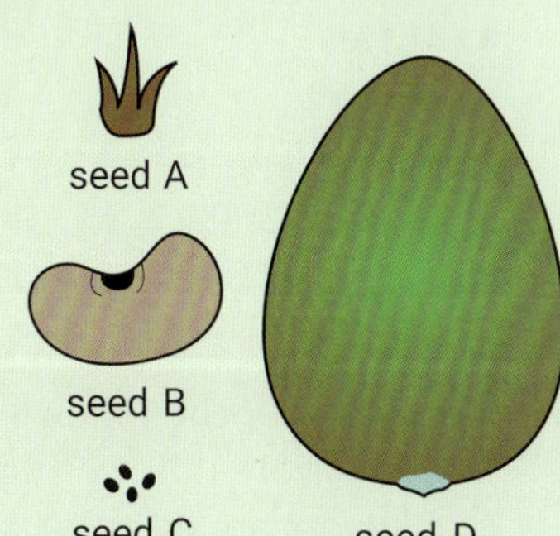

3 In most types of frog, the eggs are laid in the water together in a mass of foul-tasting jelly, whereas fish lay their eggs individually in the water. Suggest how these two reproductive behaviours help in the survival of each type of animal.

4 Many types of animals show *courtship behaviour* before they mate and produce offspring. Use the internet and other library resources to find out what courtship behaviour means. Write a report of what you find out, giving examples. How does courtship help in the survival of each animal?

5 The 'most devoted parent' award for caring for offspring should go to the male emperor penguin. Use library books or the internet to find out why the emperor penguin would win this award.

ISBN 978 1 4202 3735 1

MAIN IDEAS

Copy and complete these statements to make a summary of this chapter. The missing words are on the right.

1 All organisms are made of ______. There are many different types of cells, with different shapes and different functions.

2 An organism grows in size by making new cells in a process called ______.

3 During the stages of human growth, the ______ stage is characterised by a rapid growth spurt and the development of the reproductive organs.

4 In females, the ______ produce ova, and in males the ______ produce sperm.

5 In sexual reproduction, ______ occurs when the nucleus of a sperm cell joins with the nucleus of an ______.

6 During puberty girls begin ______ and boys start producing ______.

7 Organisms that care for their young (birds and ______) generally produce fewer eggs than those whose young are independent (fish, ______ and frogs).

8 Flowering plants use a variety of methods to ______ their seeds away from the adult plant.

reptiles
disperse
semen
ovum
cell division
cells
mammals
fertilisation
adolescence
menstruation
ovaries
testes

CH•10 REVIEW

1 Match the term in the list with the correct description below.

cells · sperm · pollination
fertilisation · nucleus · puberty

a the part of the cell that controls its activities
b the process that occurs when pollen travels to the stigma
c the process that occurs when the nucleus of a sperm joins the nucleus of an ovum
d the male sex cell
e the small building blocks that all living things are made of
f the period of time in humans when the reproductive organs start making sex cells

2 Describe the main functions of each of these parts of the human body.
a testes
b placenta
c semen
d uterus
e umbilical cord
f ovaries
g vas deferens
h fallopian tubes
i penis
j seminal vesicles
k endometrium

3 Which one of the following statements about reproduction in flowering plants is *false*?
A Flowers are brightly coloured to attract insects and birds, which help pollination.
B Pollen contains the sperm cells.
C Pollination is the same process as fertilisation.
D Pollen is made in the anthers of flowers.

ISBN 978 1 4202 3735 1

4 The partly-labelled diagram below shows the life cycle of a moth. Use the following terms to label the diagram.

adult	reproduction occurs
larva	pupa

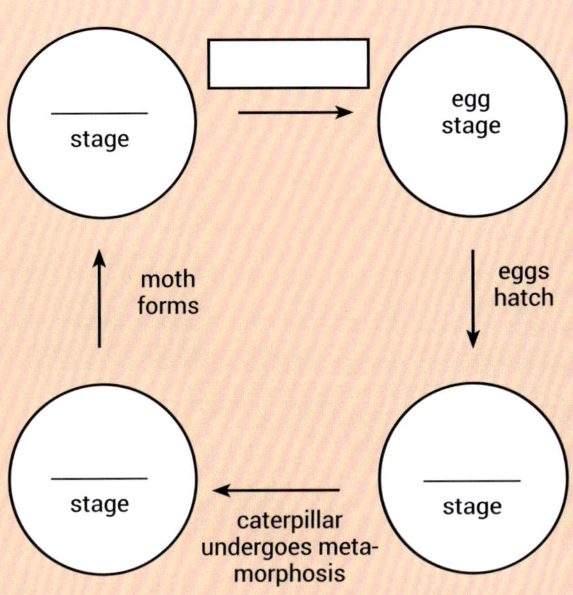

5 The graph below shows the change in the thickness of the wall of the uterus during the menstrual cycle.

a Describe in your own words what happens to the wall of the uterus during the menstrual cycle.

b In which days of the cycle does the ovary release an ovum? Why does it occur then?

c When does bleeding occur? Give a reason for your answer.

6 a A male fish externally fertilises the eggs laid by a female fish. Give two reasons why many of the eggs are never fertilised.

b Less than 0.5% of eggs laid by a frog reach adulthood, but over 60% of eggs laid by birds reach adulthood. Suggest reasons for this.

7 The fruits and seeds from various plants are shown in the diagrams below. Infer how the seeds are dispersed by each type of plant.

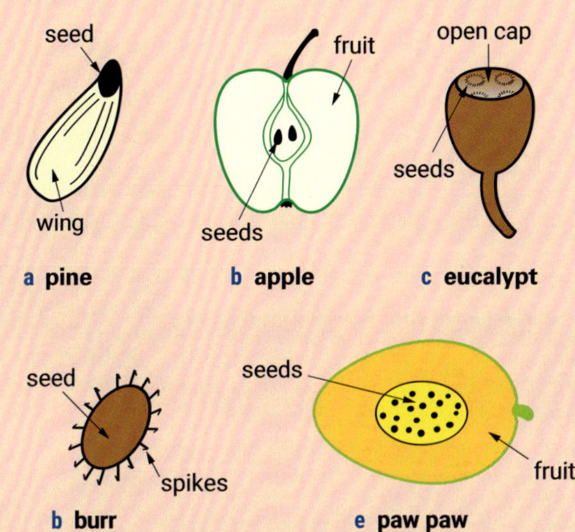

8 Write an inference for each of these observations.

a Bird's eggs are much larger that human eggs.

b Some flowers produce very strong, fragrant scents.

c In humans, the testes make millions of sperm cells.

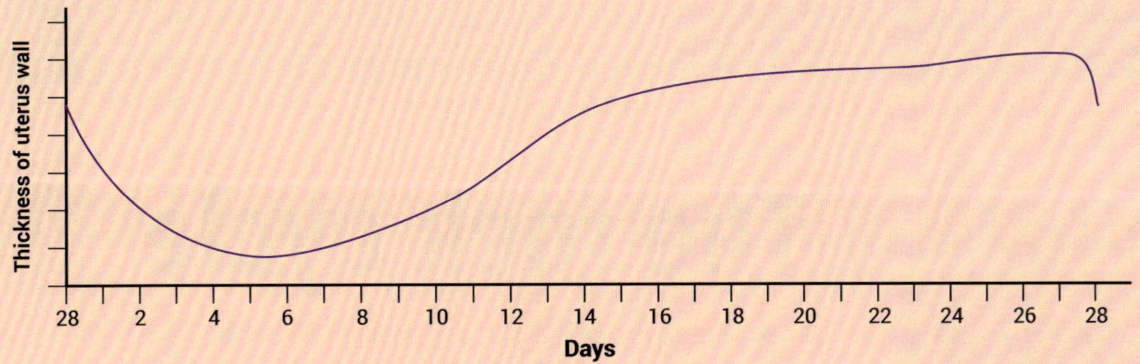

Check your answers on page 301.

ISBN 978 1 4202 3735 1

IN THIS CHAPTER

Science Understanding

- identify a range of common rocks using a key based on observable properties
- use granite to show that rocks are usually a mixture of several different minerals
- give examples of rocks and minerals that provide valuable resources
- use the rock cycle to describe relationships between igneous, metamorphic and sedimentary rocks
- investigate scientific and Aboriginal explanations of the origins of Uluru and other Australian landforms

Science Inquiry Skills

- use the internet to research the uses of rocks

CH•11 The rock cycle

ISBN 978 1 4202 3735 1

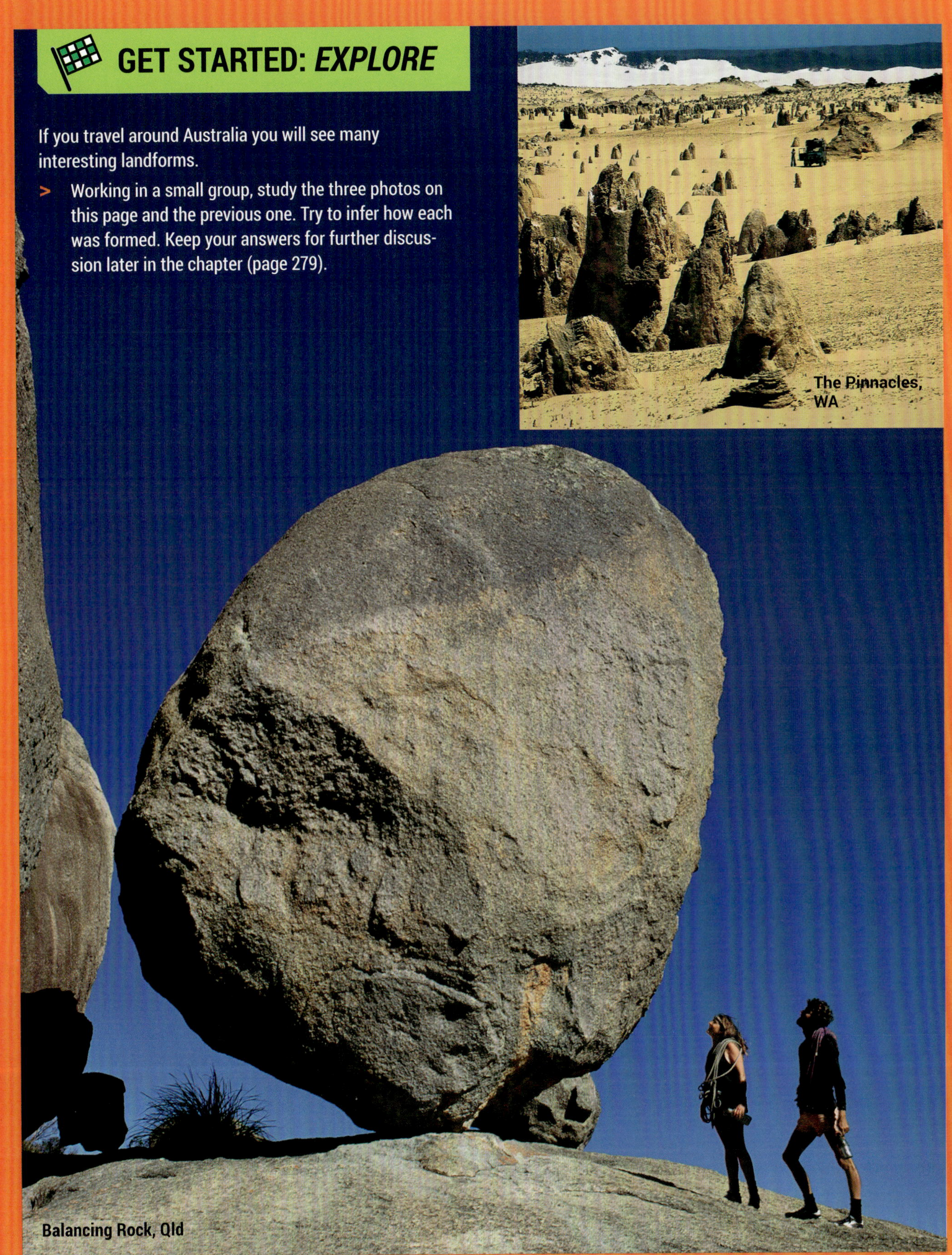

GET STARTED: *EXPLORE*

If you travel around Australia you will see many interesting landforms.

- Working in a small group, study the three photos on this page and the previous one. Try to infer how each was formed. Keep your answers for further discussion later in the chapter (page 279).

The Pinnacles, WA

Balancing Rock, Qld

11.1 Rocks from fire

Inside the Earth

Scientists infer that the Earth formed about 5000 million years ago from part of a cloud of dust and gas around the sun. As the dust and gas started to cling together, a ball of hot rock formed. Scientists think that the Earth is still cooling down. It is a bit like a baked potato: the outside crust cooled first, but the inside is still very hot.

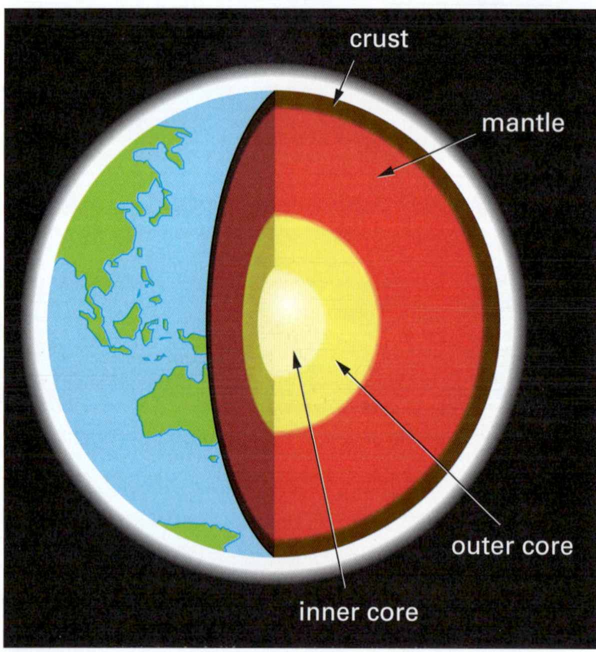

Figure 11.1 A cut-away view of the inside of the Earth

Scientists have never directly observed the centre of the Earth. They have observed molten rock from volcanoes, and have drilled holes to get rock samples from deep in the Earth's crust. They have also observed what happens to earthquake waves travelling through the Earth. From these observations they have inferred what is inside the Earth.

The evidence suggests that the Earth is not the same all the way through, and that there are layers, a bit like in a boiled egg. Scientists infer that there are four layers—the crust, the mantle, the outer core and the inner core, as shown in Figure 11.1.

The *crust* is like a thin skin, with an average thickness of about 30 kilometres. (The distance to the centre of the Earth is 6370 kilometres.) Beneath the crust is the *mantle*—about 3000 km thick. It consists of solid rock, but under certain conditions it can bend and flow like a thick paste.

The crust and the top of the mantle is called the *lithosphere* (LITH-os-fear), which means 'rocky sphere'. Below the lithosphere the rocks are under immense pressure. They are very hot, and in parts molten. This molten rock is called **magma**. It also contains dissolved gases.

At the centre of the Earth is the *core*. Scientists infer that it is made of the same materials as in most meteorites—iron and nickel. They also infer that the outer core is liquid, and the inner core is solid due to the enormous pressures.

Movements in the crust

The Earth's crust is more rigid than the mantle beneath it. In fact, the crust can be thought of as floating on the mantle, a bit like the skin on cooling custard. The mantle is constantly moving and sometimes large sections of the crust move very slowly upwards or downwards as the magma moves. When the magma underneath a pushed-up part of the crust cools, it turns into rock, forming a dome-shaped mountain. However this process is very slow, and usually takes millions of years.

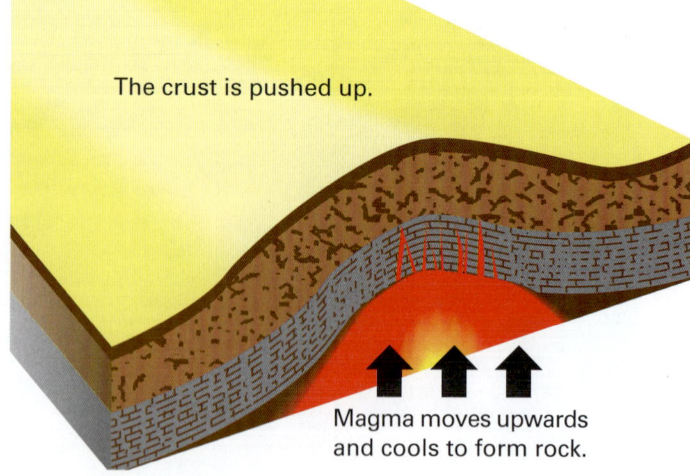

Figure 11.2 How moving magma can buckle the Earth's crust

ISBN 978 1 4202 3735 1

Volcanoes

There are sometimes weak spots in the Earth's crust that may crack and allow the magma to flow to the surface. When this happens a volcano forms. The pressure inside the Earth pushes the magma upwards. Sometimes the molten rock oozes out steadily. At other times it blasts out with incredible force.

After a volcano has erupted and the pressure has been released, the magma may harden to form a plug that blocks the vent. When this happens the eruption stops, and we say the volcano is *dormant*, or sleeping. On the other hand, if the pressure builds up, it may become active again. If it doesn't erupt again for a very long time we say it is *extinct*, or dead.

There are about 500 active volcanoes around the world. There are no active volcanoes in Australia, but there are many extinct ones; for example, Tower Hill in Victoria and Mount Gambier in South Australia.

When magma reaches the surface it is called **lava**. Lava is usually about 1000 °C, and red-hot. As it cools, it turns to solid rock. This may take weeks, or it may happen very quickly if the lava flows into water. Some volcanoes erupt quietly, with the lava spreading out to form a flat shield-shaped volcano. The volcanoes on the Hawaiian Islands are like this. Sometimes the lava is thin and runny. At other times it is thick and lumpy, like porridge, and hardly flows at all.

Volcanoes also produce gases, and some of these are poisonous. When lava contains a lot of gas it may froth violently. When this lava cools, the rock formed is full of holes where the gas bubbles used to be. *Pumice* rock is so full of holes, it is very light and floats on water. You often see it washed up on the beach.

Figure 11.4 Australia has many dormant volcanoes.

Figure 11.3 A lava fountain on an active volcano

Figure 11.5 This lava in Hawaii is very thick and slow-moving.

ISBN 978 1 4202 3735 1

Igneous rocks

Rocks that formed from the molten rock of the mantle are called **igneous rocks**. The word igneous (IG-nee-us) means 'from fire'. Examples of igneous rocks are granite (see page 273) and basalt.

Igneous rocks are made of interlocking **crystals**. These form from the substances in the magma, and have definite shapes and straight sharp edges—like sugar crystals. The crystals in granite are quite large, but the crystals in basalt are much smaller. You can usually infer how an igneous rock was formed from the size of its crystals. In Investigation 11.1 you can see how different cooling rates produce different-sized crystals.

Figure 11.6
A sample of basalt

Volcanic and plutonic rocks

There are two important groups of igneous rocks—volcanic and plutonic. **Volcanic** rocks are produced from rapidly cooling lava on or just below the surface of the Earth. The time taken for these rocks to solidify would be only days or perhaps months, depending on the thickness of the lava flow. Because the rock cools quickly it does not have time to form large crystals. Basalt is formed in this way.

If magma solidifies to form rocks well below the surface, the rocks are said to be **plutonic** (ploo-TON-ic) after Pluto, the Greek god of the underworld. The crystals in a plutonic rock are large because they have had time to grow—as long as a million years. Granite is formed in this way.

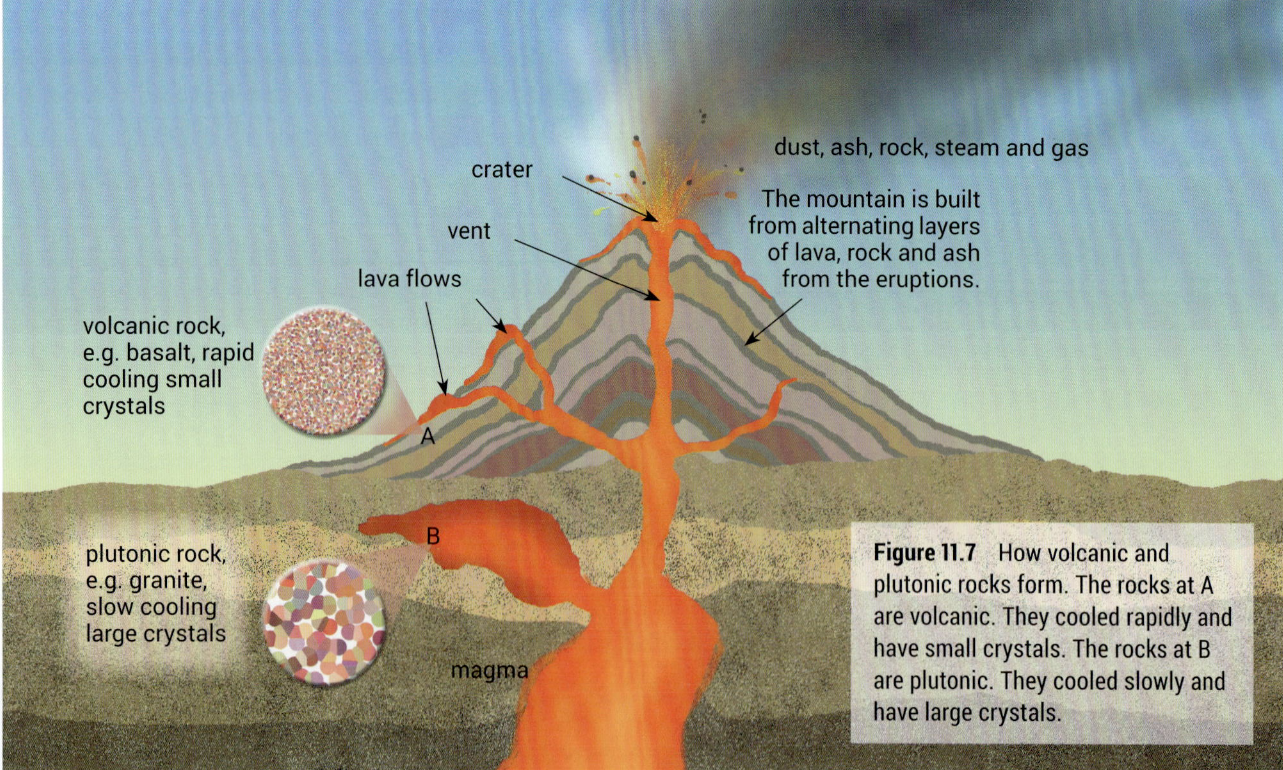

Figure 11.7 How volcanic and plutonic rocks form. The rocks at A are volcanic. They cooled rapidly and have small crystals. The rocks at B are plutonic. They cooled slowly and have large crystals.

ISBN 978 1 4202 3735 1

ACTIVITY

Do research to find out more about each of the following rock types. Classify each rock as plutonic (forming deep inside a volcano below the Earth's surface) or volcanic (forming at or near the Earth's surface). Then list any interesting facts about each rock such as its uses or other information that you discover.

Figure 11.8 Some samples of volcanic and plutonic rocks from New Zealand volcanoes

Minerals and crystals

All rocks are made of **minerals**. These are the building blocks of rocks. Some rocks contain only one mineral. For example, limestone contains only calcite (calcium carbonate). Most rocks, however, are a mixture of several different minerals. For example, granite is a mixture of three minerals—quartz, feldspar and mica. (See the activity on page 273.)

As magma solidifies, some minerals become concentrated in certain places. For example the minerals may dissolve in hot water in the cooling rock. This hot water may then seep into the surrounding rocks, carrying the minerals with it. When the water evaporates, the minerals are left behind as crystals. These minerals may contain metal compounds called **ores**, which can be mined and the metals extracted. The table on the right shows some common uses of ores mined in Australia.

Under certain conditions, minerals occur as large crystals. These crystals have definite shapes that can be used to identify the minerals. For example, quartz always forms six-sided crystals that are often transparent. Some minerals can be cut and polished to form gemstones; for example, diamonds, rubies, sapphires and emeralds.

Ore/mineral	Metal	Some uses
bauxite	aluminium	drink cans, aircraft parts
chalcopyrite (fool's gold)	copper	electrical wires, saucepans
galena	lead	batteries, fishing sinkers
haematite	iron	steel girders, railway lines

Figure 11.9 A sample of pure galena crystals, made of lead sulfide

ISBN 978 1 4202 3735 1

Fast and slow cooling

Aim

To use a model to represent what happens when crystals grow in cooling magma to form igneous rocks.

Materials

- two 250 mL beakers
- burner, tripod and gauze mat
- stirring rod
- watch glass
- styrofoam box or Esky
- hand lens (optional)
- potash alum (aluminium potassium sulfate)
- **copper sulfate**

Risk assessment and planning

Read the experiment carefully, then describe to your partner what you will be doing. Include safety precautions in your description.

Method

1. Set up the burner, tripod and gauze mat for heating.
2. One-third fill a beaker with water and heat it.
3. Add a spoonful of copper sulfate (a blue mineral) and a spoonful of potash alum (a white mineral). Continue heating, and stir until dissolved.
4. Continue adding equal amounts of copper sulfate and potash alum until no more will dissolve. The solution is then said to be *saturated*.
5. Half fill a cold watch glass with the solution and leave it to cool quickly.
6. Pour the rest of the solution into a second beaker and put it in a styrofoam box with the beakers of other groups. Leave the beaker to cool overnight.
7. The next day, pour any remaining solution off the crystals in the watch glass and the beaker. Use the hand lens to examine the crystals. Sketch the crystals in each 'rock'.

Discussion

1. Which 'rock' has the larger crystals?
2. What does the hot solution in this model represent?
3. Which mineral formed larger crystals—the copper sulfate or the potash alum?
4. Write a generalisation linking the rate of cooling to the crystal size.

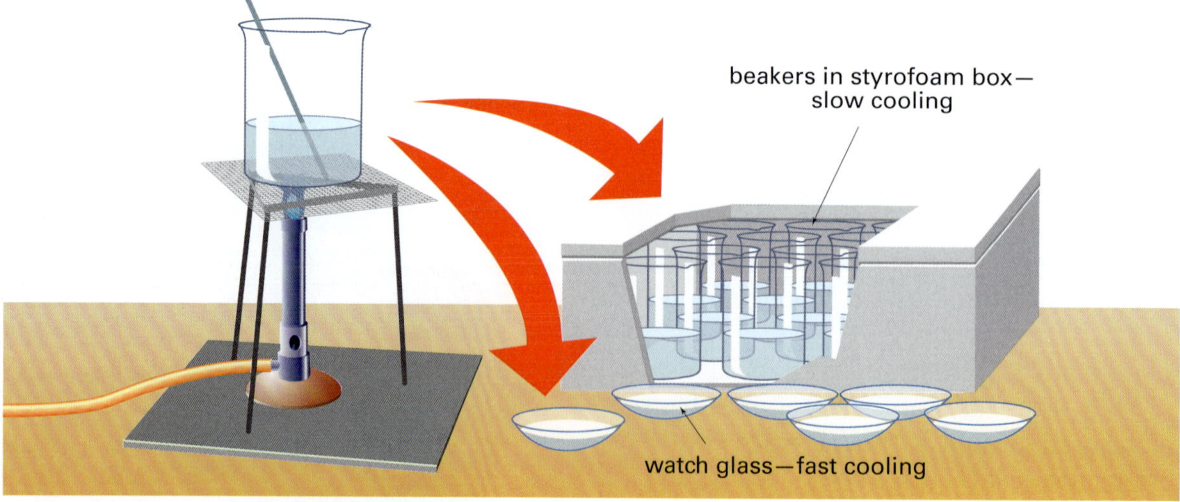

ISBN 978 1 4202 3735 1

ACTIVITY

Your teacher will give you samples of various igneous rocks, including granite and basalt. Observe them carefully and see if you can spot the different minerals in them. Use a hand lens or stereomicroscope to take a closer look.

Infer which of the igneous rocks are volcanic (formed on the surface of the Earth). Which are plutonic (formed under the ground)? Explain your answers.

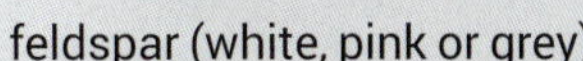

CHECK

1. How thick is the Earth's crust? Why is it called the 'crust'?
2. Draw a diagram of a volcano and put these five labels on it:

   granite formed here, magma, vent, crater, lava

3. How is basalt formed? Mark where it is formed on your diagram in Question 2.
4. Each of the following words is the correct answer to a question. Write a suitable question for each answer.
   a. crust
   b. magma
   c. igneous rocks
   d. plutonic rocks
5. What is the difference between a rock and a mineral?
6. Ask a classmate to check your spelling of the following words. Write the words in your notebook.

   basalt, lava, granite, lithosphere, igneous, magma

7. Use the table on page 271 to answer these questions.
   a. Which metal is extracted from haematite?
   b. Suggest why chalcopyrite is sometimes called fool's gold.
   c. Suggest other uses for each of the metals.
   d. Use library resources to add two other ores to the table.
8. a. How does the speed of cooling affect the size of the crystals in a rock?
   b. Why are the crystals in granite much larger than the crystals in basalt?

ISBN 978 1 4202 3735 1

CHALLENGE

1 How do volcanoes form?

2 The photo below shows a man holding up a boulder made of pumice. Why is pumice so light?

3 The table below shows the temperatures at different depths in a drill hole. Plot a line graph with depth on the horizontal axis, and temperature on the vertical axis. (If you aren't sure how to do this, see page 11.)

Depth (km)	Temperature (°C)
0	20
1	51
2	82
3	112
4	142
5	171
6	201
7	230

a Use your graph to work out how many degrees the temperature rises for each kilometre travelled into the Earth.

b Predict the temperature at the following depths: 1.5 km, 5.8 km, 8 km and 20 km.

4 Why are the crystals in the rocks on the edge of an old lava flow smaller than those in the rocks in the middle of the flow?

5 Lava is sometimes thin and runny, and sometimes thick and sticky. How would this affect the shape of the volcano formed by the lava flows?

EXPLORE

1 Make a model of the inside of the Earth showing the different layers.

2 Growing a large crystal:

- Add copper sulfate crystals to hot water until no more will dissolve. Allow the solution to cool almost to room temperature. Pour the solution into a clean beaker, leaving any undissolved solid behind.
- Ask your teacher for a well-shaped copper sulfate crystal. Carefully tie a cotton thread around it. Tie the other end of the thread to a glass rod or ice-cream stick.
- Hang the crystal in the copper sulfate solution as shown. Cover the beaker with aluminium foil or plastic wrap, and put it somewhere it won't be disturbed for a week or so.
- Check the crystal every day to observe its growth.

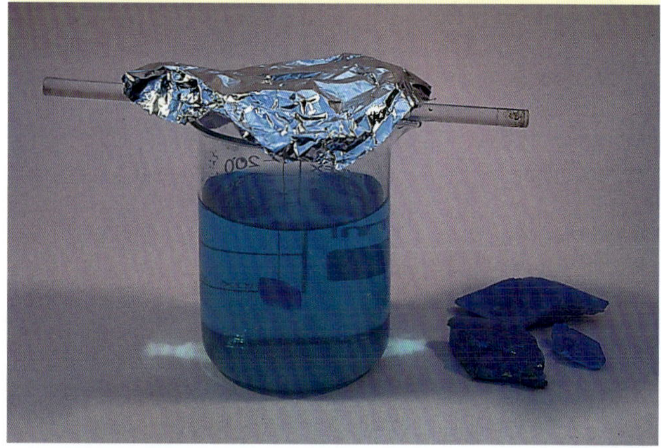

EXPLORE ONLINE

Follow the links to the websites below.

Volcano world

This site has extensive information on volcanoes around the world. It tells you which volcanoes are currently erupting, and you can see movies of eruptions.

Structure of the Earth

On this site you can explore the various layers of the Earth.

ISBN 978 1 4202 3735 1

11.2 Weathering and erosion

The drawing below shows part of a road cutting. Notice that there is solid rock at the bottom, then broken-down rock, and soil at the top. From this, scientists infer that long ago the top must have been solid rock too, but it has somehow broken down into smaller and smaller pieces.

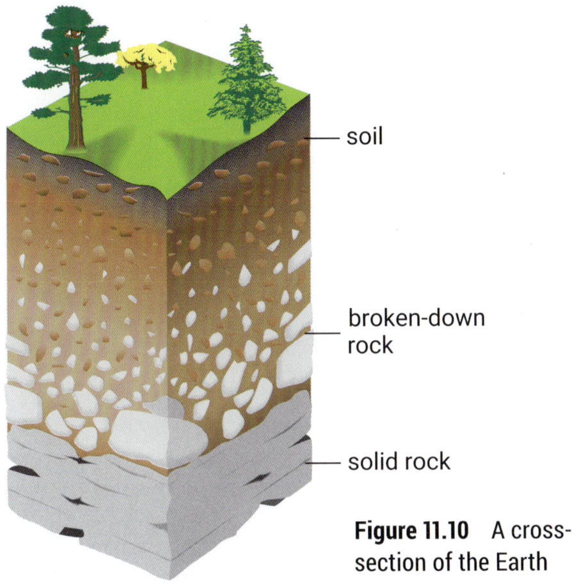

Figure 11.10 A cross-section of the Earth

This breaking down of rocks by natural processes is called **weathering**. There are two main types of weathering—*physical weathering* and *chemical weathering*. Most weathering is the result of both physical and chemical processes. Different rocks weather at different rates, depending on how hard they are; and how hard they are depends on what they are made of.

Physical weathering

If you heat a glass rod and put it into cold water, it will crack. This is because the glass on the outside of the rod contracts (gets smaller) while the inside remains the same. Similarly, rocks will crack if their temperature changes quickly. This may occur at night after a hot day, or when rain falls on hot rocks. Eventually the outer layers of the rock crack and may peel off. This process is very common in granite, and eventually rounded boulders are formed.

Figure 11.11 Hanging Rock in Victoria is made of hard volcanic rock. The surrounding softer rock has been weathered away.

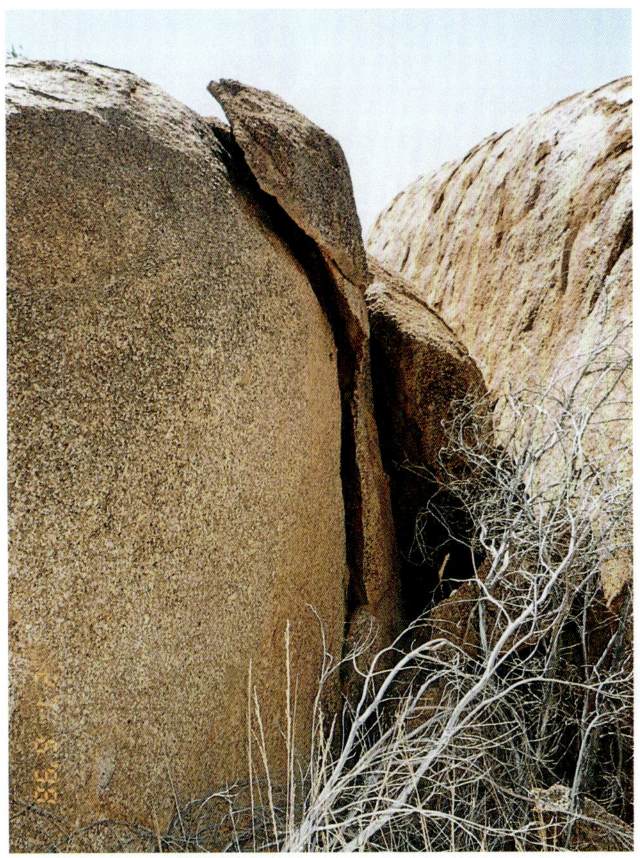

Figure 11.12 Notice that pieces are flaking off this granite, leaving a rounded boulder.

ISBN 978 1 4202 3735 1

In places where the temperature falls below 0 °C, ice can cause weathering by breaking open small cracks and holes in the rocks. How does it do this?

You may have seen a bottle in the freezer burst when the liquid inside freezes. This is because water expands when it freezes. If water in a small crack in a rock freezes, it will eventually split pieces off the rock.

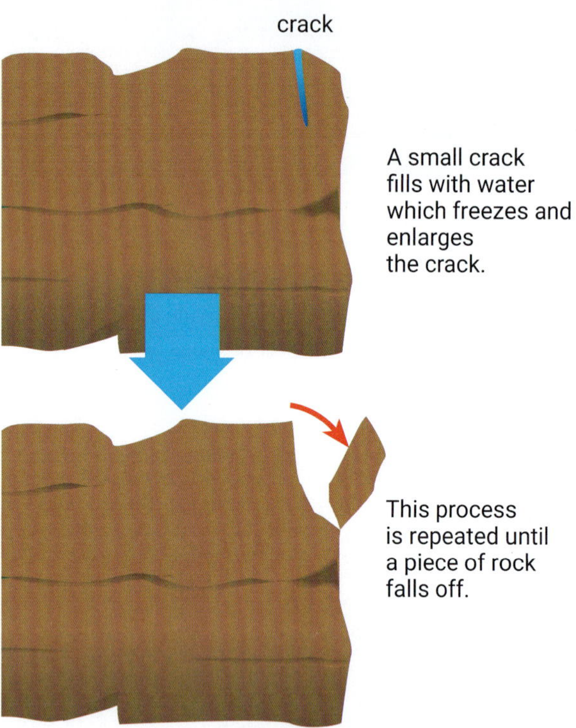

Figure 11.13 Freezing water can crack rocks.

ACTIVITY

To get an idea of how water can crack rocks when it freezes, try this activity at home.

Obtain a small plastic bottle with a tight-fitting lid. Fill it to the brim with water and close the lid tightly. Then leave it in the freezer overnight. What happens?

Plants can also cause physical weathering of rocks. Seeds from trees may fall into cracks in rocks and develop into seedlings. As the tree grows, the roots act like a crowbar, wedging the rocks apart.

Figure 11.14 Over long periods of time, tree roots can break up rocks.

Chemical weathering

Chemical weathering is the chemical breakdown of rocks. There are many different minerals in rocks, and when water and air react with these minerals, the new minerals formed are usually softer. This causes the rock to crumble.

You may have noticed that most rocks are brown on the outside. This is because iron is present in many minerals, and it reacts with the oxygen in the air to form brown iron oxide (rust), which is relatively soft.

iron + oxygen → iron oxide

If you chip a piece off some weathered rock you will notice that the unweathered rock inside is a different colour.

ISBN 978 1 4202 3735 1

Figure 11.15 Three pieces of granite. The piece on the left has not been exposed to the weather. The middle piece has been weathered (notice its brown colour). The piece on the right has been weathered to small pieces.

Igneous rocks such as granite often contain cracks and joints. Water seeps into these cracks and some of the minerals in the granite change, causing the rock to crumble. Physical weathering occurs at the same time, forming rounded boulders, and sometimes balancing rocks, as shown below.

Figure 11.16 Weathering in granite. See Balancing Rock on page 267.

A special type of chemical weathering occurs with limestone, which is made of calcium carbonate. Carbon dioxide in the air reacts with moisture in the air or with rainwater to form carbonic acid.

carbon dioxide + water → carbonic acid

Limestone reacts with carbonic acid (and other acids). So when this acidic water seeps into cracks in limestone it can dissolve quite large amounts of limestone, eventually turning the cracks into caves and arches. The Jenolan Caves near Sydney and the Chillagoe-Mungana Caves in North Queensland were formed in this way.

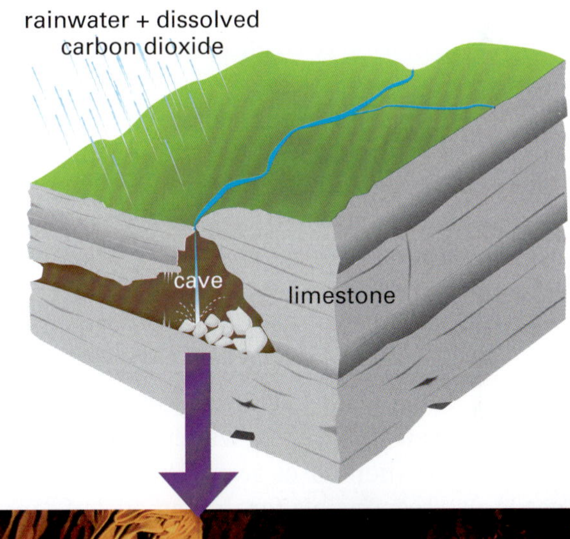

Figure 11.17 A beautiful shawl formation in the Chillagoe-Mungana Caves

ISBN 978 1 4202 3735 1

Erosion

Weathering is the process by which rocks are broken down. **Erosion** is the movement of soil and other weathered material from one place to another. The main agents of erosion are wind, water and ice. The eroded material is deposited (dumped) somewhere else as **sediment**. Erosion is a natural process, but human activity often causes it to occur more quickly.

Wind erosion

Dust storms are common in dry areas such as deserts. The sand grains carried by the wind can wear away (weather) rocks, often smoothing them and causing spectacular landforms such as the Pinnacles in Western Australia (see page 267). Fertile topsoil can be carried away by the wind if there is no vegetation to cover it and hold it in place.

Wave erosion

The action of the sea on coastal areas removes large amounts of weathered rock and sand, depositing it in other areas. The Twelve Apostles on Victoria's south-west coast (see page 266) are spectacular examples of coastal erosion. The soft rocks have been eroded by the sea, leaving the harder rocks, called sea stacks. Four of these have already collapsed and the rest will eventually fall as the sea wears their bases away. The action of the sea is also responsible for the removal of sand from beaches. The eroded sand is often carried by currents and deposited on other beaches.

Running water erosion

Running water is the most effective agent of erosion. Water running down any slope can carry soil with it. A fast-flowing river in flood can carry huge boulders, as well as pebbles, sand and soil. A spectacular example of erosion by a river is Grose Canyon in the Blue Mountains, where the Grose River has carved a groove in the Earth 10 kilometres long and 500 metres deep! It has been estimated that this has been going on for about 60 million years.

Glacier erosion

If you visit New Zealand you may see a *glacier*. This is a huge 'river of ice' moving down a valley. It moves rocks, just like a river, only very slowly.

Figure 11.18 The Werribee River in Victoria is eroding sandstone cliffs along its bank.

Figure 11.19 Grose Canyon in the Blue Mountains

ISBN 978 1 4202 3735 1

CHECK

1 Which of the following are true and which are false? Correct the ones that are false.
   a Rocks usually have cracks and holes in them.
   b Air and water cause chemical weathering.
   c Weathering is a rapid change.
   d Granite dissolves slowly in acidic water to form caves.
   e Water takes up less space when it freezes.
   f Plant roots can crack rocks.

2 How do sudden temperature changes cause weathering of rock?

3 Limestone dissolves slowly in rainwater. Explain in your own words how limestone caves are formed.

4 The diagrams below show four stages in the weathering of a granite cliff.
   a Put them in the correct order.
   b How long do you think this weathering would take—weeks, years, hundreds of years, or thousands of years?

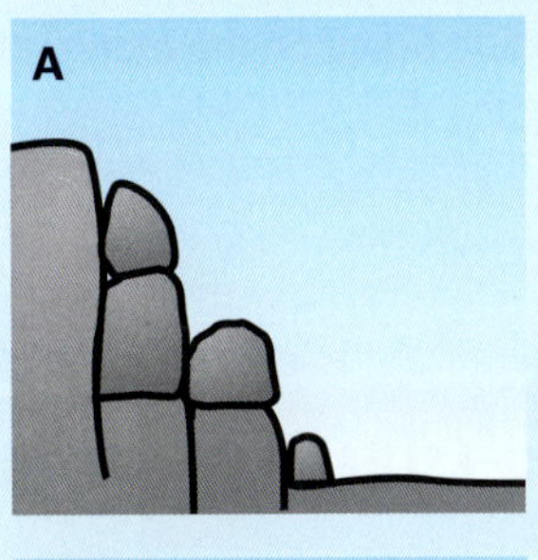

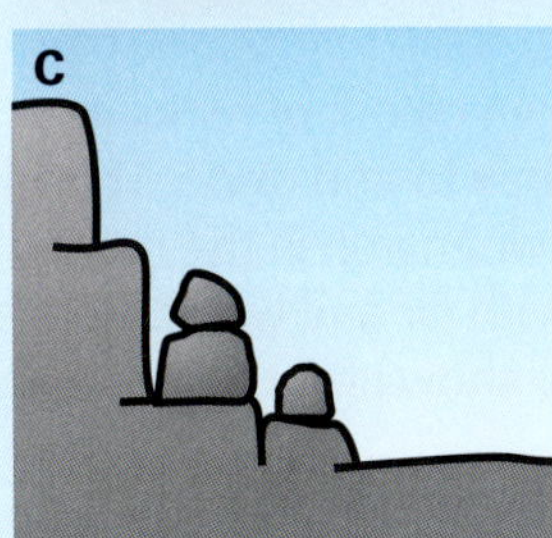

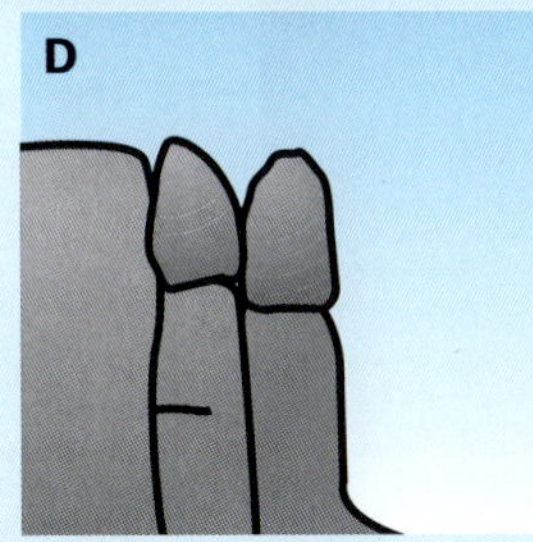

5 Clear water flowing down a river becomes muddy after rain. Why is this?

6 Granite is formed deep inside the Earth but there are many places where this rock can be seen at the surface. Suggest a reason for this.

7 Why do different rocks weather at different rates?

8 When limestone dissolves, what happens to the dissolved material?

9 A geologist has set up two measuring stations 150 km apart on a river. In a gorge in the mountains the river carries 200 million tonnes of sediment each year. Downstream it carries only 10 million tonnes per year. Infer what happens to the rest of the sediment.

10 Mt Crookneck, one of the Glasshouse Mountains in Queensland, is about 26 million years old. It is made of a volcanic rock called trachyte. Write an inference to explain its present-day shape.

Figure 11.20 Mt Crookneck, Glasshouse Mountains

11 a In what ways has human activity increased erosion by the sea?
   b How can beach erosion by wave action be prevented or reduced?

12 Look back to your inferences for 'Get started'. If necessary rewrite them using what you have learnt in this section.

ISBN 978 1 4202 3735 1

CHALLENGE

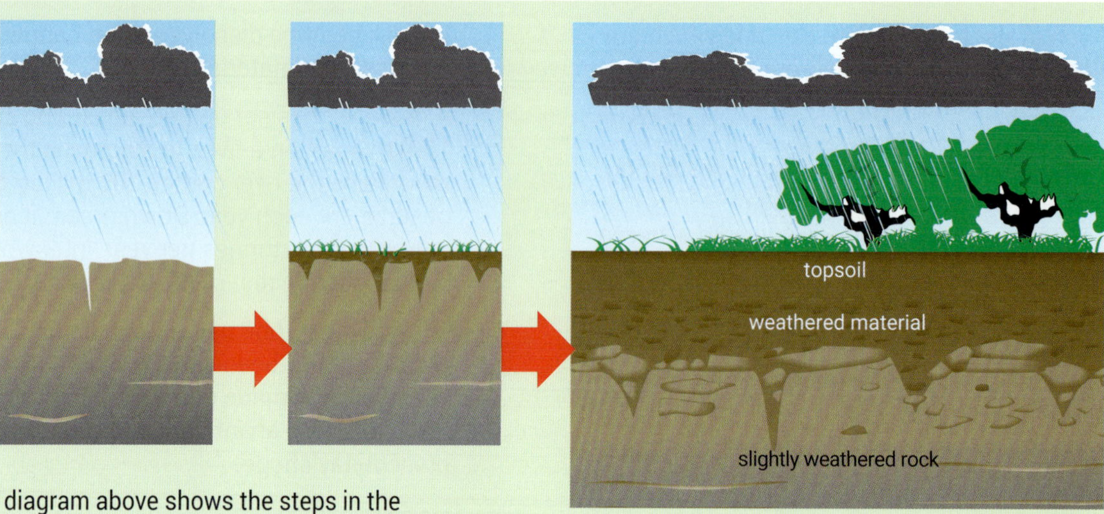

1 The diagram above shows the steps in the formation of soil. Write captions for the three parts of the diagram.

2 Why is soil in one place different from that in other places?

3 Figure 11.21 below shows Wave Rock in Western Australia. It is made of granite.
   a Write an inference to explain how it was formed.
   b Can you suggest another inference?
   c Could you test which inference is correct? If so, how?

4 How can animals such as rabbits affect the rate of weathering and erosion of soil and rocks?

5 Many water reservoirs in agricultural areas of Australia are becoming filled with sediment washed in by streams flowing into them. What action could be taken to help overcome this problem?

6 Do weathering and erosion occur on the moon? Explain your answer.

Figure 11.21 Wave Rock in Western Australia

EXPLORE

1 The brown colour of rocks is due mainly to iron oxide. This is formed when iron in the rocks reacts with air and water. Here is a way to show how this happens.

   Put some steel wool (or iron filings) in a jar, and pour in enough sand to just cover it. Sprinkle water on the sand each day. Examine the jar after about a week.

2 Look at Figure 11.17 on page 277. What are these cave formations called? Use a library to find out how they were formed.

ISBN 978 1 4202 3735 1

11.3 The rock cycle

Sedimentary rocks

Streams carry large amounts of weathered material (sediments). The size of the sediments being carried depends on the flow rate of the stream. The largest sediments are deposited first. For example, gravel quickly settles to the bottom unless the stream is flowing very quickly down a steep slope. The next to settle is the sand. Near the mouth of the stream the water moves very slowly. Here the smallest sediments settle out, making the banks muddy.

Look at Figure 11.22 below. The sediments carried by the river have entered the sea. Eventually they will settle to the bottom, the largest sediments first and the smallest sediments last. This also happens when rivers flow into lakes.

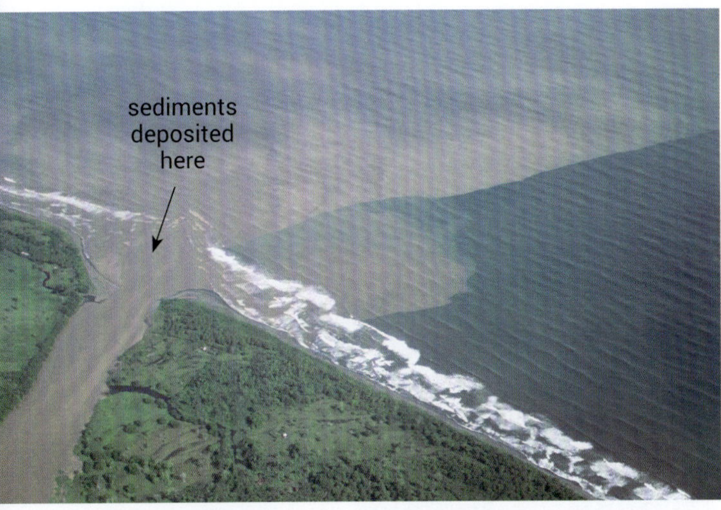

Figure 11.22 The sediments deposited at the mouth of this river will eventually turn into sedimentary rocks.

Sea water contains dissolved minerals that can soak into the sediments and cement them together. The sediments are also compressed (pressed down) by the tonnes of material on top of them. Eventually the sediments turn into **sedimentary rocks**.

Over millions of years, these sedimentary rocks can be pushed up above sea level by the movements in the lithosphere. You can usually see the sediment layers in these rocks.

Figure 11.23 Layers can be usually be seen in sedimentary rocks.

Types of sedimentary rocks

Different types of sediments produce different types of sedimentary rocks. For example, sand-sized sediments eventually produce *sandstone*. *Shale* (or mudstone) is a fine-grained sedimentary rock formed from mud. Like sandstone it is often layered, and often contains fossils. If rounded pebbles and stones are deposited and cemented together with sand, clay and other minerals, a rock called *conglomerate* (con-GLOM-er-it) is formed.

Not all sedimentary rocks are formed from deposits of weathered rocks as a result of erosion. *Limestone* is mostly formed from deposits of the remains of organisms such as shellfish, corals and certain microscopic plants. For this reason limestone usually contains fossils. *Coal* is formed from the remains of dead plants that are buried by other sediments before they can decay. Over millions of years the weight of the overlying sediments compacts the partially decayed plant material. This compacting process increases the temperature of the material and squeezes out water. Eventually coal is formed.

ISBN 978 1 4202 3735 1

Figure 11.24 Examples of sedimentary rocks

SCIENCE AS A HUMAN ENDEAVOUR

A Dreamtime legend

Different cultures have different ways of explaining the world around them. Aborigines have many legends to explain various landforms throughout Australia. For example, there is a Dreamtime legend to explain the origin of The Three Sisters in the Blue Mountains.

There were three beautiful giant sisters who fell in love with three brothers from a neighbouring tribe. Marriage was forbidden by tribal law, so the brothers decided to take the maidens by force. The large tribal battle that followed forced the medicine man to turn the sisters into stone. He intended to restore them after the danger had passed, but he was killed in the battle and nobody has been able to break the spell and turn the three sisters back to their original form.

Using what you have learnt in this chapter, how would you explain the formation of The Three Sisters?

What are the differences between your explanation and the Aboriginal explanation?

Your teacher may be able to arrange for an Aboriginal person to describe a legend about your local area.

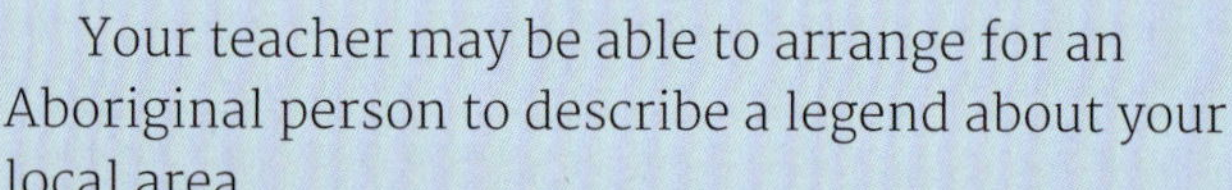

ISBN 978 1 4202 3735 1

Sedimentary rocks

Aim

To investigate the formation of sediments and sedimentary rocks.

Risk assessment and planning

Read both parts carefully and plan when you will do them. In Part B you will need to work out the details yourself.

PART A Settling of sediments

Materials

- sediment (mixture of dried mud, sand and gravel)
- clear plastic bottle with lid

Method

1 Half fill the bottle with a mixture of mud, sand and gravel. Fill the rest of the bottle with water.

2 Put the lid on the bottle and shake it. Then stand it on the bench and observe what happens.
Record the following data:
   - About how long did it take for the first sediment to settle?
   - What is the depth of sediments in the bottom of the bottle after 5 minutes? How much of the water is almost clear?
   - Can you see any layers in the sediments? Draw a sketch.

   Let the sediments settle overnight.

Discussion

1 Which sediments settled first (clay, sand or gravel)? Why?

2 Did all the sediments settle eventually? How do you know?

3 What is the relationship between the settling time and the size of the sediments? Write a generalisation.

PART B Making rocks

Materials

- sand, gravel and clay
- a range of cements, e.g. plaster of Paris, cement powder, PVA glue.

Use the Method below to make a full list of the equipment you will need.

Method

1 Make a sample of sandstone rock, as shown.

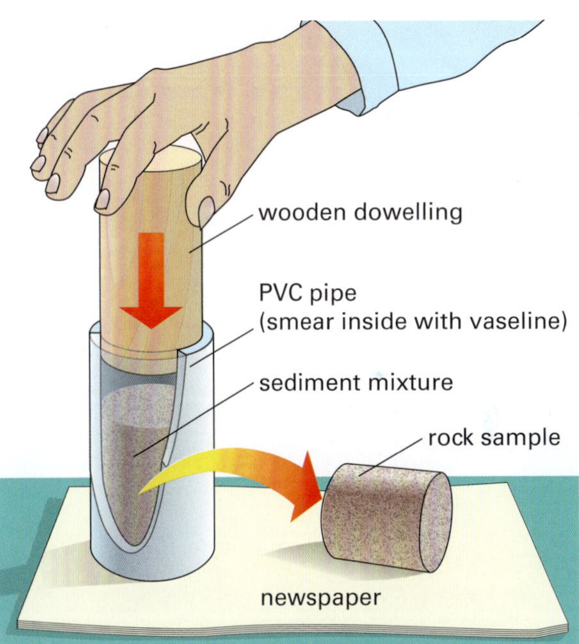

2 Experiment with the proportions of sand, clay and gravel to make different sorts of rocks. How would you make shale or conglomerate? You can also try different types of cement. For example, clay and a little water makes a good cement.

3 Leave your rock samples to dry for a few days then examine them closely. Use a hand lens if you like. How hard is the rock? How does it break?

4 Write a full report of your investigation.

ISBN 978 1 4202 3735 1

Metamorphic rocks

All rocks can be altered in some way by the effects of *heat* and *pressure*. Rocks that have been changed in this way are called **metamorphic** (met-a-MOR-fik).

1 Most metamorphic rocks are formed during mountain building, when the Earth's crust is pushed up and down by movements in the upper mantle. Enormous pressures cause the rocks to fold and twist, and tremendous heat is generated. Chemical changes produce new minerals, and the metamorphic rocks formed are harder than the original rocks.

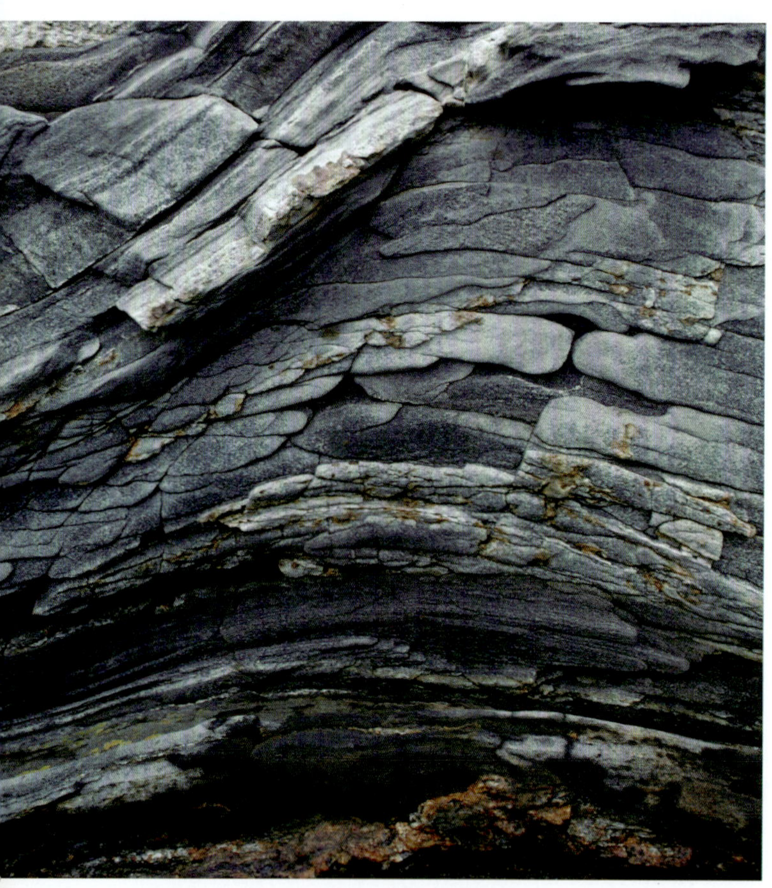

Figure 11.25 Pressure and heat produced in the folding of rocks can produce metamorphic rocks.

2 Another way metamorphic rocks form is when magma is forced up into the Earth's crust. The rocks in contact with this magma become very hot and are slowly changed to metamorphic rocks, as shown in Figure 11.26.

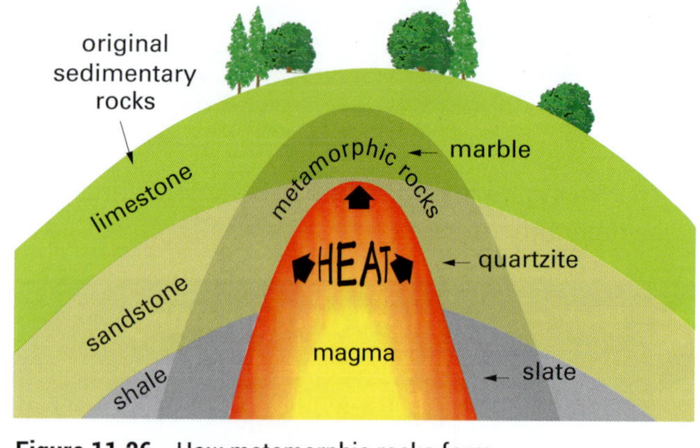

Figure 11.26 How metamorphic rocks form

Figure 11.27 Examples of metamorphic rocks

The rock cycle

To summarise, there are three types of rocks.

1 Igneous rocks are formed from magma.
2 Sedimentary rocks are formed from sediments.
3 Metamorphic rocks are formed from other rocks by heat and/or pressure.

Over millions of years, the Earth's rocks are constantly changed. Magma from deep within the Earth rises to the surface, and cools to become igneous rock. These igneous rocks are weathered and eroded to form sediments, which form sedimentary rocks.

ISBN 978 1 4202 3735 1

As layers of sedimentary rocks build up, the bottom layers sink deeper and deeper into the Earth's crust, where the temperature and pressure are greater. When this happens they can be changed into metamorphic rocks. If the temperature is high enough, they may melt to produce magma again. They may also be uplifted by huge Earth forces pushing the rocks upwards to form mountains.

The Earth's surface is constantly being worn down and uplifted by mountain-building forces. As this happens the rocks are changing from one form to another. The whole process is called the **rock cycle**.

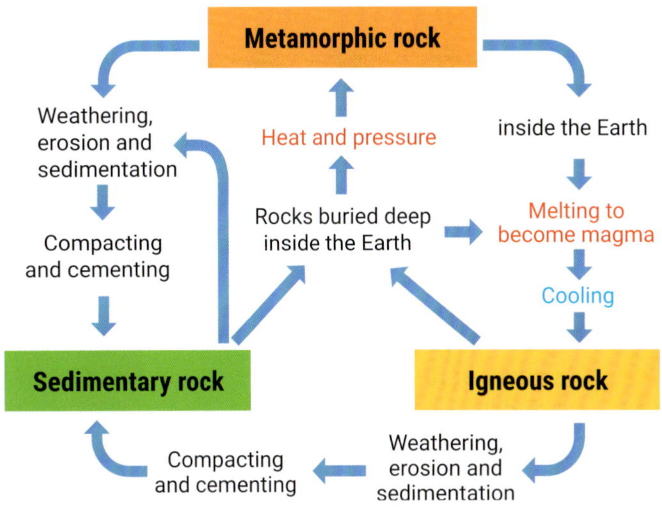

Figure 11.29 Simplified rock cycle diagram

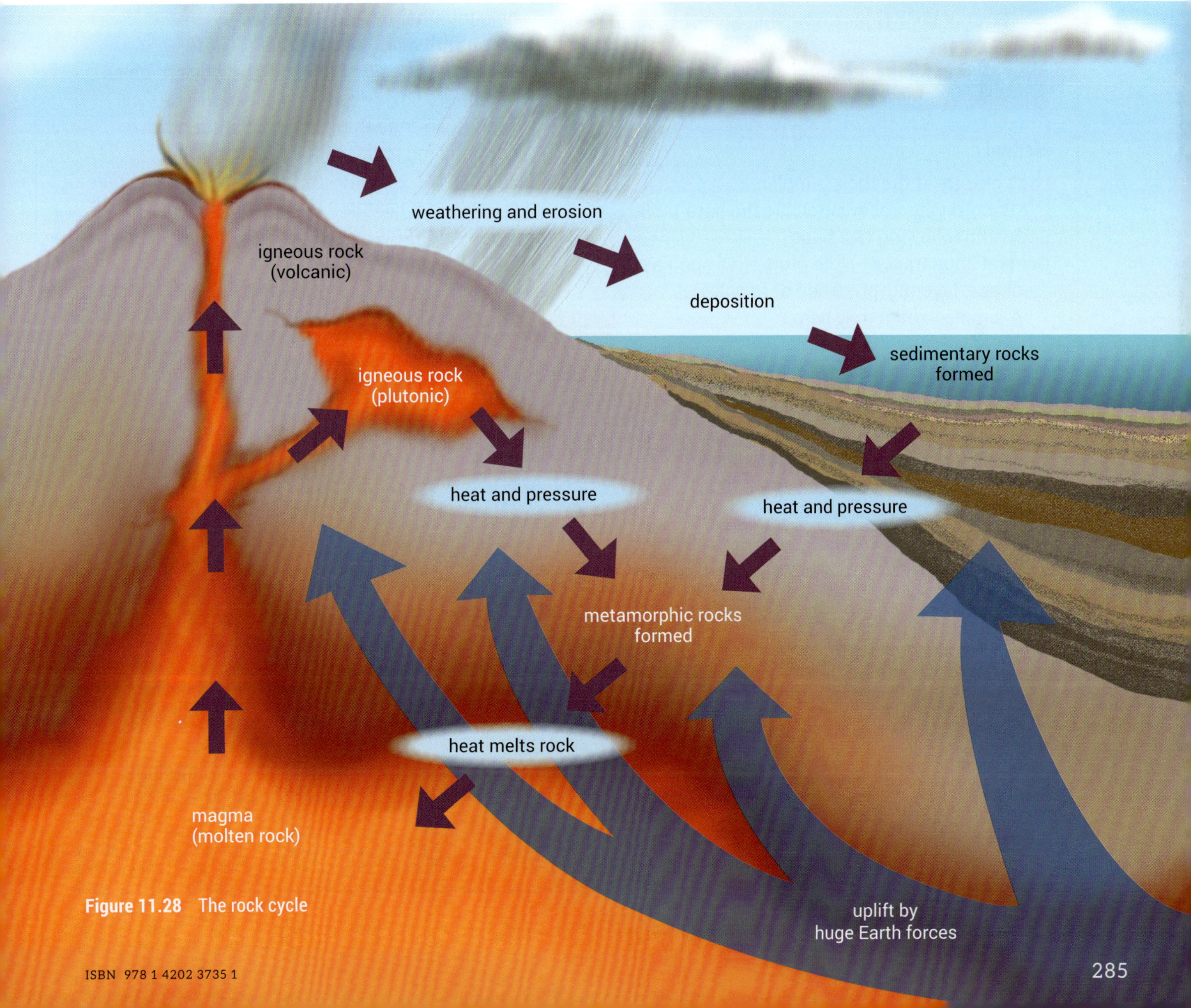

weathering and erosion
igneous rock (volcanic)
deposition
sedimentary rocks formed
igneous rock (plutonic)
heat and pressure
heat and pressure
metamorphic rocks formed
heat melts rock
magma (molten rock)
uplift by huge Earth forces

Figure 11.28 The rock cycle

ISBN 978 1 4202 3735 1

Looking at rocks

Aim

To observe a variety of igneous, sedimentary and metamorphic rocks.

Materials

- variety of rocks or rock kit
- hand lens
- table knife or metal spatula
- dilute **hydrochloric acid**

Risk assessment and planning

Draw up a data table in which to record the properties of the rocks.

What safety precautions will be necessary?

Method

Look closely at the rock samples.

Record your observations in the data table, under the following headings.

General appearance—Note properties such as shape, layering, presence of fossils. A simple sketch may help.

Colour—Note the colour.

Hardness—Use a table knife or metal spatula to see if you can scratch or split the rock.

Grain size—This is the size of the rock fragments (in sedimentary rocks) or the size of the crystals (in igneous rocks). The terms usually used to describe grain size are *fine*, *medium* and *coarse*.

Texture—The size, shape and arrangement of the fragments or crystals in the rock. Note whether they interlock as in granite (see the photo on page 273) or whether they are separate grains held together by finer material.

Reaction with acid—Add a drop of dilute hydrochloric acid to see if it produces bubbles of carbon dioxide gas. (*Teacher note:* You will need to arrange things so that students are not handling rocks with acid on them.)

Classification—Use the key below to see if you can classify each sample as igneous, sedimentary or metamorphic. You may also be able to classify rocks you have collected yourself.

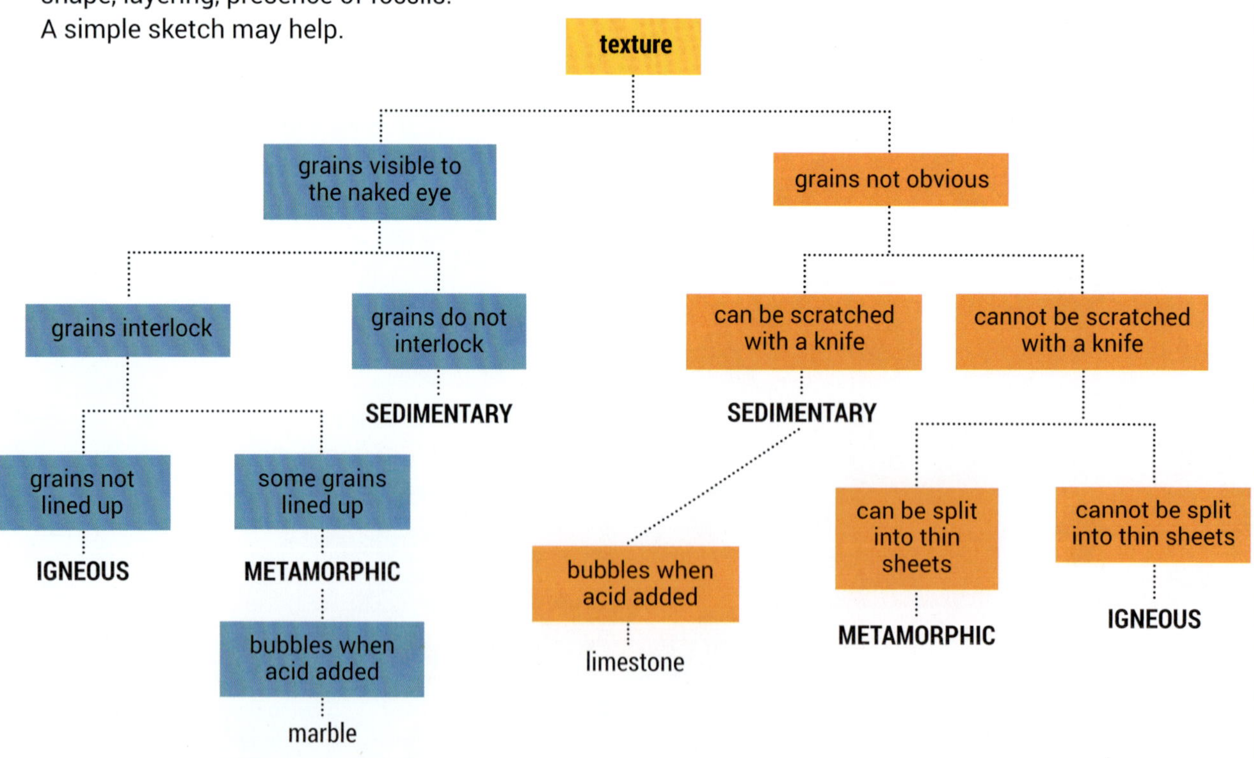

ISBN 978 1 4202 3735 1

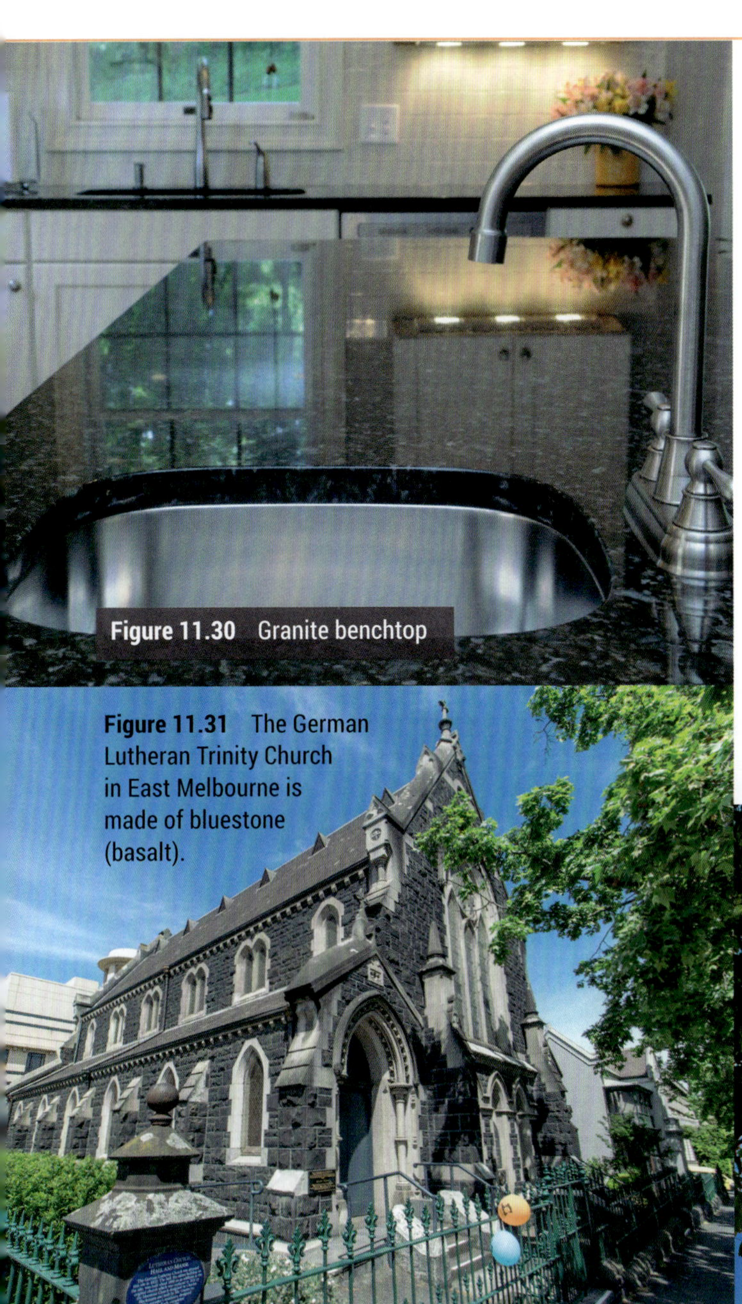

Figure 11.30 Granite benchtop

Figure 11.31 The German Lutheran Trinity Church in East Melbourne is made of bluestone (basalt).

Figure 11.32 Cutting limestone slabs in a quarry (Mount Gambier)

Use the internet to research the following commonly used rocks:

- dolerite
- granite
- limestone
- marble
- pumice
- quartzite
- sandstone
- shale
- slate

Record your information in a table under these headings:

- type of rock (sedimentary, igneous or metamorphic)
- properties
- how it was formed
- how it is used.

For useful websites follow the links to:

List of rocks

The rock cycle

Figure 11.33 St Paul's Cathedral in Melbourne is made from local sandstone.

SCIENCE AS A HUMAN ENDEAVOUR

Explaining Australia's landforms

In central Australia there are two famous rock formations called Uluru/Ayers Rock and Kata Tjuta/Mount Olga. Geologists have inferred how these landscapes were formed, and Aboriginal people have their own explanation.

According to geologists, about 550 million years ago, granite mountains to the south were eroded and sedimentary rocks were deposited in a low area in central Australia. Then about 400 million years ago these rocks were pushed upwards and folded to form a large mountain range. Since then this mountain range has been eroded to form a series of smaller mountains, including Uluru and Kata Tjuta (KAH-tah-JOO-tah), made up of near vertical layers of sedimentary rocks. These rocks are much harder than the surrounding rocks, which have been weathered and eroded. Look at the diagram.

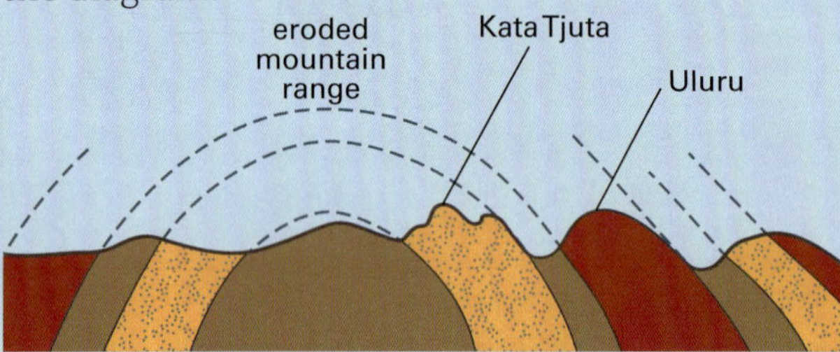

According to Aboriginal people, the world was once a flat featureless plain. During the Dreamtime, creation beings in the form of people and animals travelled widely across the land and formed the landscapes we know today (see the Dreamtime legend on page 282). The spirits of these ancestral beings are still here today, and that is why places like Uluru and Kata Tjuta are considered sacred sites.

There are several legends about the formation of Uluru. One is that it was built during the creation by two boys playing in the mud after rain. Their bodies are preserved as boulders on the top of Mount Connor to the east of Uluru. Another legend is that there was a great battle at Uluru at the end of the Dreamtime. The rock python people were attacked by a party of poisonous snake warriors. The marks on the south-west face of the rock are the scars left by the warriors' spears.

The local Aboriginal peoples say that Kata Tjuta (many heads) is the home of the snake Wanambi who, during the rainy season, stays curled up in a waterhole on the summit of Mt Olga, the highest of the many dome-shaped rocks. The dark lines on the side of the mountain are the hairs of his beard and the wind that blows through the gorge is his breath. Other rocks of Kata Tjuta represent Aboriginal ancestors known as mice women, kangaroo man, lizard woman and giant cannibals.

Question

1. In your own words, describe how geologists infer Uluru and Kata Tjuta were formed.
2. What was the Aboriginal Dreamtime?
3. Why do the Aboriginal people consider Uluru and Kata Tjuta to be sacred sites?
4. How do you think geologists would explain the dome-shaped rocks of Kata Tjuta?
5. How is the Aboriginal explanation of the creation of Uluru–Kata Tjuta different from the geologists' explanation?

Figure 11.34 Kata Tjuta

ISBN 978 1 4202 3735 1

CHECK

1 These four stages in the formation of a sedimentary rock are mixed up:

cementing and compressing
deposition (settling)
weathering
erosion

Put them in the correct order.

2 a List the following sediments in the order in which they would settle in water that has been vigorously shaken in a bottle:

coarse sand
clay
fine sand
gravel

b Suppose you found sediments that had been deposited at A, B and C in the diagram below. Which sediments would be gravel, which mud, and which sand? Explain your answer.

3 How could you tell the difference between sandstone and conglomerate? Draw a diagram to show the difference between the two rocks.

4 Which type of rock is formed as a result of:
a weathering and erosion?
b heat and pressure?
c melting and cooling?

5 Where does all the force come from to squeeze sediments together to form sedimentary rocks?

6 Copy the rock cycle diagram below. Then use the diagram on page 285 to add the names of the rock types and label the arrows.

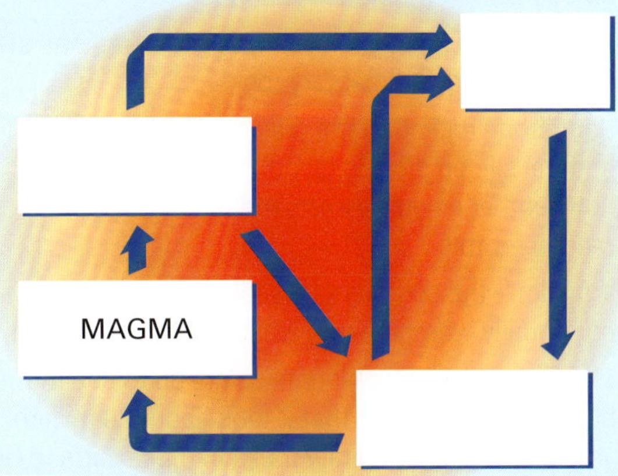

7 Name three igneous, three sedimentary and three metamorphic rocks.

8 a How can granite be changed into a sedimentary rock?
b Can a metamorphic rock be changed into an igneous rock? How?

9 The photo below shows an open-cut coal mine.
a Why is coal called a fossil fuel?
b What effect does coal mining have on the environment?

ISBN 978 1 4202 3735 1

CHALLENGE

1 Many buildings were once built from stone. Why are buildings rarely made of stone now?

2 Which rock would be best to build a small bridge across a stream? Explain your choice.

3 Suggest why marble is usually found in regions of heavily folded sedimentary rocks.

4 Sandstone and quartzite are both made of the mineral silicon dioxide (sand), but quartzite does not weather as quickly as sandstone. Suggest a reason for this.

5 Why don't the continents disappear as they are continually worn down by weathering and erosion?

6 The cliffs in the Blue Mountains are up to 500 metres high. Geologists infer that it took about 20 million years for the sedimentary rocks in these cliffs to be formed on the ocean floor. Calculate the average depth of sediment deposited each year.

7 Mary found a cliff-face with four different layers, as shown below.

a Suggest why the sediments in the layers are different sizes and colours.

b Which layer was most likely formed during a period of floods? Explain your inference.

c Which layer was most likely formed when the area was swampy? Explain your inference.

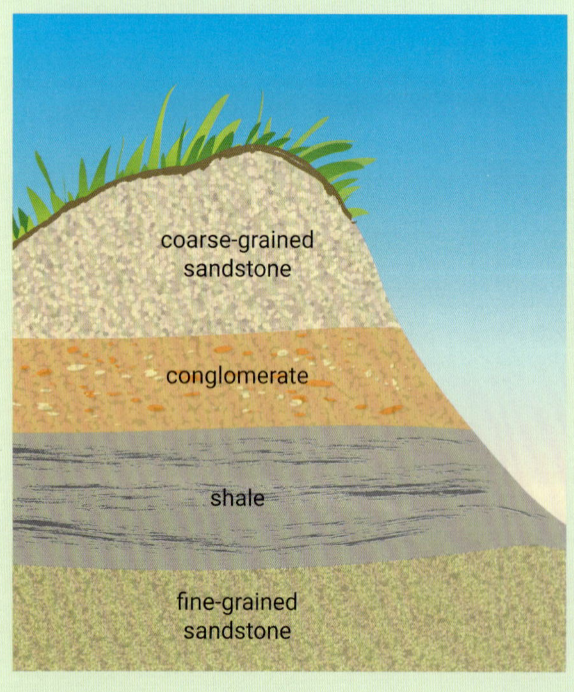

8 A, B and C in the diagram below are sedimentary rocks and D is a plutonic igneous rock. Which of these four rocks would be the oldest? Which would be the youngest? Explain your answer.

9 Imagine you are a piece of rock that has recently come from a volcano. Using the rock cycle on page 285 describe your 'life' over the next few million years.

EXPLORE

Collect as many different kinds of rocks as you can. The best places to look are where bare rocks are exposed—on hillsides, cliffs, quarries, building excavation sites, rocky gullies, stream beds and road cuttings. To get a good specimen you will need to break off a piece of rock about the size of a clenched fist with a geological hammer. Warning: When chipping rocks always wear safety glasses to protect your eyes.

Note the difference between the weathered surface of the specimen and the unweathered rock inside. Label each specimen, and make a note of where you found it. You could use the key on page 286 to help you identify your rocks.

Did you find any sedimentary rocks? Can you tell whether they are sandstone, conglomerate or shale? If you think a sedimentary rock may contain fossils, try to split it along the layers. Fossils are common in limestone.

Did you find any igneous rocks? Basalt is very common. It is grey-coloured and its crystals are too small to see. Granite (page 273) is also common and often forms round boulders. It has large crystals in it.

ISBN 978 1 4202 3735 1

MAIN IDEAS

Copy and complete these statements to make a summary of this chapter. The missing words are on the right.

1 The inside of the Earth is very hot, and in parts it is _______.

2 Molten material below the Earth's surface is called _______. Where the magma breaks through the surface in a volcano, it is called _______.

3 _______ rocks are formed by the cooling and hardening of molten rock.

4 Rocks are formed from minerals. _______ are minerals from which metals can be extracted.

5 _______ is the slow breaking down of rocks into smaller and smaller pieces. It may be physical or chemical.

6 _______ is the process by which weathered materials are carried away, mainly by water.

7 Sedimentary rocks are formed by the processes of weathering, erosion and _______, followed by the cementing together and hardening of _______.

8 _______ rocks are formed from other rocks by heat and/or pressure.

9 Rocks can be changed from one type to another over millions of years. This is called the rock _______.

molten
weathering
metamorphic
erosion
cycle
ores
igneous
sediments
lava
magma
deposition

CH•11 REVIEW

1 Which of the following are rocks and which are minerals?
- **a** basalt
- **b** bauxite
- **c** calcite
- **d** marble
- **e** pumice
- **f** quartz
- **g** sapphire

2 Granite is a rock:
- **A** containing crystals big enough to see.
- **B** made up mostly of dark minerals.
- **C** formed by the cooling of a lava flow.
- **D** made up of layers of different colours.

3 A soil sample containing different-sized grains was mixed with water in a jar. After the mixture settled, four layers could be seen. Which one of the diagrams shows the correct order in which the layers probably formed?

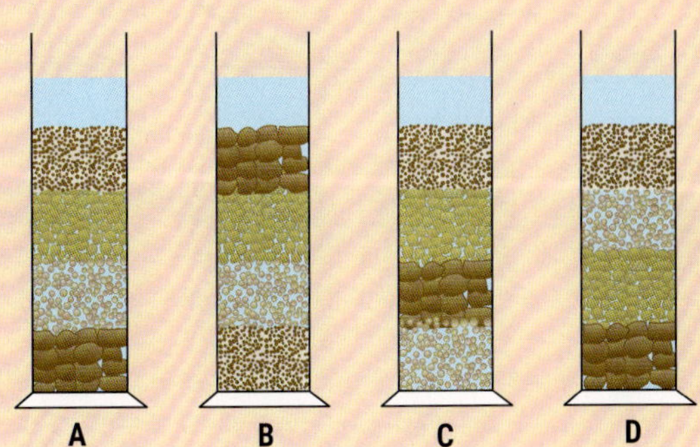

ISBN 978 1 4202 3735 1

4 Just below the crust of the Earth is the:
 A mantle.
 B inner core.
 C outer core.
 D atmosphere.

5 You want to find out whether an igneous rock has been formed by slow or rapid cooling. Which of these properties of the rock would you need to observe?
 A hardness
 B shape
 C colour
 D crystal size

6 The diagram below is a cross-section of an extinct volcano.
 a Where would you expect to find igneous rocks with the smallest crystals? Why?
 b Which of the following rocks would you most likely find at 4?
  A granite
  B pumice
  C basalt
  D sandstone
 c Are the rocks at 3 volcanic or plutonic?
 d Copy the diagram and mark on it where you are likely to find metamorphic rocks.
 e Why are the layers of sedimentary rocks not horizontal?

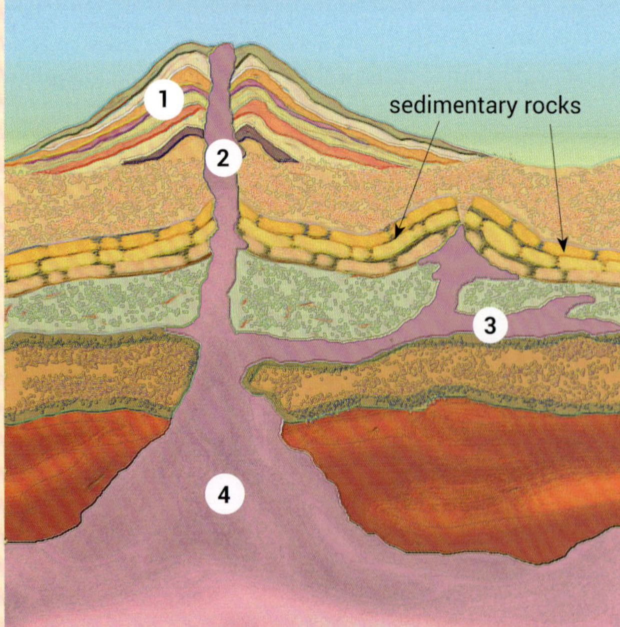

7 The surface of the Earth is covered by many different landforms. These were formed by:
 A erosion.
 B movements inside the Earth.
 C volcanoes.
 D all of the above.

8 Put these five stages of the weathering and erosion of granite into the correct order:

boulder	gravel	fine sand
coarse sand	solid rock	

9 Explain in your own words the difference between lava and magma.

10 You are a rock sculptor and have been given the job of making an outdoor sculpture of Australia's first president.
 a What properties would the rock you use for the sculpture need to have?
 b Suggest which rock you would use.

11 Use your knowledge of the rock cycle to explain the sequence of events that could over millions of years change a rock exposed near the top of a mountain into metamorphic rock. Describe the changes needed and draw a diagram.

12 A mesa is a flat-topped land form. Many mesas in Australia are capped with basalt or other hard rock, as in the photo below. Use diagrams to explain the flat-topped shape.

Figure 11.35 Mt Connor in Central Australia is a 'mesa'.

Check your answers on page 302.

ISBN 978 1 4202 3735 1

Answers to Reviews

If your answer does not agree with the answer given here, go back to the chapter and read the relevant section again. Your answers may be slightly different from the answers given here. If in doubt, check with your teacher.

CH•1 Let's experiment

1 **B**

2 **D**—One-quarter of the candle burns in 2 hours, so you can predict that the whole candle will burn in $4 \times 2 = 8$ hours.

3 **A**—It is difficult to control how hard you hit (**B**) or throw (**C**) the balls.

4 **a** Steel is a metal and conducts electricity.
b Metals conduct electricity, but non-metals don't conduct electricity.
c You would not expect carbon (a non-metal) to conduct electricity. You could modify the hypothesis as follows: Metals conduct electricity, but *most* non-metals don't.

5 **C** (**D** is incorrect because the solubility increases with temperature at a uniform rate. In other words, the graph is a straight line.)

6 **a** about 1 pm
b about 11.30 am and between 12.30 pm and 1.30 pm
c at the top of the high range
d Most probably there were clouds around 12 noon that blocked some of the UVB.

7 **a** Cut a piece of dirty cloth into equal-sized pieces. Using the same quantities of soap powder and water, wash one piece of cloth in Sudso and the others in other types of washing powders for the same time. Compare the results. Do the experiment in both hot and cold water.
b The variables to control are:
- how big and how dirty each piece of cloth is
- amount of washing powder you use
- volume of water you use
- temperature of water
- method of washing the cloth
- how long you wash the cloth.

c You are purposely changing the type of washing powder.
d You will measure the cleanness of the cloth.

ISBN 978 1 4202 3735 1

CH•2 Solids, liquids and gases

1 **D**—Because the statement says *all matter*, it is a generalisation rather than an inference.

2 See the diagram below.

3 **C**

4 There are several examples on page 30 but you have probably thought of others.

5 *Aluminium* (density 2.7 g/cm^3) and *lead* (density 11.3 g/cm^3) will both sink in water. And because their densities are less than the density of mercury (14 g/cm^3), they will float in mercury.

6 **a** Density of pure gold $= \frac{\text{mass of pure gold}}{\text{volume of pure gold}}$

$= \frac{1930\text{ g}}{1000\text{ cm}^3}$

$= \mathbf{19.3\ g/cm^3}$

b Density of crown $= \frac{\text{mass of crown}}{\text{volume of crown}}$

$= \frac{1500\text{ g}}{1000\text{ cm}^3}$

$= \mathbf{15.0\ g/cm^3}$

c No—the density of the crown is less than 19.3 g/cm^3, therefore it is not of pure gold.

7 Your answer should be something like this:

Shopping bags are normally made of LPDE plastic. These are cheap but are difficult to dispose of and at present cannot be recycled. However, the green shopping bags made of polypropylene can be reused many times. Plastic is a *synthetic* material made from coal. It is therefore *non-renewable*. The bags could be made from paper, a *processed* material made from wood, which is *renewable*. However, paper bags may not be as strong as plastic ones. Bags could also be made from cotton canvas. These are more expensive but can also be reused.

8 **a** **A** particles vibrating or moving very slowly
D particles very close together, almost touching
G very strong bonds between particles

b **C** particles moving freely and rapidly
F wide spaces between particles
I very weak bonds between particles

c **B** particles moving around freely but slowly
E particles fairly close together
H particles held together to some extent but free to move around

d **D** particles very close together, almost touching

e **D** particles very close together, almost touching
G very strong bonds between particles

f **A**

9 **a** The particles in gases are much more spread out than the particles in liquids or solids. There are fewer particles packed into each cubic centimetre. Hence gases have lower densities than liquids or solids.

b When a gas is cooled, its particles lose energy and don't move as quickly. They move closer together and attract each other more strongly. As a result the gas condenses to a liquid.

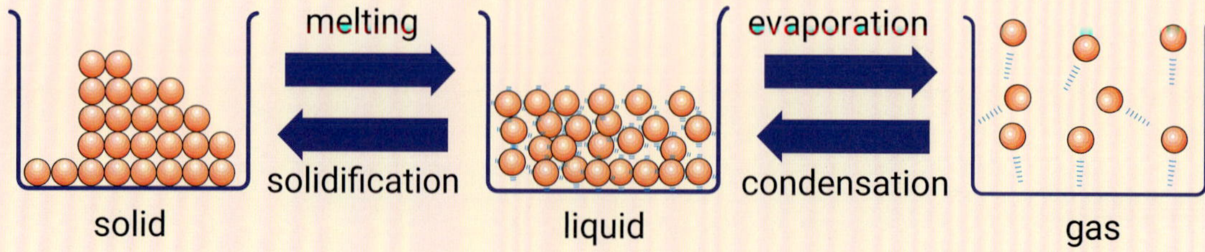

ISBN 978 1 4202 3735 1

CH•3 Introducing energy

1 **D**

2 **C**—This is against the law of conservation of energy (page 69).

3 **B**

4 **A**—Heat energy is transferred from the hot tea to the cup.

5 **a** **B**
b **C**

6 **a** The other 95 joules of energy are wasted as heat energy. This is why the bulb becomes so hot.

b $$\text{efficiency} = \frac{\text{energy}}{\text{input energy}} \times 100$$

$$= \frac{5 \text{ joules}}{100 \text{ joules}} \times 100$$

$$= 5\%$$

7

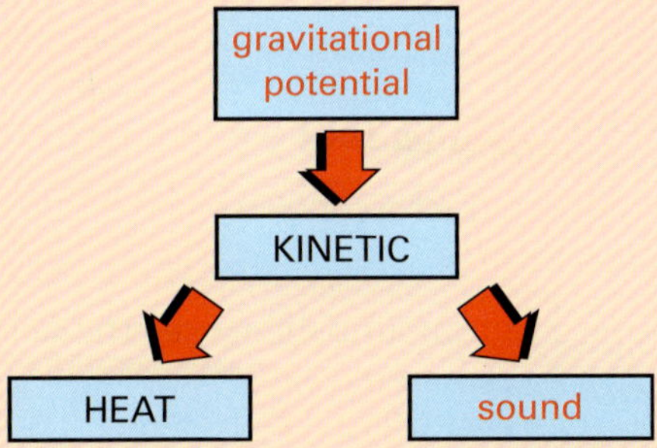

8 People often say that electrical energy is made in power stations, but this is not scientifically correct. According to the law of conservation of energy, energy cannot be made—you can only change it from one form to another. In a power station the chemical energy in coal or the kinetic energy of falling water is converted into electrical energy.

9 **a**

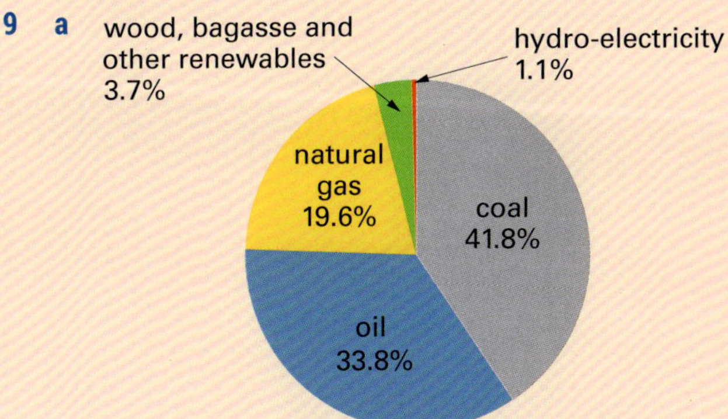

b coal, oil and natural gas

c 4.8% (hydro-electricity + wood, bagasse and other renewables)

d Bagasse is the crushed, juiceless remains of sugar cane left after extracting the sugar. It is used as a fuel and for making wallboard etc.

10 gravitational potential energy of water in dam

kinetic energy of water in pipe

kinetic energy of spinning turbine

electrical energy in electrical generator

11 The more efficiently a ball changes its kinetic energy into elastic potential energy (then back to kinetic energy), the higher it bounces. To compare different balls, you could drop them from the same height onto the same surface and measure how high they bounce. To make the test completely fair, the balls would need to be the same mass and size.

ISBN 978 1 4202 3735 1

CH•4 Cells of life

1 **C**—Plant and animal cells both have a nucleus, cytoplasm and organelles.

2 **C**—Cells have many and varied shapes.

3 **a** chloroplast
b cytoplasm
c nucleus
d cell membrane
e cell wall
f vacuoles

4 A ×10 objective and a ×4 eyepiece lens gives a total magnifying power of ×40. An object 0.05 mm in diameter would appear to be 40 × 0.05 = 2 mm in diameter.

5 **a** During photosynthesis carbon dioxide and water combine to produce glucose and oxygen. This chemical reaction requires energy, which is provided when chlorophyll absorbs sunlight. Respiration is the reverse of photosynthesis. Glucose combines with oxygen to produce carbon dioxide and water. During this reaction energy is released.

b Cell respiration produces the energy needed for:
- cells to make proteins
- muscle movement
- nerves to send nerve impulses (messages)
- cells to divide.

See page 90.

6 A unicellular organism consists of a single cell that contains all the structures necessary to live an independent life. A multicellular organism contains many different types of cells that work together for the survival of the organism.

7 The stomach is an organ because it is made up of many different types of *tissues*, e.g. gland tissue, muscle tissue and connective tissue. These tissues contain specialised *cells* that work together to digest food.

8 Tissue A could be found in the lining of the gut where its function would be to produce mucus that is slippery and allows the food to move smoothly through the gut.

The cells of tissue B could form a flat surface like paving stones, and this tissue could be found in the skin.

9 **a** The liquid rose because water from the beaker passed through the semipermeable membrane and into the funnel. This occurred because the water concentration was greater in the beaker than in the liquid in the funnel.

b More water would have diffused into the funnel and the water level would have risen further.

c If the two liquids were reversed, the level in the funnel would fall and the level in the beaker would rise slightly.

d The water concentration is greater in the celery than in the salty water surrounding it. So water diffuses out of the celery into the salty water. As a result the cells in the celery lose water, making them soft and not as firm as before. When this happens the stick of celery becomes limp.

MICROSCOPE LICENCE TEST

1 Setting up a microscope—see the Skillbuilder on page 81.

2 Making a wet-mount slide—see the Activity on page 82.

ISBN 978 1 4202 3735 1

CH•5 Chemical reactions

1 **B**

2 **C** Some reactions are slow, e.g. rusting.

3 **a** precipitate formed
b heat produced
c gas produced
d colour change

4 **B**

5 **c** gas burning in a stove (fastest)
d a cake baking
a paint drying
b fruit rotting
e a metal can rusting (slowest)

6 sugar + HEAT → carbon + water

7 **D**

8 The food rotted. This was a chemical reaction that produced new substances. Some of these new substances were gases. It was these gases that caused the lid to bulge.

9 **A** Hydrogen is used as a rocket fuel *because it burns explosively with oxygen*. It is true that hydrogen is the lightest gas known, but this is not the reason it is used as a rocket fuel.

10 **B**

11

1 amount of water added to the bread	B made special allowance
2 temperature in each jar	C did not allow for it
3 type of glass each jar is made of	D not important
4 amount of light falling on the jars	A deliberately varied

CH•6 Heat energy

1 **D**

2 **C**—see page 145

3 **a** **b**

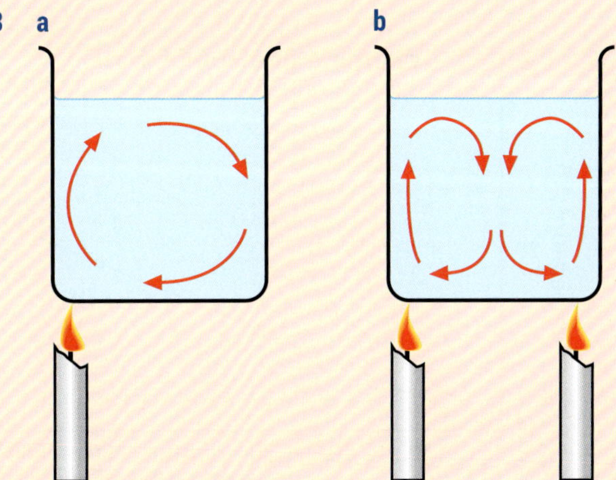

4 **a** true
b false. Conduction is fast in *conductors* (or *slow* in insulators).
c true
d false. The sun transfers heat energy to the Earth by the process of *radiation.*
e false. The hotter an object is, the *more* radiation it emits.
f true
g true

5 According to the particle theory (page 135), the particles in a hot object move more rapidly than the particles in a cooler object.

6 The amount of heat in an object depends on its mass, its temperature and what it is made of. So a bucket of water has more heat energy than a teaspoon of water at the same temperature.

7 When your hand is above the candle, heat rises through the air by *convection*. When your hand is beside the flame, some heat travels to your hand by *conduction*, but air is a poor conductor of heat. The candle is not hot enough to produce a lot of radiation.

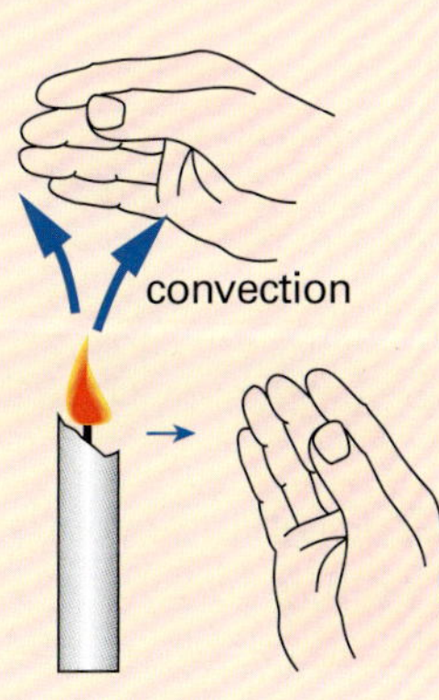

ISBN 978 1 4202 3735 1

8 **a** Heat travels from the heating element to the sandwich by *radiation*.

b Heat cannot travel *downwards* by convection, and conduction through the air would be very slow because air is a poor conductor.

9 **a**

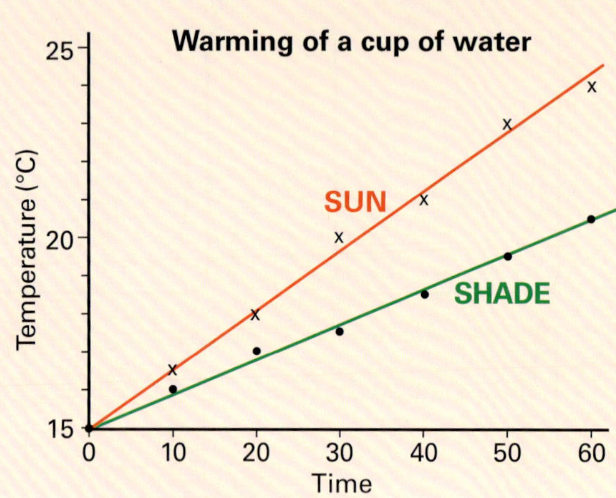

b The cup of water in the sun warms up more rapidly than the cup of water in the shade.

c same volume of water in each cup
identical cups
same initial water temperature
identical thermometers

d Heat was transferred to the cups by radiation from the sun.

e If you painted the cup black, it would absorb more radiation. However, as it warmed up it would probably lose heat more rapidly than an ordinary styrofoam cup.

10 Heat flows from warm places to cold places. So the insulation is to slow down the movement of heat from warm to cold. It is therefore better to say 'to keep the warm in' rather than 'to keep the cold out'.

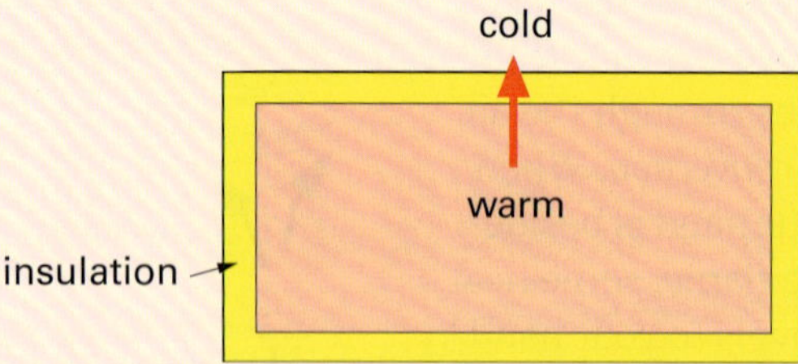

11 You can base your experiment on Experiment 6.1 on page 147.

1 Make two model sheep, e.g. by wrapping wool around soft drink cans.
2 Wet the wool on one of the cans.
3 Fill both cans with warm water at the same temperature, and put a thermometer in each.
4 Record the temperature in each can every minute for 15 minutes.
5 Plot the results on a graph and decide which 'sheep' cools more rapidly. This should give you some idea of whether sheep get colder when it is raining.

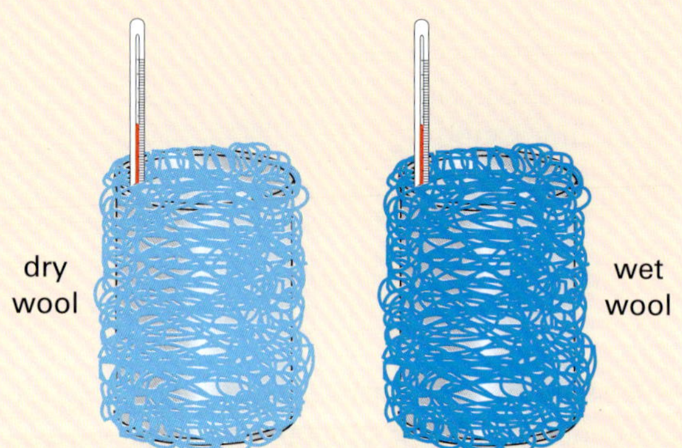

ISBN 978 1 4202 3735 1

CH•7 The human body

1 **a** urine **b** arteries
c stomach **d** trachea
e small intestine **f** ventricle

2 **B**—see page 163

3 **D**

4 **a** 3—the small intestine (see page 166)
b 1—enzymes start digesting starch in the mouth
c 2—the stomach
d 1—the mouth
e 4—the large intestine

5 Solid wastes (faeces) pass out of the gut through the anus. Liquid wastes are removed by the kidneys through the bladder. Gaseous wastes are removed by the lungs and breathed out through the mouth. (See pages 175 and 176.)

6 **a** Vessel B contains blood that is being pumped *away* from the heart. Therefore, it is an artery and would have thicker walls.
b The blood in chamber 1 would have less oxygen since the blood has come from the body and is being pumped to the lungs to receive more oxygen.
c The blood flows through vessel A into chamber 1, then into chamber 2. It is pumped from chamber 2 to the lungs in vessel B. It then returns from the lungs through vessel C to chamber 3, and finally is pumped from chamber 4 to the body in vessel D.

7 **a** smallest amount of protein—food C
largest amount of water—food C
b Food A is likely to be eggs, since it is high in protein and contains a lot of water but no carbohydrate.
Food B is likely to be fried beef sausages, since it is high in fat.
Food C is likely to be oranges, since it is mostly water, but contains a fair bit of carbohydrate (sugar).
Food D is likely to be wholemeal bread, since it is high in carbohydrate (starch).
c The same mass of food was used in each of the tests to make it easy to compare the results.

CH•8 Elements and compounds

1 **B**—Only elements, e.g. the substances copper, mercury and chlorine, can normally be seen without a microscope.

2 **C**

3 **A**—There are only 90 elements found naturally, but they can combine to form many thousands of different compounds.

4 **C**—Sugar is a compound of carbon, hydrogen and oxygen.

5 **C**—Only NO_2 has nitrogen and oxygen in the ratio 1:2. The ratios of nitrogen to oxygen are:
A NO 1:1
B N_2O 2:1
C NO_2 1:2
D N_2O_2 2:2 or 1:1

6 **B** represents an element—atoms of one type only. (**A** represents a mixture of elements and compounds, **C** represents a mixture of elements and **D** represents a compound.)

7 Four (three hydrogen and one nitrogen)—there are four atoms although there are only two different *types* of atoms.

8 **a** 1 and 4
b 2 and 4
c 4—it gives a yellow flame *and* produces a purple gas.

9 **B**—The tests show that the elements hydrogen, chlorine and zinc are present. The zinc reacted with the acid, so the acid contains only hydrogen and chlorine.

10 Your answer should be something like this: Elements and compounds are substances but atoms and molecules are invisible particles. A compound is a substance containing two or more elements combined together. A molecule is two or more atoms joined together.

11 The electric current causes the water to decompose into the elements hydrogen and oxygen (see page 198).

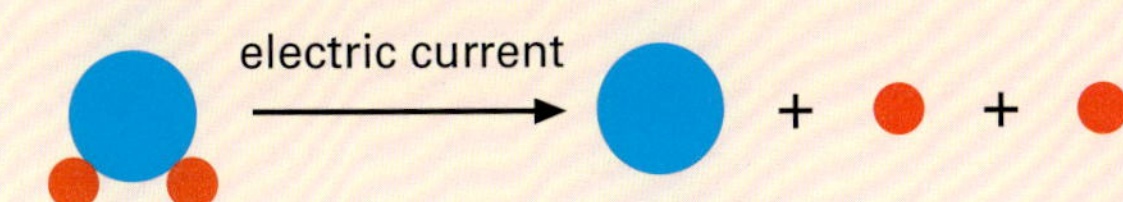

ISBN 978 1 4202 3735 1

When hydrogen and oxygen are mixed, a lighted match causes them to combine again. In both cases a chemical reaction occurs.

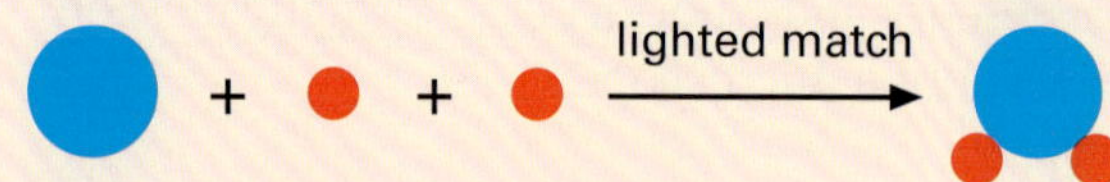

12 The scientist needs to use chemical reactions to break the compounds into their elements, as in Question 11. If she does this, the first compound ●■ will give her equal amounts of ● and ■. The second compound ●■● will give twice as much ● as ■.

CH•9 Light and sound

1 **B**—Only for this ray (AX) is the angle of incidence equal to the angle of reflection.

2 **D**—See page 233.

3 **A**

4 **D**—The light rays are refracted through a rectangular glass block as shown in **A**, but the rays that come out the other side of the block are always parallel to the incoming rays.

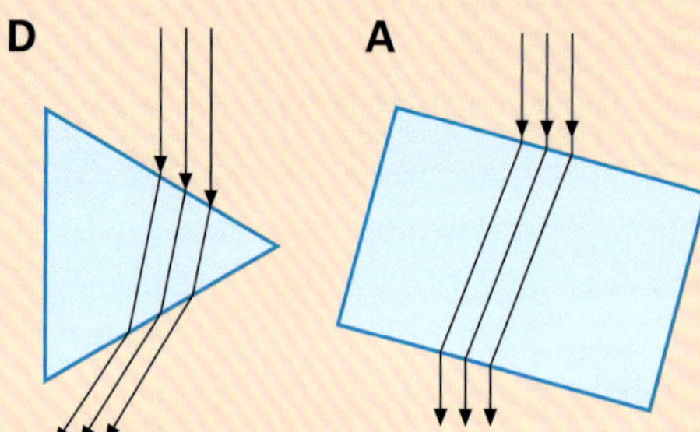

5 **a** 15 cm
b Lens B will be less curved and thinner.

6 **a** green light only
b red and violet light
c The filter is green because a green filter will transmit its own colour and absorb the others.

7 White light is shining in your eye because it contains green, blue and red light, affecting all three types of colour vision receptors.

8 When different coloured lights pass through a prism, red light is refracted least, blue the most, with green light in between. So if beam C is green, then B must be yellow (between red and green) and A must be red.

9 **a** The sound wave that returned in 0.10 seconds came from the school of fish.
b The sound wave took 0.05 seconds to travel to the fish. If sound travels 1450 m each second, then the fish are 1450 × 0.05 = 72.5 m below the ship.
c The table on page 230 shows that the speed of sound in water is less at lower temperatures. Therefore as the temperature decreases the speed of sound also decreases. The sound wave will take longer to return, and the fishermen may therefore think that the water is deeper than it is.

ISBN 978 1 4202 3735 1

10 See Figure 9.48.

11 See Figures 9.35 and 9.36.

12 Correctly identify the following wave diagrams as high pitch and low pitch.

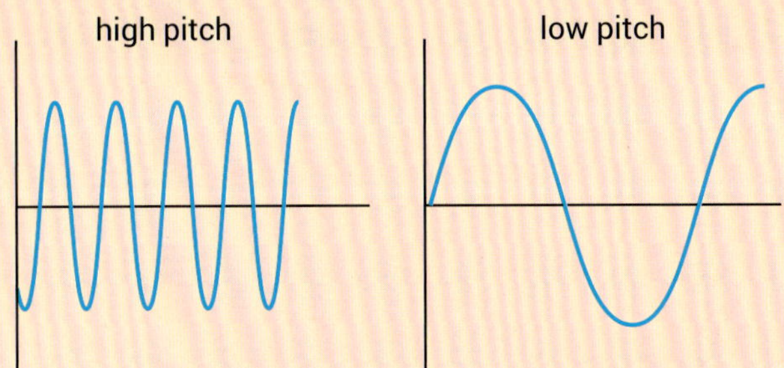

13 The player moves their fingers up and down the neck of the violin, pushing on the strings at different points. This effectively makes the strings shorter or longer, depending on where they press. A shorter string makes a higher-pitch note, and a longer string makes a lower-pitch note.

14 **a** approximately 100 dB
b Listening to a loud repetitive sound for a long time can damage a person's hearing, even if the sound is below 100 dB.
c Wear headphones or ear plugs. Stay inside where noise is lower. Take breaks from the noise by moving to a quiet area.
d An in-the-ear hearing aid could be used, which collects the sound and amplifies it, making it louder so that you can hear it. Alternatively, a cochlear implant could collect sound and transmit it directly into the nerves of the ear.

15 See Figures 9.14 and 9.15. Long-sightedness causes light to focus behind the retina; it is corrected with a converging lens. Short-sightedness causes light to focus in front of the retina; it is corrected with a diverging lens.

CH•10 Growth and reproduction

1 **a** nucleus **b** pollination **c** fertilisation
d sperm **e** cells **f** puberty

2 **a** testes—make sperm
b placenta—passes food and oxygen from the mother to the unborn baby
c semen—liquid that contains sperm
d uterus—hollow organ in female in which baby develops
e umbilical cord—joins the foetus to the placenta inside the mother's uterus
f ovaries—make ova (eggs) in the female
g vas deferens—tube that carries semen from the testicles to the urethra
h fallopian tubes—carry eggs from ovaries to uterus; often where fertilisation occurs
i penis—transfers semen containing sperm into the vagina during sexual intercourse
j seminal vesicles—produce fluid that mixes with sperm to make semen
k endometrium—the lining of the uterus; it builds up and is shed during menstruation

3 C

4

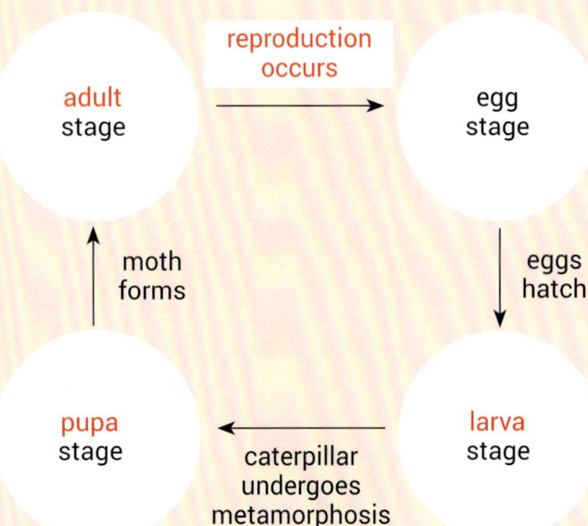

5 **a** At the beginning of the menstrual cycle, the uterus lining is shed, causing bleeding. The lining is thinnest about day 5, then quickly builds up during ovulation (days 11–15). The lining remains thick until bleeding occurs again at the end of the cycle.

ISBN 978 1 4202 3735 1

- b The ovary releases an ovum between days 11 and 15 when the lining of the uterus is ready to receive a fertilised egg.
- c Bleeding occurs at the end/beginning of the cycle if the egg has not been fertilised.

6 a First, the eggs may be eaten by other animals, and second sperm from the male may not reach the eggs to fertilise them.
- b Fewer frogs' eggs reach adulthood than birds' eggs do because (1) the frogs' eggs are fertilised externally, which means that many eggs may not be fertilised, and (2) the young tadpoles are not cared for by the adult frog, so many young may be eaten by other animals.

7 a The pine seed has a wing that allows it to be carried in the wind.
- b The apple has a sweet, edible fruit that is eaten by animals. The seeds pass through the gut of the animal and out in its droppings, and are spread this way.
- c The eucalypt has very small, light seeds which fall out of the 'gumnut' and are carried away by the wind.
- d The burr seed has spikes that stick to the fur, hair or feathers of animals. The animal may pick off the burr some distance from the plant.
- e The paw paw has edible flesh around its seeds. The seeds are spread in animal droppings in the same way as apple seeds.

8 a Because the fertilised egg develops outside the mother's body, the egg must contain enough food to sustain the developing chicken.
- b Strong, fragrant scents attract insects, which pollinate the flowers.
- c Millions of sperm cells are needed because very few survive the journey through the female reproductive system to fertilise the egg.

CH•11 The rock cycle

1

Rocks	Minerals
basalt marble pumice	bauxite calcite quartz sapphire

2 A

3 D

4 A

5 D

6 a 1—as the rocks here cooled most rapidly
- b C
- c **plutonic**—as they formed below the surface
- d The rocks around the magma would be changed to metamorphic rocks as shown below.

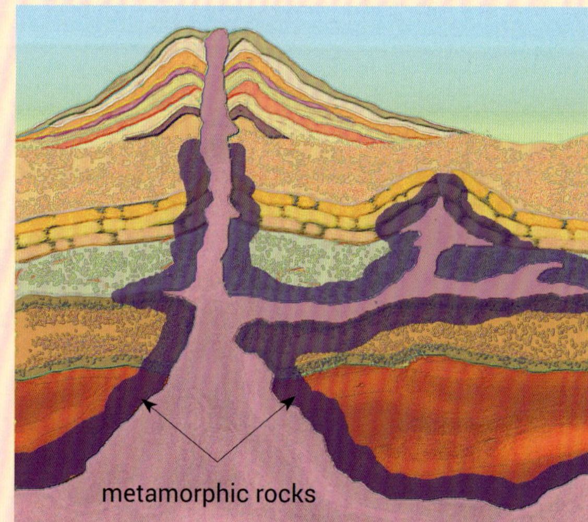

- e The layers of sedimentary rocks are not horizontal because they have been pushed up by the magma.

7 D

8 solid rock → boulder → gravel → coarse sand → fine sand

9 Magma is molten rock under the Earth's crust before it reaches the surface. Lava is molten rock after it reaches the surface.

10 a As you have to carve the rock you would not want it to be too hard. As the sculpture has to sit out in the weather you would not want it to weather too easily.

ISBN 978 1 4202 3735 1

b Using information from your internet research on page 287, the best rock to use is probably **marble**. (Granite would be too hard, and limestone would probably weather too easily.)

11 First, the rock would be weathered and eroded. The sediments would then be deposited in the ocean or a lake and changed into sedimentary rock. As this sedimentary rock is buried under other rock layers, it could be changed by heat and pressure into a metamorphic rock. See the diagram below.

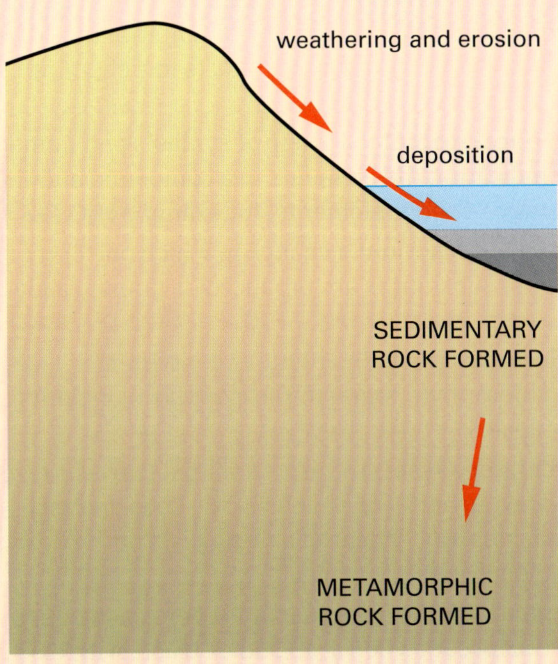

12 Basalt does not weather as easily as most sedimentary rocks. It also prevents the rocks under it from being weathered and eroded. This is why a flat-topped mesa is sometimes formed, as shown.

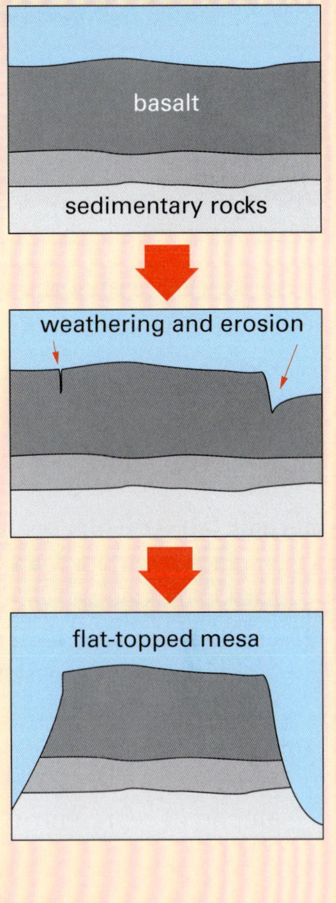

ISBN 978 1 4202 3735 1

Glossary

The words in this list occur in **dark type** throughout the book. The number after each entry gives the page where you will find more information. For some words the pronunciation is given. The syllable in capitals should be stressed; for example, foetus (FEE-tus).

addition (of colour): making colours by adding different coloured lights together 223

aerobic (air-OH-bic) respiration: the breakdown of glucose with oxygen to produce energy plus carbon dioxide and water 92

alveoli (AL-vee-OH-lee): minute air sacs in the lungs which allow gases to pass into and out of the blood capillaries 174

amplitude: the height of a wave which represents how much energy it carries; in a sound wave amplitude represents the loudness 229

arteries: thick-walled blood vessels that carry blood away from the heart 170

atmosphere: the thin layer of gases surrounding the Earth (or any other planet) 119

atoms: particles too small to see, that make up all matter 184

atria (singular atrium): two thin-walled chambers of the heart, above the ventricles 172

capillaries: microscopic blood vessels with very thin walls that allow substances in the blood to pass to and from the body cells 170

carbohydrates: a food type that supplies energy for the body; carbohydrates include sugars and starches 163

cartilage: material that covers the ends of bones and stops them rubbing on each other 160

cell division: the process by which a cell divides to make two new cells 246

cell membrane: the thin covering surrounding a cell which controls the movement of substances into and out of the cell 83

cell wall: the tough outside layer of a plant cell 83

cells: the building blocks of all living things; cells are usually microscopic 80

change of state: a change from one state of matter to another, for example from solid to liquid 39

chemical bonds: attractive forces between atoms 41

chemical energy: the form of energy stored in chemicals, for example foods and fuels 61

chemical equation: a chemical sentence that tells you the reactants and products in a chemical reaction 115

chemical formula: a group of symbols and numbers indicating the elements in a compound and the ratio of those elements; for example H_2O 192

chemical reaction: a change in which one or more new substances are produced; it cannot easily be reversed 111

chloroplasts: small structures containing chlorophyll, found in the cytoplasm of plants and algae 83

circulatory system: the system that carries oxygen and nutrients to all cells in the body and takes wastes away from cells: it consists of the heart, blood vessels and blood 170

combustion (burning): a rapid chemical reaction that occurs when a substance reacts with oxygen in the air, producing heat and light energy 112

compound: a pure substance that contains atoms of two or more elements combined in a fixed ratio; it can be broken down into its elements by chemical reactions 192

conduction: the transfer of heat through a solid, or the passing of an electric current through a solid or liquid 141

conductor: a substance that allows heat or electricity to move through it easily 141

conservation of energy: this law says that energy cannot be made or destroyed—it can only be changed from one form to another 69

control the variables: to keep all the variables the same, except the one you are purposely changing in an experiment 8

ISBN 978 1 4202 3735 1

convection: the transfer of heat in a liquid (or gas) by the movement of particles, when less dense liquid rises and more dense liquid flows in to take its place 142

crystal: a solid with naturally straight edges, flat sides and regular angles 270

cytoplasm (SIGH-toe-plaz-um): a jelly-like substance that fills most of a cell 83

data: information gathered by observation, experiment or library research; it may be qualitative or quantitative 4

decibel: dB, the unit for measuring the loudness of sound 229

decomposition: a chemical reaction where a substance breaks down to simpler substances 112

density: a measure of how much matter is packed into a given volume; it is measured in grams per cubic centimetre (g/cm^3) 32

dependent variable: a variable that changes in response to changes in the independent variable; values for this variable are graphed on the vertical axis 11

diffusion: the gradual mixing of substances caused by the random movement of particles from a region of high concentration to a region of lower concentration 48, 97

digestion: the physical and chemical breakdown of food into soluble materials 166

digestive system: the organs or parts of the body that work together to break down food into a smaller, more usable form 166

dispersion: the splitting up of white light into the colours of the spectrum 220

efficiency: a measure of how well a machine uses the energy put into it; a perfect machine would have an efficiency of 100% 69

elastic potential energy: the energy stored in compressed or stretched springs or other elastic devices 60

electrolysis: splitting up a chemical into its elements by passing an electric current through it 199

electromagnetic waves (radiation): waves that can pass through a vacuum; they include visible light, ultraviolet and infrared rays, microwaves, X-rays, gamma rays and radio waves 231

electromagnetic spectrum: the full range of electromagnetic radiation, such as heat, light, ultraviolet and X-rays 231

element: a pure substance made up of only one type of atom; it cannot be broken down into simpler substances by chemical reactions 186

embryo (EM-bree-oh): a developing baby, from the time of fertilisation to birth 255

energy: the ability to do work or cause changes; there are many different forms of energy 56

energy chain: a series of steps in which energy changes form; for example, chemical energy → heat energy → kinetic energy 68

enzymes (EN-zimes): substances made by special cells in the body to speed up chemical reactions 166

erosion: the process by which weathered material is carried away by water, wind or glaciers. 278

evaporation (e-VAP-or-AY-tion): the process in which a liquid turns into a vapour and seems to disappear 39

excretion (ex-KREE-shun): the process of removing wastes from the body by the liver and kidneys 197

excretory system: system of organs (including the kidneys and bladder) which removes waste materials from the body 175

experiment: a well thought-out scientific test, designed to answer a question or solve a problem 4

faeces (FEE-seas): solid waste produced by the body and removed through the anus 176

fair test: an experiment in which you change something, measure something and keep everything else the same 8

fats: a food type that supplies a large amount of energy and which can be stored in the body 163

fertilisation (FUR-til-eyes-AY-shun): the process in which the nuclei of a sperm and ovum join to make a new living thing 252

focus: the point at which rays of light meet after reflection from a curved mirror, or refraction by a lens 209

foetus (FEE-tus): a human embryo between about 14 weeks and birth 256

fossil fuels: fuels obtained from material that was once living; for example oil, coal and natural gas 71

ISBN 978 1 4202 3735 1

frequency: the number of wavelengths in a given period of time; also known as the pitch of sound. 229

fuel cell: a device for generating electricity by the chemical combination of a fuel (for example hydrogen) and oxygen 200

generalisation: a statement or conclusion, based on many observations, that holds true for most cases: for example, the heavier a truck is, the longer it takes to stop 5

global warming: an increase in the global temperature of the Earth thought to be due to the build-up of greenhouse gases in the atmosphere 126

gravitational potential energy: the energy stored in a raised object 59

greenhouse effect: the trapping of heat energy by gases in the atmosphere, causing its temperature to rise; carbon dioxide is the main greenhouse gas 123

heat: a type of energy that can raise the temperature of things; it is measured in joules 134

hypothesis (high-POTH-e-sis): a generalisation that explains a set of observations or gives a possible answer to a question; it can be tested by experimenting 9

igneous rock (IG-nee-us): a type of rock formed by the cooling and hardening of magma or lava 270

independent variable: a variable that is purposely changed in an experiment; values for this variable are graphed on the horizontal axis 11

insulator: a substance that does not allow heat or electricity to move through it easily 141

joule (J): the unit for measuring work and energy 56

kidneys: organs that filter and remove waste materials from the blood 175

kinetic (kin-ET-ic) energy: the energy that a moving object has 59

law of reflection: the angle of incidence of a light ray is equal to the angle of reflection 209

lava: hot molten material that flows from beneath the Earth's crust onto the surface, usually from volcanoes; it cools to form volcanic rock 269

ligaments: strong fibres that hold bones together 160

liver: a large body organ that stores and distributes digested food, and breaks down harmful substances 175

longitudinal wave: a wave in which the movement of particles within it is in the same direction as that in which the wave is moving, for example a sound wave 229

long-sightedness: when the lens of the eye focuses light behind the retina; it is corrected with a converging lens 217

lungs: large organs that absorb oxygen from the air and remove carbon dioxide from the body 175

magma: molten rock within the Earth 268

matter: a term used to include anything that has mass and occupies space (has volume) 30

menstrual cycle: a cycle of about 28 days, which includes ovulation (release of an egg) and menstruation (bleeding) 253

menstruation (MEN-strew-AY-shun): the 3–5 day period of the menstrual cycle during which bleeding occurs 253

metamorphic (meta-MOR-fik) rock: a type of rock formed from other rocks by the action of heat and pressure 284

metamorphosis (meta-MOR-fo-sis): a dramatic change in an organism's appearance and habits during its life cycle 249

minerals: substances that occur naturally in the Earth; they are the building blocks of rocks 271

mitochondria (might-oh-KON-dree-a): organelles in which respiration occurs in a cell 90

model: a way of representing something that cannot be observed directly because it is too small, too large or too complicated; for example, a model of a molecule 41

molecule: a tiny particle containing two or more atoms in a fixed ratio and joined by chemical bonds 184

muscle contraction: the process in which a muscle becomes shorter and pulls on bones or other parts of the body 159

non-renewable energy: resources that are not replaced as they are used; for example, coal and oil 71

nuclear energy: the energy stored inside the nuclei of atoms 61

nucleus (NEW-klee-us): the round dark-coloured structure that controls the activities of a living cell 83

ISBN 978 1 4202 3735 1

oesophagus (uh-SOF-a-gus): the tube leading from the mouth to the stomach 166

optical fibres: fibres made of thin pure glass that allow the transmission of digital light pulses over long distances 215

ores: minerals or rocks from which metals can be extracted 271

organ: a collection of tissues that has a particular function in the body, for example heart, kidney 87

organelles (OR-gan-els): small structures found in the cytoplasm of cells, for example chloroplasts 83

osmosis (oss-MOW-sis): the movement of water through a semipermeable membrane (such as a cell membrane) from a region of high concentration to a region of lower concentration 99

ovary (OH-var-ee): the female reproductive organ that makes ova (eggs) 251

oviducts (OH-vee-ducts): the tubes in a female that link ovaries with the uterus 255

ovulation (OV-you-LAY-shun): the time when an egg is released from an ovary 254

ovum (plural ova): the female sex cell, also called an egg 257

particle theory: the theory that all matter is made up of particles (atoms or molecules) that are too small to see and that are always moving 40

penis (PEE-nis): the male reproductive organ used to pass sperm into the female; urine also passes out though the penis 254

photosynthesis (foe-toe-SIN-thu-sis): the process in which the energy of sunlight is absorbed by chlorophyll in green plants and is used to make food and oxygen 117

physical change: a change where the properties of a substance change, but the substance is still the same; for example, when water freezes 111

placenta (pla-SEN-ta): a spongy pad at the end of the umbilical cord that grows on the mother's uterus and passes food and oxygen from the mother to the unborn baby 255

plasma: fourth state of matter that exists at very high temperatures; it consists of charged particles even further apart than the particles in a gas 42

plutonic (ploo-TON-ic) rock: igneous rock that formed from magma that hardened beneath the Earth's surface; for example, granite 271

pollination (pol-in-AY-shun): the movement of pollen from an anther to a stigma of a flower 259

potential energy: stored energy, available to be converted to other forms of energy 59

precipitate (pre-SIP-it-ate): a solid that forms as a result of mixing two solutions 112

products: the new substances produced in a chemical reaction 115

properties: the characteristics of materials, for example strength, colour, hardness, flexibility 108

proteins: a food type that provides the materials for the growth and repair of cells 163

puberty (PEW-ber-tee): the period of time (usually between the ages of 10 and 15) during which sexual development occurs 249

qualitative: a type of observation using words without measurement 4

quantitive: a type of observation that involves measurement 4

radiation: the transfer of heat from a hot object through space (or air) to a cold object 143

reactants: the substances you start with and that react with each other in a chemical reaction 115

reaction rate: the speed of a chemical reaction 116

refraction: the bending of light that occurs when light passes from one transparent substance to another 209

reliable: results are reliable if they are the same when the experiment is repeated many times 147

renewable energy: resources that can be replaced as they are used; for example, solar energy 71

respiration (RES-per-AY-shun): the process in living things of getting energy from foods 90

respiratory system: system of organs in the body (consisting of the mouth, trachea, bronchi and lungs), which takes in oxygen from the air and releases carbon dioxide 174

rock cycle: the slow process by which rocks are constantly being formed and changed from one type to another 285

rusting: a chemical reaction when iron reacts with water and oxygen in the air to form iron dioxide 112

scattering (of light): the bouncing of light from particles such as dust or smoke 226

ISBN 978 1 4202 3735 1

sediment: weathered material such as mud, sand and gravel carried mainly by running water and deposited somewhere else 278

sedimentary rock: a type of rock formed by the cementing together and hardening of sediments 281

semen (SEE-men): a liquid containing the sperm made in the male reproductive organs 254

sex cell: a special cell for reproduction. The male sex cell is a sperm and the female cell is an ovum (egg) 251

short-sightedness: when the lens of the eye focuses light in front of the retina; it is corrected with a diverging lens 217

solar energy: energy from the sun; it can be converted into heat or electrical energy 61

spectrum: the rainbow colours produced when white light is split up after passing through a prism or raindrops 220

sperm cell: the male sex cell 251

states of matter: there are three states of matter—solid, liquid and gas; a substance can exist in any of these three states 30

stem cells: unspecialised cells that can develop into any one of many different types of cells in the body 102

subtraction (of colour): making colours by mixing different paints or pigments together 223

symbols: signs, markings or letters that represent something else; for example, the symbol for copper is Cu 186

system: something that has parts that work together as a whole; for example the solar system, the digestive system 158

temperature: how hot or cold something is; it is measured in degrees Celsius 134

tendons: strong fibres that attach muscles to bones 159

testes (TES-teez): the two male reproductive organs that make sperm 251

theory: what a hypothesis becomes after it has been supported again and again by experimental results 40

tissue: a group of similar cells organised to do a particular job in the body; for example muscle tissue 87

total internal reflection: occurs when light hits a boundary between two transparent substances at a large angle of incidence and is reflected, with none transmitted 215

trachea (track-EE-a): a cartilage-banded pipe that takes air from the throat to the lungs 174

uterus (YOU-ter-us): the thick-walled organ in a female in which the embryo develops 255

vacuole (VAK-you-ole): a liquid-filled space found mainly in plant cells, which is used to store water and dissolved food 83

vagina (va-JI-na): the female reproductive organ that accepts the penis. It is also the birth canal through which the baby passes to the outside during birth 253

variable: any changeable factor that may influence the results of an experiment 8

veins: vessels or tubes that carry blood towards the heart; these tubes contain valves, which allow blood to flow in one direction only 170

ventricles: the thick-walled chambers of the heart that pump blood around the body 172

vitamins and minerals: substances needed in very small amounts by your body to keep it healthy 163

volcanic rock: igneous rock formed from lava that hardened on the Earth's surface; for example, basalt 270

weathering: the slow physical and chemical breakdown of rocks by the action of rain, cold etc. 275

work: the result of a force moving an object a certain distance; energy is needed to do work 56

ISBN 978 1 4202 3735 1

Index

ISBN 978 1 4202 3735 1

 ISBN 978 1 4202 3735 1

ISBN 978 1 4202 3735 1

ISBN 978 1 4202 3735 1

ISBN 978 1 4202 3735 1

ISBN 978 1 4202 3735 1

Acknowledgements

The author and publisher are grateful to the following for permission to reproduce copyright material:

PHOTOGRAPHS

Anteater Publications/Peter Stannard, **6**, **8**, **12**, **51**, **67**, **69**, **83**, **93**, **145**, **173**, **190**, **210**, **211**, **218**, **222**, **228**, **229**, **232**, **246**, **248**, **260**, **261**, **262**, **263**, **267**, **273**, **274**, **277**, **279**; AUSCAPE/Clive Bromhal, **259**, /Jean Paul Ferrero, **258**, C. Andrew Henley, **258**; Citizen of the Planet/Peter Bennett, **202**; BHP Billiton Science and Engineering Awards, **20**; Coo-ee Picture Library, **120**; CSIRO/M W Mules/CC BY 3.0, **23**, /Mark Talbot/CC BY 3.0, **25**; Dr J. A. Campbell, University of Canterbury, New Zealand, **150**, **151**; Dreamstime/Carlos Caetano, **149**, /Neal Cooper, **21**, /Johanna Goodyear, **237**, /Anthony Hathaway, **148**; GETTY IMAGES /Auscape/UIG, **23**, /Capture, **220**, /Bloomberg/Duncan Chard, **39**, /Corbis/Robbie Jack, **225**, /Galen Rowell, **281**, /Steve Raymer, **225**, /Paul Seheult, **194**, /Education Images/UIG, **271**, /GCA GCA/Science Photo Library, **247**, /Hulton Archive, **45**, /The Image Bank/Andy Sacks, **128**, /Gabriel M. Coven, **133**, /Iconica/ Art Wolfe, **263**, /Phillip Hayson, **278**, /Juice Images/Ian Lishman, **84**, /Lonely Planet Images/Richard I'Anson, **267**, /Mark Metcalfe, **25**, /Andrew Watson, **191**, /Photodisc, **251**, /Photolibrary/ Gary Braasch, **284**, /David Messent, **282**, /M Phillips David, **174**, /Paul Souders, **46**, /Photo Researchers/Science Source, **185**, /Science Photo Library, **80**, **256**, / SPL/Prof. P. Motta, **251**, /SCIEPRO, **176**, /Science Source, **187**, /Biophoto Associates, **247**, /M Phillips David, **174**, /Time & Life Pictures, **45**, /Joseph Van Os, **282**, ISTOCK/Arpad Benedek, **292**, /Samuel Clarke, **61**, /George Clerk, **123**, /Linda Epstein, **244**, /gece**33**, **127**, /g-miner, **271**, /gordonsaunders, **119**, /Highwaystarz-Photography, **146**, /Björn Kindler, **73**, /kjorgen, **245**, /Fedor Kondratenko, **111**, /meyes, **148**, /Stephen Morris, **281**, /Christian Nasca, **119**, /nikkytok, **129**, /nzphotonz, **177**, /pxhidalgo, **269**, /Jorge Salcedo, **287**, /sarkophoto, **127**, /sturti, **177**, /StefaNikolic, **208**, /stocknroll, **139**, /Wendy Townrow, **276**, /wScottLoy, **60**, /ZambeziShark, **280**, FAIRFAX/Robert Rough, **289**, /Dean Sewell, **116**, REUTERS/Mike Segar, **102**; SHUTTERSTOCK/abd, **217**, /Adisa, **148**, /Khakimullin Aleksandr, **106**, /anyaivanova, **4**, /Aprilphoto, **30**, /Banana Power, **17**, BioMedical, **159**, /BMJ, **245**, /BlueRingMedia, **252**, /Tyler Boyes, **282**, **284**, /Massimo Cavallo, **70**, /Nino Cavalier, **175**, /Robert Cicchetti, **183**, /Ziga Cetrtic, **127**, /Marcel Clemens, **186**, /Neale Cousland, **266**, /Crepesoles, **35**, /dantess, **231**, /Designua, **237**, /die Fotosynthese, **111**, /Everett Collection, **2**, /FiledIMAGE, **288**, /Fotokostic, **118**, /Foto-Ruhrgebiet, **186**, /Deyan Georgiev, **30**, /GrashAlex **263**, /Georgejmclittle, **115**, /hecke**61**, **29**, /Shawn Hempel, **36**, /Jubal Harshaw, **78**, /Ben Heys, **269**, /I AM NIKOM, **36**, /iatlo, **246**, /ifong, **163**, /IxMaster, **157**, /JonMilnes, **146**, /Juancat, **28**, /Sakdinon Kadchiangsaen, **270**, / Evgeny Karandaev, **61**, /Sebastian Kaulitzki, **156**, **172**, /Volodymyr Krasyuk, **61**, /Dmitry Kalinovsky, **146**, /Kuznetcov_Konstantin, **62**, /Olga Kovalenko, **186**, /Lev Kropotov, **246**, /D. Kucharski K. Kucharska, **85**, /kuehdi, **250**, /Lebendkulturen.de, **80**, /Petr Malyshev, **58**, /Steve Mann, **203**, /MarcelClemens, **35**, /Maridav, **40**, **269**, /mantis, **80**, **282**, /michaeljung, **193**, /MichaelaPerryman, **278**, /Lindsey Moore, **112**, /NikoNomad, **40**, /NoPainNoGain, **126**, /OGphoto, **112**, /Tyler Olson, **177**, / optimarc, **282**, /Pablo Prat, **127**, /pedrosala, **127**, /PHB.cz (Richard Semik), **258**, /phloem, **39**, /Photodiem, **3**, /photoinnovation, **62**, /Leigh Prather, **61**, /posteriori, **168**, /Alexander Raths, **5**, /ra**2**studio, **206**, /r.classen, **107**, /Oleksiy Rezin, **61**, /Romariolen, **13**, /RugliG, /Dario Sabljak, **132**, /Samarskaya, **62**, /Yuri Samsonov, **134**, /Joseph Sohm, **55**, /sonsam, **282**, **287**, /Taina Sohlman, **66**, /South**12**th Photography, **38**, /Clint Spencer, **59**, /stihii, **159**, /stocksolutions, **146**, Syda Productions, **165**, /Ferenc Szelepcsenyi, **263**, /Tomsickova Tatyana, **186**, /Leah-Anne Thompson, **62**, /Venus Angel, **161**, /Nils Versemann, **287**, /vm**2002**, **13**, /Volt Collection, **54**, /vvoe, **282**, **284**, /wavebreakmedia, **207**, /Angela Waye, **257**, /WHITE RABBIT83, **42**, /Peter J. Wilson, **117**, /Olena Yakobchuk, **127**, /ymgerman, **59**; SCIENCE PHOTO LIBRARY, **182**, /Blaine Image, **275**, /Ian Boddy, **216**, /Martin Bond, **200**, /Dr. Jeremy Burgess, **94**, /Trevor Clifford Photography, **114**, /Thomas Deerinck, NCMIR, **80**, /Edelmann, **256**, /Eye of Science, **91**, **184**, /Steve Gschmeissner, **86**, /Alfred Pasieka, **201**, /Claude Nuridsany & Marie Perennou, **186**, /David Scharf, **214**, /Sheila Terry, **138**; Dr Michael Tyler, **22**; Courtesy of South Australia Department of State Development, **287**; Ken Williamson, **274**, **275**.

The author and publisher would like to acknowledge the following:

NASA, **61**, NASA/ESA/Hubble Heritage Team (STScI / AURA), **128**.

While every care has been taken to trace and acknowledge copyright, the publisher tenders their apologies for any accidental infringement where copyright has proved untraceable. They would be pleased to come to a suitable arrangement with the rightful owner in each case.